CONTROL AND SYSTEMS THEORY

A Series of Monographs and Textbooks

Editor
JERRY M. MENDEL
University of Southern California
Los Angeles, California

Associate Editors

Karl J. Åström
Lund Institute of Technology
Lund, Sweden

Michael Athans
Massachusetts Institute of Technology
Cambridge, Massachusetts

David G. Luenberger
Stanford University
Stanford, California

Volume 1 Discrete Techniques of Parameter Estimation: The Equation Error Formulation, *Jerry M. Mendel*

Volume 2 Mathematical Description of Linear Systems, *Wilson J. Rugh*

Volume 3 The Qualitative Theory of Optimal Processes, *R. Gabasov and F. Kirillova*

Volume 4 Self-Organizing Control of Stochastic Systems, *George N. Saridis*

Volume 5 An Introduction to Linear Control Systems, *Thomas E. Fortmann and Konrad L. Hitz*

Volume 6 Distributed Parameter Systems: Identification, Estimation, and Control, *edited by W. Harmon Ray and Demetrios G. Lainiotis*

Volume 7 Principles of Dynamic Programming. Part I. Basic Analytic and Computational Methods, *Robert E. Larson and John L. Casti*

Volume 8 Adaptive Control: The Model Reference Approach, *Yoan D. Landau*

Additional Volumes in Preparation

ADAPTI
CONTRO

[illegible]

Acc't 21MØØ

201 Karman 301-46

ADAPTIVE CONTROL

The Model Reference Approach

YOAN D. LANDAU

Laboratoire d'Automatique
Institut National Polytechnique de Grenoble
Grenoble, France

MARCEL DEKKER, INC. NEW YORK AND BASEL

Library of Congress Cataloging in Publication Data

Landau, Yoan D.   [Date]
Adaptive control.

(Control and systems theory; v. 8)
Includes bibliographical references and index.
1. Adaptive control systems--Mathematical models.
I. Title.
TJ217.L37   629.8'312   79-62
ISBN 0-8247-6548-6

MARCEL DEKKER, INC.

270 Madison Avenue, New York, New York 10016

Current printing (last digit):
10 9 8 7 6 5 4

PRINTED IN THE UNITED STATES OF AMERICA

To Lina and Vlad

Ce qui est simple est toujours faux.

Ce qui ne l'est pas est inutilisable.

Paul Valéry

Mauvaises Pensées

CONTENTS

PREFACE

The design of high-performance control systems generally requires the use of adaptive control techniques when the parameters of the controlled process either are poorly known or vary during normal operation. Among various alternative methods, the use of the technique known as *model reference adaptive systems* (MRAS's) seems to be one of the most feasible approaches possible for the implementation of adaptive control systems. The prime characteristic of such an adaptive system is the presence of a so-called *reference model* as part of the system, which can appear under various forms.

This book presents a detailed coverage of various types of model reference adaptive systems, the corresponding design methods (from a unified point of view), and their application to relevant situations.

The basic control problems for which model reference adaptive systems might be used are (1) adaptive model-following control, (2) on-line and real-time parameter identification, and (3) adaptive state observation. In the first case, a reference model specifies the desired control performance. In the second and third cases, the process whose parameters are to be identified or whose states are to be observed represents the reference model.

In any case, the various design problems can be solved from a unified point of view when stability considerations are used as a basis for the problem formulation. In solving these problems, emphasis has been placed upon the use of such relatively new and successful concepts in systems theory as positivity and hyperstability. To keep the unified presentation in perspective, other approaches to the design of model reference adaptive systems are briefly reviewed and evaluated. When dealing with concrete problems, the author has attempted to share with the reader his practical experience concerning both the interpretation and the soundness of various designs.

It is hoped that this book will be of special interest to practicing engineers facing the problem of designing adaptive control systems and that the design examples and the case studies which are presented in detail will be of assistance to them. The book is also intended as a textbook for graduate students taking a course on techniques of adaptive control. Since the book covers most of the recent developments in the field of model reference adaptive systems, it is hoped that researchers will find it useful as a basic reference work and as a starting point for further investigation in the field of adaptive control.

The book is written assuming that the reader is familiar with linear control system theory, state-space representation of linear systems, linear differential and difference equations, and vector and matrix operations. Some acquaintance with nonlinear and time-varying system theory as well as with basic stability concepts would be helpful although not absolutely necessary. The specific stability, positivity, and hyperstability background necessary for understanding the book is included in the appendices for completeness.

To present the results contained in this book, we were obliged to choose between the *inductive* approach, that is, starting with simple examples and then stating the not always obvious generalization for high-dimension problems, and the *deductive* approach in which general results are first derived in a formal manner and then illustrated by examples. Finally, a middle ground was chosen. Whenever it was considered necessary, examples were discussed first in order to illustrate the major steps and the motivations behind them. Many of the book's important results are presented as formal theorems or summarized in tables. In this way, they can be easily identified from the body of the text.

The problems included at the end of each chapter (except the first) complete the book material in the sense that they allow the reader (1) to gain familiarity with the design techniques by solving numerical examples and (2) to extend concepts and design methods to less standard but practical situations.

The emphasis placed throughout the book on analytic design methods for model reference adaptive systems does not mean (1) that they are self-sufficient in solving practical problems or (2) that ad hoc simplification of adaptation algorithms is not possible for particular applications. To guide the reader, the book contains a detailed review of various pragmatic applications of model reference adaptive techniques. However, without a solid analytical understanding of the adaptation techniques, a user will not be able to modify an adaptation algorithm in attempting to apply it to difficult and new situations.

For the most part this book is an outgrowth of the several survey papers on model reference adaptive systems published by the author since 1971[1] and also of the graduate courses on nonlinear and adaptive systems delivered by the author since 1970 at the Institut National Polytechnique de Grenoble, Section Spéciale d'Automatique. In addition, important parts of the material presented in the book are based on the author's research activities since 1969 at ALSTHOM--Directions des Rescherches (Département d'Automatique et d'Electronique), Grenoble, and at the Institut National Polytechnique de Grenoble (Laboratoire d'Automatique), both under the partial support of Délégation Générale pour la Recherche Scientifique et Technique. Similarly, fundamental preparatory research was done at NASA--Ames Research Center (System Analysis Branch), in 1971-1972 under support from the National Research Council and the National Academy of Sciences. This work was accomplished at Laboratoire d'Automatique, Institut National Polytechnique de Grenoble, which also provided invaluable assistance in preparing the manuscript.

Yoan D. Landau

[1](1) "Les systèmes adaptatifs avec modèle (théorie, mise en oeuvre, applications)," *Automatisme*, Vol. 15, no. 5, pp. 272-292, May 1971. (2) "Model Reference Adaptive Systems. A Survey (MRAS--What Is Possible and Why?)," *Trans. ASME, Dym. Syst. Meas. Control*, Vol. 94, series G, pp. 119-132, June 1972. (3) "A Survey of Model Reference Adaptive Techniques (Theory and Applications)," *Automatica*, Vol. 10, pp. 353-379, July 1974.

ACKNOWLEDGMENTS

If it is true that the initial decision and the continuing determination of an author are necessary conditions for producing a book, it is equally true that the help of many others is an additional necessary condition.

The author is indebted to Professor René Perret, Director of the Laboratoire d'Automatique, Institut National Polytechnique de Grenoble, and Professor Rolland Rouxel, Head of the Département d'Automatique et d'Electronique, ALSTHOM--Direction des Recherches. They welcomed him to Grenoble on a rainy day in December 1968 and since then have given him encouragement and support in his research in the field of model reference adaptive systems.

The author particularly thanks Professor Yasundo Takahashi and Professor Michael Rabins, Editor Emeritus and past Editor, respectively, of the *Transactions of the American Society of Mechanical Engineers* (*Journal of Dynamic Systems, Measurements and Control*), who carefully reviewed the first version of the manuscript. Their suggestions provided both encouragement and guidance for improving and completing the final version of this work.

It is a pleasure for the author to acknowledge his many associates and students who throughout the years have contributed directly or indirectly to this work, in particular: E. Sinner, B. Courtiol, G. Béthoux, L. Muller, H. Medeiros-Silveira, A. Bénéjean, A. Gauthier, and J. M. Dion. Their work forms the basis for diverse parts of the material presented in the book.

The author also wishes to acknowledge his parents, who motivated his interest in research in general.

Finally, but perhaps most importantly, the author wishes to acknowledge an incalculable debt of time to his wife and son. The patience and assistance of his wife made the writing of this book possible.

INTRODUCTION

Chapter 1 is an introduction to the *model reference adaptive systems*. Its first intention is to familiarize the reader with the various concepts, terminology, and block diagrams which will appear throughout the book. The second intention of this chapter is to provide motivation to the reader for studying model reference adaptive systems by indicating conceptually how these systems have evolved and where they have already been applied.

The design problem for model reference adaptive systems is stated in Chap. 2, which starts with the presentation of the mathematical description of various types of model reference adaptive systems. This is followed by a discussion of the design hypothesis and of the equivalent representation of the model reference adaptive systems as a nonlinear time-varying feedback system.

Chapter 3 presents briefly, with the help of examples, three basic approaches to the design of model reference adaptive systems in the chronological order of their appearance [designs based on the use of (1) local parametric-optimization theory, (2) Lyapunov functions, and (3) positivity and hyperstability concepts].

The use of positivity and hyperstability concepts for the design of continuous-time model reference adaptive systems is then discussed in detail in Chap. 4. The design of various configurations when the state vector is available or when only input and output measurements are available is presented. A comparison of this approach with those based on the use of Lyapunov functions and of local parametric-optimization theory is included.

Because model reference adaptive systems are time-varying nonlinear systems, the extrapolation of the designs for continuous type to discrete type model reference adaptive systems is not evident.

In addition, the potential offered by the use of digital computers gives a challenge to develop specific adaptation algorithms which do not in practice have a counterpart in continuous-time model reference adaptive systems. For these reasons, Chap. 5 is devoted to the design of discrete-time model reference adaptive systems also using the positivity and hyperstability concepts. The class of adaptation algorithms with time-varying gains is discussed here in detail because of their practical appeal for solving adaptation problems in the presence of measurement noise.

The basic designs for various types of model reference adaptive systems having been presented, we proceed to the presentation of their application to various control problems.

In Chap. 6 we discuss the design of adaptive model-following control systems. Because adaptive model-following control is viewed as a natural extension of linear model-following control, when the parameters of the controlled process are poorly known or vary during operation, one starts by reviewing briefly the linear problem for every case. The cases of adaptive tracking when the state vector of the controlled process is available or when only input and output measurements are available are discussed as well as the case of adaptive state regulation and the extension of the designs to the discrete case. The implementation aspects and two case studies concerning an aircraft control problem and an adaptive speed controller for electrical drives are also examined.

The use of model reference adaptive techniques for parametric identification is the topic of Chap. 7. Identification of continuous-time and discrete-time models of dynamic processes is discussed. The problem of input requirements for parameter convergence and the effect of measurement noise are examined in detail. Comparisons of various designs from the identification point of view are included. Two case studies are presented in this chapter.

Chapter 8 is devoted to the design of adaptive state observers. These adaptive observers are an extension of the linear asymptotic observer, and they can also simultaneously provide identification of the parameters. Several designs are presented and evaluated.

The book also contains four appendices. Appendix A briefly reviews some stability concepts and Lyapunov functions. Appendix B reviews the basic results of the theory of positive dynamic systems, and Appendix C reviews some basic results of the theory of hyperstable systems. The presentation of the solutions for an integral inequality which is a cornerstone for deriving adaptation algorithms is given in Appendix D.

Only the references directly pertaining to the text are indicated at the end of each chapter (enclosed in square brackets in the text).

Since the book was written as a thorough coverage of the field, all the portions of the text need not be read by someone interested in a specific problem. In fact, starting with Chap. 3, every chapter can be read independently, even if references are made, for example, in Chaps. 6, 7, and 8 to the results of Chap. 4 or 5.

For those interested in applications, for a first reading, we recommend the following sequence: Chap. 1; the case studies of Chaps. 6 and 7; then Chap. 2; then Chap. 3 (Sec. 3.3) followed either by Chap. 4 or 5, depending on one's interest in continuous-type or discrete-type problems; and then Chap. 6, 7, or 8, depending on the practical control problem which the reader wants to solve.

For a one-quarter graduate- or senior-level course, we recommend the following sequence: Chaps. 1, 2, and 3 (omitting part of the material); Secs. 4.1, 4.2, 5.1, 5.5, 6.1, 6.2, 6.3, 7.1, 7.2-1, 7.3-1, and 8.3 (the last as an application of Sec. 5.5).

For a research course in adaptive control, Chaps. 4, 5, 6, 7, and 8 can serve as reference.

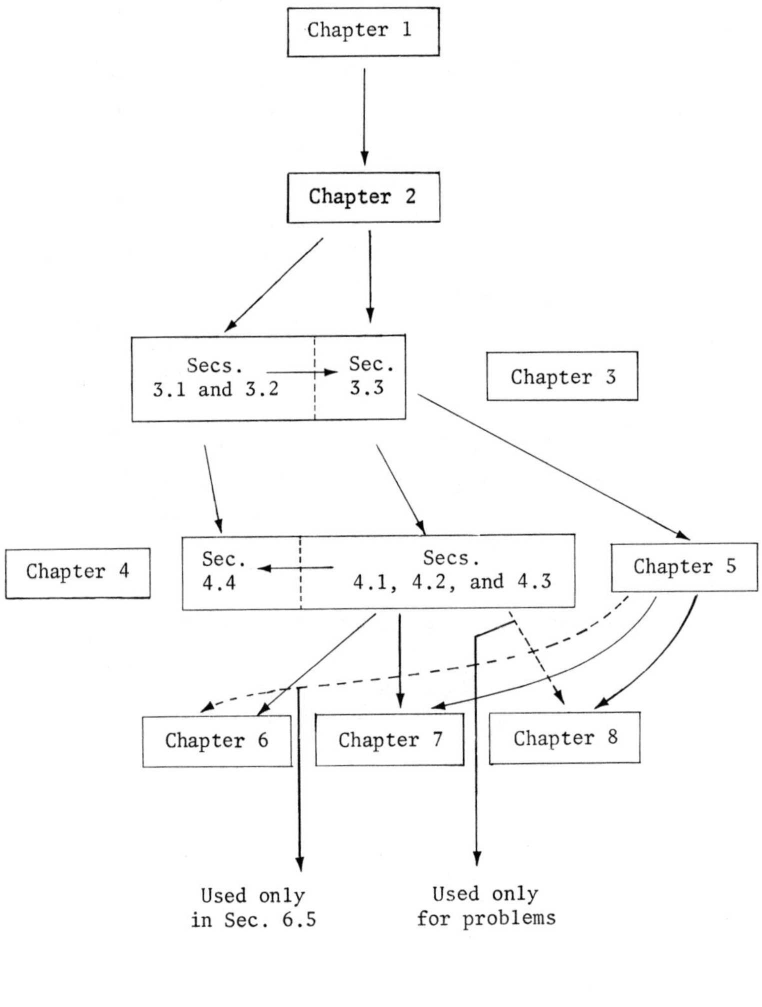

Logical dependence of the chapters and sections.

A, B, P	capital letters denote matrices (except letter V)
$\underline{x}$, $\underline{y}$, $\underline{u}$, $\underline{\theta}_p$	lowercase underlined letters denote vectors
x, y, u, θ_p	lowercase letters denote scalars
L	Laplace transform
V	Lyapunov function
$P > 0$, $P \geq 0$	positive definite (or semidefinite) matrix
B^+	left Penrose pseudo-inverse
$\|.\|$	modulus of a scalar
$\|\|\underline{x}\|\|$	the norm of the vector $\underline{x}$
R^n	the n-dimensional vector space
$R^{n\times m}$	the $(n \times m)$-dimensional space
C^m	the m-dimensional piecewise continuous vector function space
tr H	trace of $n \times m$ matrix H $(\mathrm{tr}\, H = \Sigma_{i=1}^{n} h_{ii})$
$\Longrightarrow$	implies
I	unit matrix
E{.}	expectation
ε	belonging to
MRAS	model reference adaptive system
LMFC	linear model-following control
AMFC	adaptive model-following control
SVF	state-variable filter
SISO	single-input/single-output
IP	index of performance
$(IP)_{RM}$	index of performance expressing the differences between the reference model and the adjustable model

ADAPTIVE CONTROL

Chapter One

INTRODUCTION TO MODEL REFERENCE ADAPTIVE SYSTEMS

The adaptive control technique we are about to study can be introduced either as a particular class of adaptive techniques or as a development of the conventional feedback control in order to be able to satisfactorily solve concrete practical problems. From the author's point of view, the second approach has more appeal. Why and how the concepts of *reference model* and *model reference adaptive systems* have evolved are stressed in Sec. 1.1. Then, in order to be able to place the model reference adaptive system within the broad context of adaptive control systems, we shall very briefly review adaptive control systems in Sec. 1.2. We shall emphasize the similarities and differences between the conventional feedback control systems and adaptive control systems.

We shall then return, in Sec. 1.3, to a more detailed presentation of model reference adaptive systems. Basic definitions, configurations, and classifications will be given. The duality existing between model reference adaptive systems used for control and those used for parameter identification or adaptive state estimation will be emphasized. The potentiality of model reference adaptive systems for solving adaptive control problems is illustrated by a brief review of several applications covering various fields.

1.1 THE NEED FOR MODEL REFERENCE ADAPTIVE SYSTEMS

For a long time the automatic control of physical processes, in spite of the use of the principle of *feedback*, has been an experimental technique deriving more from art than from scientific bases. The requirement for more complex and higher-performance control systems has been the impulse for the development of a systematic control theory. Even with the development of such a systematic control theory, there is still usually something else missing when a practical control system must be designed: a good knowledge of the dynamic characteristics of the controlled plant (eventually in a statistical sense).

To implement high-performance control systems when the plant dynamic characteristics are poorly known or when large and unpredictable variations occur, a new class of control systems called *adaptive control systems* has evolved which provides potential solutions. The adaptive control concept seems to be old--perhaps as old as the feedback concept itself--but significant interest in this type of system has arisen only since the early 1950s.

Many solutions have been proposed in order to make a control system "adaptive" (i.e., to assure high performance when large and unpredictable variations of the plant dynamic characteristics occur). Among them a special class of adaptive systems, *model reference adaptive systems* (MRAS's), evolved in the late 1950s. One of the main innovations of such systems is the presence of a *reference model* which specifies the desired performance. For several years, this was a very controversial concept. Nevertheless, a careful examination of the linear control system design problem and of the adaptive control problem leads naturally to the concept of a reference model, a concept which appears, in fact, to be very useful in a variety of situations. To appreciate these aspects, let us now briefly review the development of control theory.

In its first stages, control theory was developed (along with the mathematical description of dynamical phenomena) through linear, time-invariant, differential equations derived in a deterministic environment. Earlier, significant results were obtained by transposing the control problem into the frequency domain through linear integral transform techniques. Later, brilliant theoretical results were obtained in determining *optimal control policy*, and at this stage the *state-variable* concept appeared to be very useful in describing dynamical processes.

To briefly introduce optimal control theory, note that if a controlled plant is described by a state-vector differential equation

$$\dot{\underline{x}} = A\underline{x} + B\underline{u} \tag{1.1-1}$$

where $\underline{x}$ is the state vector, $\underline{u}$ is the control input, and A and B are constant matrices of appropriate dimension, it is possible under certain assumptions to find the optimal control $\underline{u}$ (as a function of $\underline{x}$) which minimizes the quadratic index of performance:

$$J = \int_0^{t_1} (\underline{x}^T Q\underline{x} + \underline{u}^T R\underline{u})\, dt \tag{1.1-2}$$

where Q and R are positive definite matrices (eventually Q could be nonnegative) and superscript T denotes transposition. Several comments must be made immediately:

1. The number of dynamic processes which can be described correctly by linear, time-invariant differential equations is small. In fact, this hypothesis is more the exception than the general rule. The linear, time-invariant description only gives a valuable description of the system dynamical behavior around a steady-state value of the state vector.
2. It is supposed that the desired performances of the control system can be completely specified by a quadratic index of performance and that the matrices Q and R in Eq. (1.1-2) are given. These assumptions are far from being evident in many cases.
3. The description of Eq. (1.1-1) introduces the concept of *dynamical parameters* defined by the terms of the matrices A and B. These parameters are very rarely measurable directly.

4. The construction of the optimal control law requires access to the entire state vector $\underline{x}$. In many situations, some components of the state vector $\underline{x}$ are not directly available (or the cost of these measurements is prohibitive).

To discuss the consequences of the above observations in control system design, let us first assume that the linearity and invariant hypotheses are in force, that the plant parameters are known, and that the entire state vector is measurable. The control system designer, having available all the mathematical tools necessary for computing the optimal control law, must first solve an important problem: how to choose Q and R in the performance index of Eq. (1.1-2). This problem becomes more and more complex as the dimension of the system increases. Furthermore, in most realistic cases in addition to the minimization of an index of performance of the type given in Eq. (1.1-2), one is obliged to assure a certain dynamic behavior for the evolution of the inputs and the states of the controlled plant. Because the quadratic index of performance of Eq. (1.1-2) does not contain any explicit information about the evolution at each instant of the states and inputs of the controlled plant, it is difficult to specify in terms of Q and R such widely accepted indices of performance as rise time, overshoot, and damping. This type of difficulty can be avoided if the desired indices of performance are specified by the performances of an *idealized* control system called the *reference model*.[1]

If $\underline{x}_M$ is the state vector of the reference model which receives the control signal $\underline{u}_M$, one must design a control system which minimizes a quadratic index of performance:

$$J = \int_0^{t_1} [(\underline{x}_M - \underline{x})^T Q(\underline{x}_M - \underline{x}) + \underline{u}^T R\underline{u}]\, dt \qquad (1.1\text{-}3)$$

With this formulation of the index of performance, the significance of the weighting matrices Q and R becomes clearer. Depending on the

[1]The first available reference using a reference model for implementing a linear optimal control system seems to be Ref. 1.

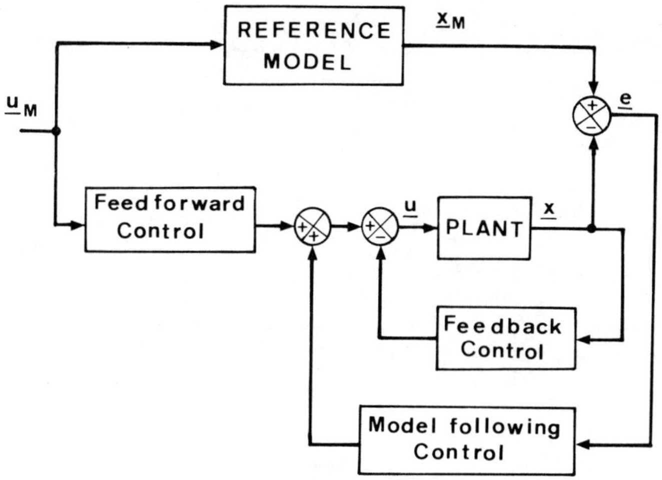

Fig. 1.1-1 Linear model-following control (LMFC) system.

choice of Q and R, the states of the model will be followed with more or less precision, but the dynamic response of the plant states will be defined by the model.

The reference model can be used implicitly or explicitly. In the first case, it is used only for the computation of the control law; in the second case (see Fig. 1.1-1), the reference model is a part of the control system itself. One can see that in the last case the new state vector defined as $\underline{e} = \underline{x}_M - \underline{x}$ allows one to directly measure at each instant the difference between the desired performance specified explicitly by the dynamic response of the states of the reference model and the real performance of the controlled plant. Even at this early stage in the design of a linear time-invariant control system, the concept of a reference model appears to be a useful tool for solving a practical problem within the framework of the optimal control theory for linear systems.

Let us now turn our attention to the problem of state measurements. It is known, given the hypothesis that the plant parameters are known, that the states which are not directly measurable can be

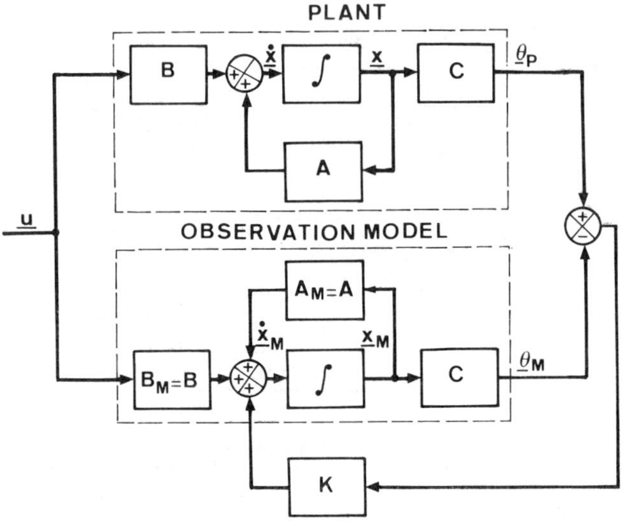

Fig. 1.1-2 Linear state observer.

reconstructed using the *observer* techniques. Assuming that the available measurements (outputs) are related to the plant states through the relation

$$\underline{\theta}_p = C\underline{x} \tag{1.1-4}$$

where $\underline{\theta}_p$ is the output vector (of lower dimension than $\underline{x}$) and C is a nonsquare matrix, the basic scheme for an observer is given by Fig. 1.1-2. In fact, in observer techniques a *model* of the plant is built, and it is excited by the same input as the plant. The difference between the output of the real plant and the output of the observation model is fed through a convenient gain as an additional input to the *observation model* in order to assure that

$$\lim_{t\to\infty}(\underline{x}_M - \underline{x}) = \underline{0} \tag{1.1-5}$$

At this point one can observe the *duality* which exists between the linear model-following control (LMFC) system represented in Fig. 1.1-1 and the linear observer represented in Fig. 1.1-2. *In spite of their different objectives, their structure is similar.* In both cases one can distinguish two subsystems: a real plant and a model created artificially. In both cases, it is desired to have a similar dynamic behavior of the two state vectors. To achieve this, the *difference between the state vectors or the output vectors is used as the main source of information.*

Now we turn to one of the major a priori problems in designing a control system: how to determine the parameter matrices A and B of the controlled plant. This is the well-known problem of *parameter identification.*

Assuming that the structure of A and B is known, the objective of the identification is to find the values of A and B such that a model having these parameters behaves (as nearly as possible) like the real plant for a given set of inputs. One easily recognizes that the reference model concept appears in the identification problem; the plant plays the role of the reference model for the identified model.

From the above considerations, it is clear that even when we want to design an optimal linear control system, the *model reference concept is very useful either for specifying desired performance or for the observation of unaccessible states.* Furthermore, the model reference concept appears basic to system identification.

In fact, as indicated earlier, in many practical situations, large rapid and unpredictable variations of the plant parameters occur. When the mean values of the parameters are known with enough precision and the possible variations fall within a known stochastic model, stochastic control theory can be used to design the control system. Nevertheless, in practice this approach is often difficult to apply because of

A poor knowledge of the mean value of the parameters

Difficulty in finding the stochastic model of the parameter variations and variations of the parameters which are too large, in which case the stochastic approach leads to poor performance

Let us now investigate the problem of parameter variations from the designer's point of view. The objective is to design a control system which maintains its nominal performance despite the variations of the dynamic parameters. To do this, it will be necessary either

To design a feedback control which makes the system performance insensitive to the parameters variations, or

To measure on line (and without delay) the plant parameters and to accordingly modify the parameters of the control law, or

To compare the desired index of performance (corresponding, for example, to the nominal values of the parameters) with the real index of performance and from this information to modify the feedback control law

The first approach has led to a great deal of work in the area called *sensitivity of control systems*. Valuable results can be obtained only if small variations of the plant parameters from their nominal values occur. The second and third approaches lead to the realization of *adaptive* control systems.

Without going into details at this point, one must mention that the second approach has several limitations, one being the fact that a delay exists in determining the plant parameters, which will make the adaptation slow (in most of the cases).

The third approach allows one to easily understand the sense of adaptive control and its particularities versus a conventional feedback control. In fact, both are feedback systems, but the adaptive control system has a supplementary feedback loop acting not upon the state variables but upon an abstract variable which is the *system performance* evaluated by the means of an *index of performance*. The difference between the *desired index of performance* and the *measured index of performance* will act through an adaptation law upon the parameters of the feedback control loop or the input to the controlled plant in order to keep the *real index of performance* close to the *desired index of performance*.

The problem is then to determine the *adaption law* (as the "control law" in feedback control) which allows one to maintain the index of performance close to the desired one in the presence of environment variations.

The behavior of an adaptive control system will depend on the index of performance used to measure the performance of the system.

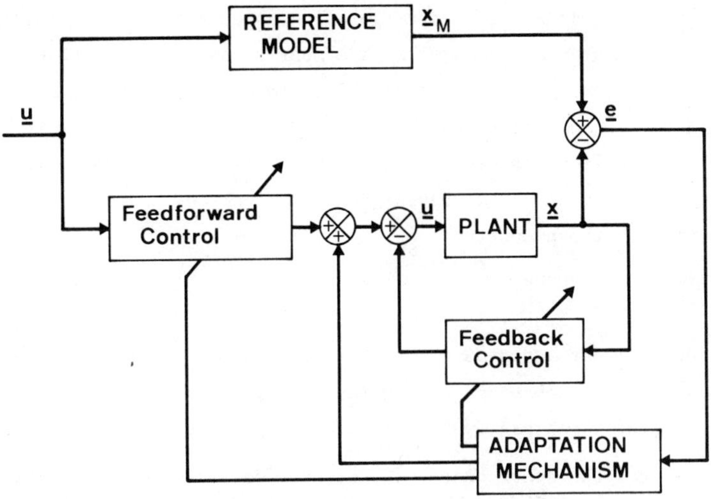

Fig. 1.1-3 Adaptive control system including a reference model (adaptive model-following control system).

For example, if the quadratic integral criterion is used as the index of performance, the adaptation will only be accomplished after some delay ΔT necessary to evaluate the index of performance. At this point, a question arises: Is it possible to instantaneously measure the performance? The affirmative answer to this question is given by the use of an adaptive control system including a reference model.

Such an adaptive system is represented in Fig. 1.1-3. We have previously seen that the differences between the states of the reference model (which is the explicit realization of an optimal system) and those of the plant are a measure at each instant of the difference between the desired performance and the real performances. In this case, the *error vector* $\underline{e} = \underline{x}_M - \underline{x}$ is not only used to eliminate the difference between the state variables of the plant and those of the model, as in the linear case. *It is also used to modify the parameters of the control law, or to generate an auxiliary control signal, when the values of the plant parameters differ from the nominal ones* used in the design of the linear control system.

The need for an adaptive configuration appears also in building observers when the plant parameters are poorly known or are subject to large variations. In this case, the *difference between the output of the plant and that of the observation model is used* not only to construct a supplementary input to the observation model, as in the linear case, but also *for adjusting the parameters of the observation model* such that the property

$$\lim_{t\to\infty}(\underline{x}_M - \underline{x}) = \underline{0} \qquad (1.1\text{-}5)$$

of the ideal linear asymptotic observer is maintained. In this case, one has an *adaptive observer* having the structure of a model reference adaptive system which has two interesting properties:

It assures the convergence of the observed state vector to those of the plant, independently of the initial values of the parameters of the observation model.

It allows one to determine, under certain conditions, the dynamic parameters of the plant. (The parameters of the observation model will converge to the values of the plant parameters under the effect of adaptation.)

This type of adaptive observer is called *identifier and observer with an adjustable model*. One concludes that the model reference adaptive systems represent a natural complement to the optimal linear control systems, with an integral quadratic index of performance allowing one to cope more efficiently with practical problems.

In fact, model reference adaptive systems appeared before the theory of linear optimal control with an integral quadratic index of performance was fully developed.

We believe that model reference adaptive systems have their origin in the transposition in the control field of the adaptive reaction of human behavior and of the concept of *causality models*. The latter seem to be a general mode of reasoning where the adaptive character of human beings clearly appeared.

For example, to describe physical, social, or biological phenomena, one often starts with a phenomenological model structure which characterizes the qualitative cause and effect links. Starting from this structure, one tries to determine the parameters which allow a quantitative definition of the cause and effect relations. One compares these models to the real phenomena and man, who plays the role of the adaptation mechanism, modifies the values of these parameters until a model which corresponds to reality is obtained. This operation (general model validation) is equivalent to *identification with an adjustable model*. When one wishes to utilize the phenomena for which a causality model--more or less exact--has been established, or to act upon it, one defines a desired behavior of the model and determines a control strategy for these phenomena in order to obtain the desired behavior. Fully aware of the imperfection of these models and of the perturbations which can modify the validity of the model considered, one compares the real behavior obtained and that desired. From this observation, one tries to adapt the control strategy in order to obtain a behavior close to the one desired. This kind of operation corresponds to an *adaptive model-following control system*.

1.2 ADAPTIVE CONTROL SYSTEMS. AN OVERVIEW (Some Examples and Definitions)

Adaptive control systems have evolved as an attempt to avoid degradation of the dynamic performance of a control system when environmental variations occur. While a feedback control system is oriented toward the elimination of the effect of state perturbations, the *adaptive control system is oriented toward the elimination of the effect of structural perturbations upon the performances of the control system*. These structural perturbations are essentially caused by the variations of the dynamic parameters of the controlled plant. Let us consider some examples of this situation.

Aeronautical Situation

The dynamic behavior of an aircraft depends on the altitude, speed, and configuration of the craft. The ratio of variation of some parameters lies between 10 to 50 for minimum and maximum values.

Nautical Situation

The dynamic characteristics of an oil tanker, because of its large dimensions, vary drastically from deep water to shallow water where the depth and volume of the ship become comparable to those of the water basin in which it navigates. Its dynamic characteristics vary also with the load.

Electromechanical Systems

The dynamic behavior of a dc motor varies with the moment of inertia and the friction of the load [2]. This is a typical situation occurring in a variety of applications such as rolling mills, wiring machines, machine tools, tracking telescopes, etc. The maximum to minimum ratio of parameter variation ranges from about 3 to 100.

The thyristor bridges which are now used almost exclusively for motor control due to their advantages (energetic efficiency, price, volume) introduce structural perturbations in the control system when the firing angle varies. This affects both the equivalent gain and time constant of the process.

Metallurgical Processes

The parameters of the dynamic model which characterize the various metallurgical processes and which are very useful for their control vary from batch to batch. Further, the working conditions are seldom constant (e.g., the reactor characteristics vary during their life, the material introduced is never exactly the same, the dynamic characteristic depends on the starting conditions, etc.).

After seeing above some of the situations encountered by a control engineer in practice, let us next examine his design problems. From the practical point of view, his objective is to synthesize the simplest control which assures acceptable performance in spite of the variations of the dynamic parameters. If small parameter variations occur, a fixed control law designed using sensitivity considerations will be satisfactory, but when acceptable performance cannot be obtained in this way because of large parameter variations an adaptive approach should be considered. In other words, the choice between a simple control law issued from basic control theory and an adaptive control law depends on the acceptability of the control system performance over a set of given environmental situations. The *acceptable performing system* is then an important concept in adaptive control. It allows one to specify under what conditions the control system operates satisfactorily [3].

We can now introduce the following definition of an adaptive system.[2]

DEFINITION 1.2-1 [4]. An adaptive system measures a certain index of performance (IP) using the inputs, the states, and the outputs of the adjustable system. From the comparison of the measured index of performance and a set of given ones, the adaptation mechanism modifies the parameters of the adjustable system or generates an auxiliary input in order to maintain the index of performance close to the set of given ones (i.e., within the set of acceptable ones).

This definition is illustrated in Fig. 1.2-1.

In the above definition, the *adjustable* system must be understood to be a system capable of adjusting its performance either by modifying its parameters (internal structure) or by modifying its input signals.

The measurement of the *index of performance* can be done in several ways, sometimes directly, sometimes indirectly, as, for example, through the identification of the dynamic parameters of the system.

The *comparison-decision block* makes the comparison between the given set of IP's and the measured IP and decides if the measured IP is within the *acceptable* IP set. If not, the adaptation mechanism will act accordingly in order to modify the system performance either by modifying the parameters of the adjustable system or by changing the system control inputs. In the general case, the comparison-

---

[2]Agreement over definitions for adaptive systems remains a very controversial area. For other definitions, see [3-7].

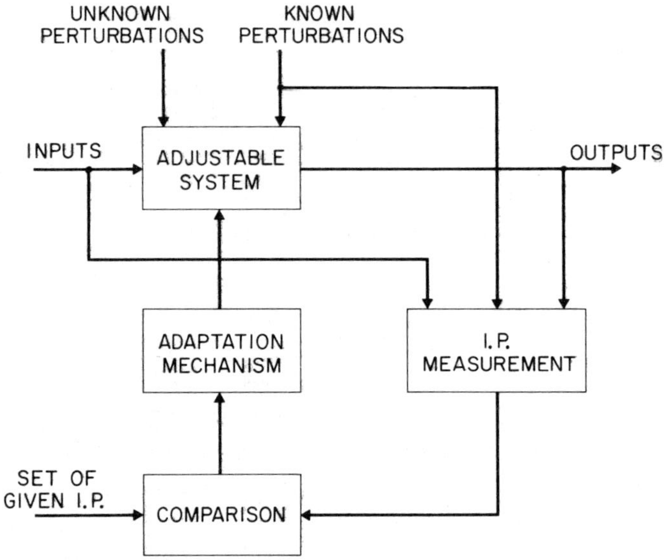

Fig. 1.2-1 Basic configuration of an adaptive system.

decision block makes a class separation as well as an evaluation of the distances from the separation boundary.

One should note that the implementation of the three fundamental blocks (IP measurement, comparison-decision, adaptation mechanism) is very intricate, and in some cases it is not an easy task to decompose the adaptive system according to the block diagram represented in Fig. 1.2-1. Despite this problem, *one of the fundamental characteristics which allows one to decide whether or not a system is truly "adaptive" is the presence or absence of a closed-loop control on the IP*. There are many control systems which are designed to achieve acceptable performance in the presence of parameter variations which are called "adaptive," but they do not assure a closed-loop control of the IP and as such are not truly adaptive.

We shall next give such an example. Consider the *open-loop adaptive system* represented in Fig. 1.2-2. It assumes a rigid relationship between some measurable variables of the environment and

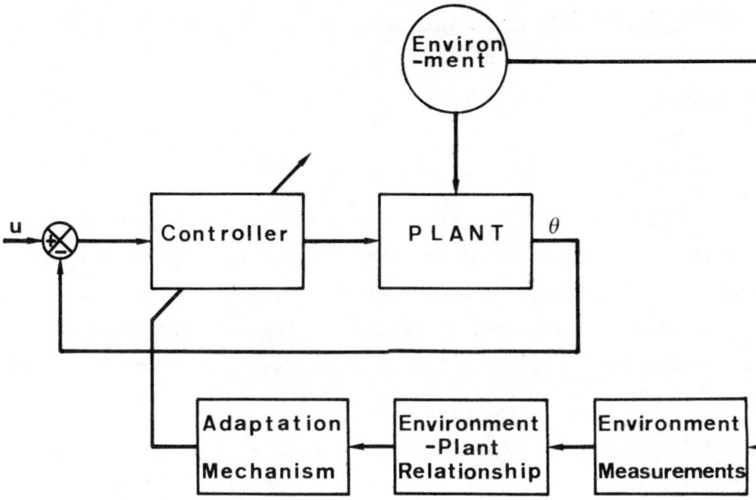

Fig. 1.2-2 Open-loop adaptive system.

the dynamic parameters of the system. Using this relationship, one can choose between the modifications of the system parameters or its inputs. One has here an *open-loop adaptive control*, because the modifications on system performances operated by means of the adaptive mechanism are not checked (measured) and fed back to the comparison-decision block. A typical application of such a principle is the modification of the characteristics of an autopilot of an airplane using speed and altitude measurements for each flight configuration. For the remainder of the book, the term *adaptive system* will mean that such a system fulfills Definition 1.2-1.

There is a variety of criteria which can be used for the classification of adaptive systems. Basically, one can consider the classification with regard to the characteristics of the various components of the adaptive loop and the tactics used in order to achieve the adaptation.

First, one can classify the adaptive systems with regard to the type of index of performance used, which will determine, of course, the characteristics of the IP measurement block. A nonexhaustive classification list is given below.

Index of Performance

Static

Dynamic

Parametric

Functional of state variables and inputs

A static IP is, for example, the efficiency of a combustion engine. A dynamic IP is, for example, the shape of the response of a system to a step input. A (dynamic) parametric IP is, for example, the damping factor of a closed-loop control system. A functional of state variables and inputs is the quadratic integral criterion.

Characteristics of the Comparison-Decision Block

Substractor

Determination of the *min* or *max* of a variable specifying the IP

Belonging to a certain domain of values

The first type is used when the given index of performance is specified uniquely for a certain situation and can be represented by a certain signal. The second type is used in so-called *extremal* adaptive systems where one wishes to maximize or minimize a certain IP (e.g., the maximization of the efficiency of an internal combustion engine) [7]. The third type is used when, for example, one wishes to keep the damping factor of a closed-loop control system between, say, 0.6 and 0.8.

Classification with regard to the adaptation mechanism is done in relation to its action upon the adjustable system. One can distinguish two situations.

Adaptation Mechanism with

Parametric adaptation

Signal synthesis adaptation

For example, in a control loop one can modify either the parameters of the controller or, directly, the control signal applied to the plant by adding a supplementary signal to the output of the controller.

Still another type of classification can be made with regard to the adaptation tactics, i.e., the operational mode of the various components. One can distinguish three basic situations.

Adaptation Tactics

Deterministic

Stochastic

Learning (evolutionary)

The first type does not need any comments. The second one assumes that either the IP measurement block or the comparison-decision block has been designed using stochastic concepts. The learning approach supposes the inclusion of a memory and pattern-recognition features for the various blocks. Such a *learning* system is able to memorize its previous experience and to recognize already-seen situations. This allows the system to improve its adaptive action based on its previous experience. *If a learning system is an adaptive system, an adaptive system does not necessarily feature learning properties.* Learning systems constitute an important and separate class of systems and will not be discussed in this book.

A final classification can be made regarding the condition of operation for the adaptive loop.

Conditions of Operation

With test signals applied (1) to the system inputs or (2) to the adjustable parameters of the systems

Without test signals

The need for test signals is determined by the requirement of assuring a good and rapid measure of the IP. When the signals which already exist are not sufficient to satisfactorily measure the IP, test signals are used (for example, to identify the dynamic characteristics of a plant when the reference input is constant).

We conclude this section by emphasizing the fact that in order for a control system to be adaptive, it must contain, in addition to the classical feedback loop which acts upon the states of the output, a supplementary "feedback" loop acting upon the IP of the control system.

1.3 MODEL REFERENCE ADAPTIVE SYSTEMS

1.3-1 Model Reference Adaptive Systems: A Special Class of Adaptive Systems

Among the various types of adaptive system configurations, *model reference adaptive systems* are important since they lead to relatively easy-to-implement systems with a high speed of adaptation which can be used in a variety of situations. In a model reference adaptive system, the given set of IP's appearing in Fig. 1.2-1 is replaced by one dyanmic-type IP, which becomes the *reference* IP. To generate this reference IP, one uses an auxiliary dynamic system called the *reference model*, which is excited by the same external inputs as the adjustable system. The *reference model specifies in terms of input and model states a given index of performance*. In this case, the comparison between the given IP and the measured IP is obtained directly by comparing the outputs (or the states) of the adjustable system and of the reference model using a typical feedback comparator (substractor). The difference between the outputs of the reference model and those of the adjustable system is used by the adaptation mechanism either to modify the parameters of the adjustable system or to generate an auxiliary input signal in order to minimize the difference between the two IP's expressed as a functional of the difference between the outputs or the states of the adjustable system and those of the model (i.e., in order to maintain the measured IP in the neighborhood of the reference IP, which constitutes a set of acceptable IP's).

The basic scheme of a model reference adaptive system is given in Fig. 1.3-1. The disturbances represented in Fig. 1.3-1 are parameter disturbances modifying either the reference model or the adjustable system. One of the most important advantages of this type of adaptive system is its high speed of adaptation. This is due to the fact that a measure of the difference between the given index of performance specified by the reference model and the index of performance of the adjustable system is obtained directly by the comparison of the states (or outputs) of the model with those of the

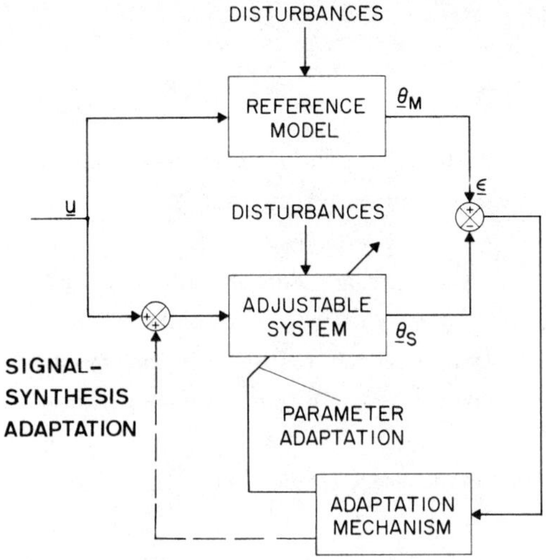

Fig. 1.3-1 Basic configuration of a model reference adaptive system (MRAS).

adjustable system. As a counterpart, a certain a priori knowledge of the structures (sometimes of the model, sometimes of the adjustable system) is necessary for the implementation of this type of adaptive system.

The basic scheme of the model reference adaptive system given in Fig. 1.3-1 is called a *parallel* model reference adaptive system in order to differentiate it from other model reference adaptive system configurations where the relative placement of the reference model and of the adjustable system is not the same.

1.3-2 Some Definitions

Consider the basic scheme of a model reference adaptive system, represented in Fig. 1.3-1, and a state-space representation for the reference model and the adjustable system. The reference model is described by

$$\dot{\underline{x}} = A_M \underline{x} + B_M \underline{u} \qquad (1.3\text{-}1)$$

$$\underline{\theta}_M = C\underline{x} \qquad (1.3\text{-}2)$$

The parallel adjustable system is described by

$$\dot{\underline{y}} = A_S(t)\underline{y} + B_S(t)\underline{u} \tag{1.3-3}$$

$$\underline{\theta}_S = C\underline{y} \tag{1.3-4}$$

where $\underline{x}$ and $\underline{y}$ are state vectors (n-dimensional), $\underline{u}$ is the input vector (m-dimensional), A_M and B_M are constant matrices, $A_s(t)$ and $B_s(t)$ are time-varying matrices (all of appropriate dimension), $\underline{\theta}_M$ and $\underline{\theta}_S$ are output vectors (r-dimensional), and C is the output matrix of appropriate dimension. Note that the same matrix C is considered for both the model and the adjustable system without loss of generality, but it fixes the definition of the state vector of one system relative to that of the other.

We now introduce some important definitions [4].

DEFINITION 1.3-1. State generalized error ($\underline{e}$): the variable vector that represents the difference between the state vector of the model ($\underline{x}$) and the state vector of the adjustable system ($\underline{y}$):

$$\underline{e} = \underline{x} - \underline{y} \tag{1.3-5}$$

DEFINITION 1.3-2. Output generalized error ($\underline{\varepsilon}$): the variable vector that represents the difference between the output of the model ($\underline{\theta}_M$) and the output of the adjustable system ($\underline{\theta}_S$):

$$\underline{\varepsilon} = \underline{\theta}_M - \underline{\theta}_S = C\underline{e} \tag{1.3-6}$$

DEFINITION 1.3-3. State distance: any norm of the difference between the state vector of the reference model ($\underline{x}$) and the state vector of the adjustable system ($\underline{y}$).

DEFINITION 1.3-4. Parametric distance: any norm of the difference between the parameter vector (or matrix) of the reference model and the parameter vector (or matrix) of the adjustable system.

DEFINITION 1.3-5. Adaptation law (algorithm): the relation between the generalized error and the corresponding modifications of the parameters or of the input to the adjustable system.

DEFINITION 1.3-6. Adaptation mechanism: set of interconnected linear, nonlinear, or time-varying blocks used to implement the adaptation law.

DEFINITION 1.3-7. Model reference adaptive system (MRAS): Given an IP specified by the input, the output, and the state of the reference model,

$$\underline{\theta}_M = \underline{f}_M(\underline{u}, P_{M_i}, \underline{x}, t) \tag{1.3-7}$$

where P_{M_i} are parameters of the model; also, given an adjustable system (S),

$$\underline{\theta}_S = \underline{f}_S(\underline{u}, P_{S_i}, \underline{y}\ t) \tag{1.3-8}$$

where P_{S_i} are parameters of the adjustable system, and given an index of performance,

$$(IP)_{RM} = F(\varepsilon, P_{M_i} - P_{S_i}, \underline{e}, t) \tag{1.3-9}$$

which expresses the difference between the given IP specified by the model and that of the adjustable system; the MRAS minimizes the $(IP)_{RM}$ defined above by parameter adaptation or by signal-synthesis adaptation via the adaptation mechanism that has the generalized error as one of its inputs.

1.3-3 Typical Configurations

It was indicated briefly in Sec. 1.1 that in many control problems, especially for multivariable systems, it is difficult to select the weighting matrices of a quadratic performance index in order to specify the desired transient characteristics of the control system. In such cases, one uses a *linear model-following control* (LMFC) system, which is represented in Fig. 1.3-2(a), where the model specifies the desired performance of the control system. The same system is represented in an arrangement slightly different from the usual one in Fig. 1.3-2(b), where the role played by the reference model is more evident [1, 4].

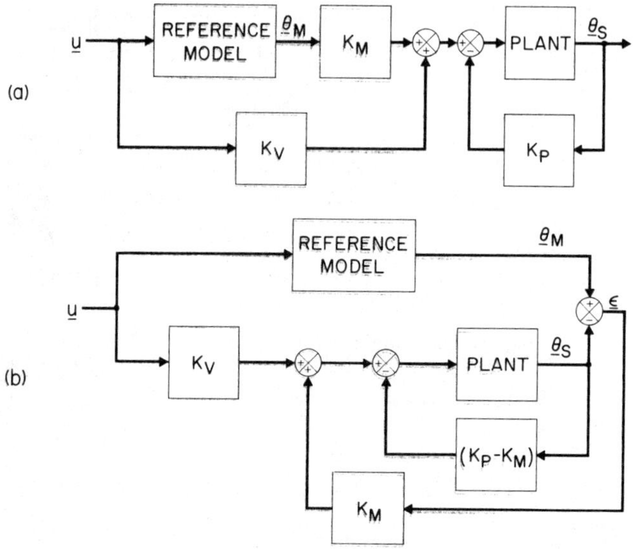

Fig. 1.3-2 Linear model-following control (LMFC) system. (a) Typical representation. (b) Equivalent representation.

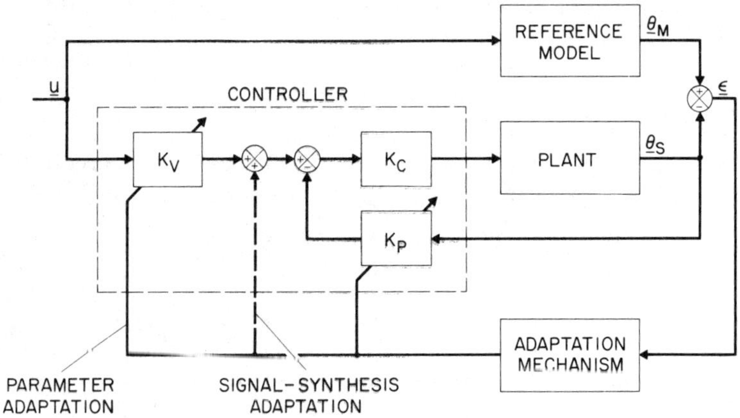

Fig. 1.3-3 Adaptive model-following control (AMFC) system.

For the class of applications in which the parameters of the plant and of the model are well known, a linear approach to the design of the control matrices is possible (such a control assures the minimization of a quadratic index of performance in terms of the difference between the states of the model and those of the plant). When the parameters of the plant are unknown or important variations from their nominal values occur an *adaptive model-following control* (AMFC) system can be implemented (Fig. 1.3-3).

The adaptation mechanism modifies the parameters of the control matrices K_V and K_P or synthesizes an auxiliary input signal to assure "good model following" even in the above-specified situations. This type of scheme is also often called a *model reference adaptive control* (MRAC) system [4].

This type of configuration can be used, for example, in aeronautics, where an in-flight simulator (the adjustable system) is used to simulate the behavior of a future airplane (specified by the reference model). It can also be used to realize an adaptive model-following servomechanism, as used, for example, to build adaptive speed controllers for electrical drives. In the last case, the dc motor is subject to important dynamic variations because of the variations, for example, of the load moment of inertia, and the reference model specifies the desired transient behavior.

Another variant of the AMFC configuration is represented in Fig. 1.3-4. The feedforward and feedback matrices are constant and are designed by supposing that the controlled plant always behaves as the reference model. To obtain such behavior when parameter variations occur, an adaptive loop (signal-synthesis adaptation) is built in order to assure that the controlled plant behaves always as specified by the reference model. This configuration is sometimes used when only part of the controlled plant is subject to parameter variations or when pure integration is inherent (as in a positioning system [Fig. 1.3-4(b)]). Such a system is used for positioning an optical tracking telescope [8], for rudder control for ships [9], and for some adaptive speed control of electrical drives. In the

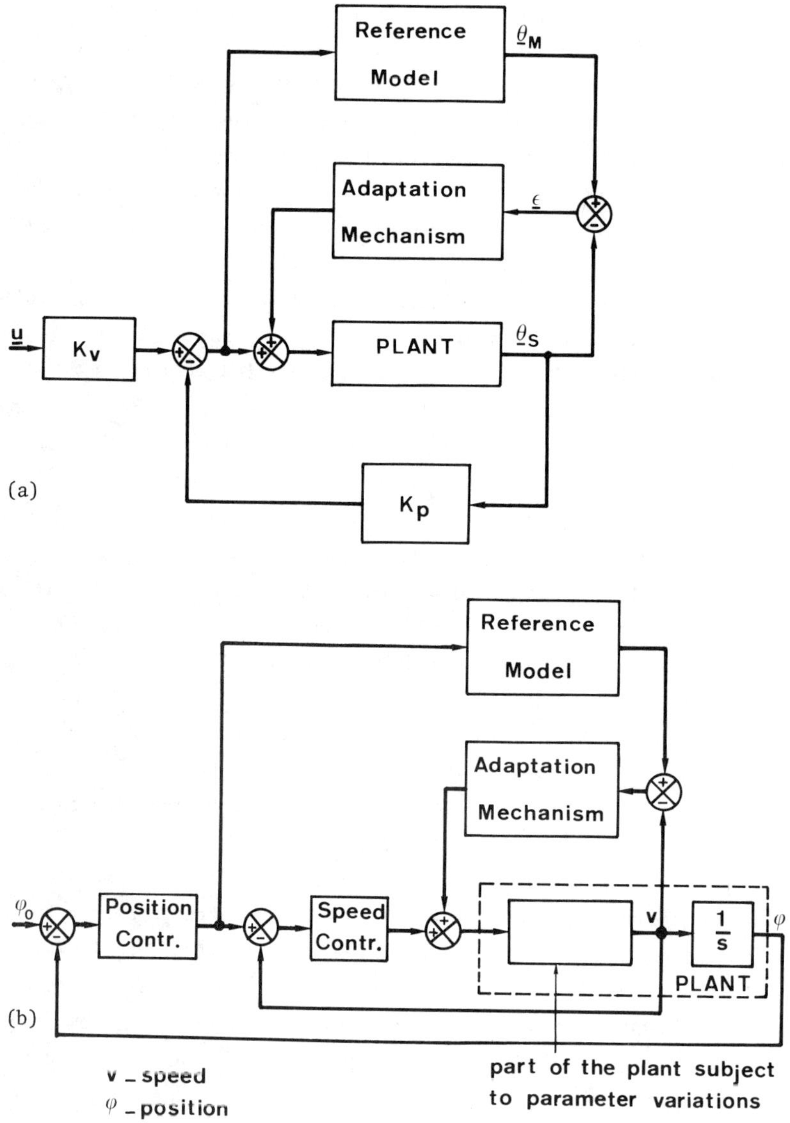

Fig. 1.3-4 Another configuration of an AMFC system. (a) Basic configuration. (b) An example; a position control system.

last case, the controller is a conventional proportional + integral (PI) controller, adjusted for the parameters of the reference model (the adaptive loop keeps the dynamic behavior of the electrical motor close to that of a reference model).

A dual problem to that of the LMFC system is the problem of the reconstruction of the unmeasurable states. Such a state reconstruction (or state observer) is represented in Fig. 1.3-5. For the implementation of the state observer, it is assumed that the plant parameters are known. The feedback matrix K is designed such that in the absence of noise

$$\lim_{t\to\infty}(\underline{x} - \hat{\underline{x}}) = 0 \quad \text{for all } \underline{x}(0),\ \underline{x}(0) \in R^n \tag{1.3-10}$$

When the parameters of the observer differ from those of the plant, the asymptotic convergence of the observed state to the real one

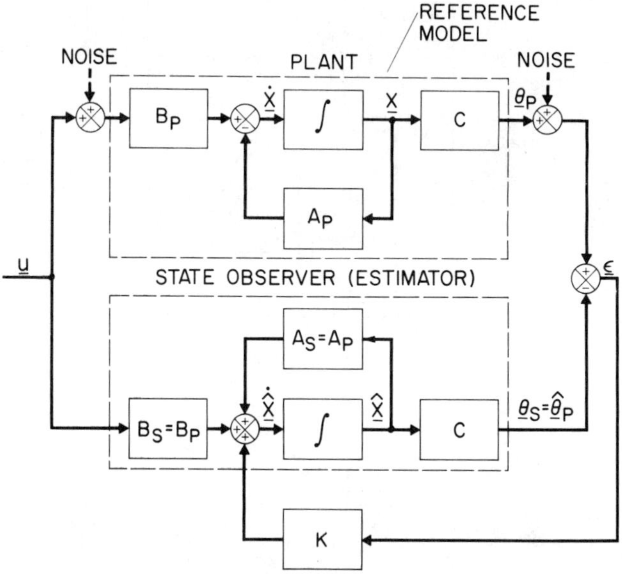

Fig. 1.3-5 Linear state observer.

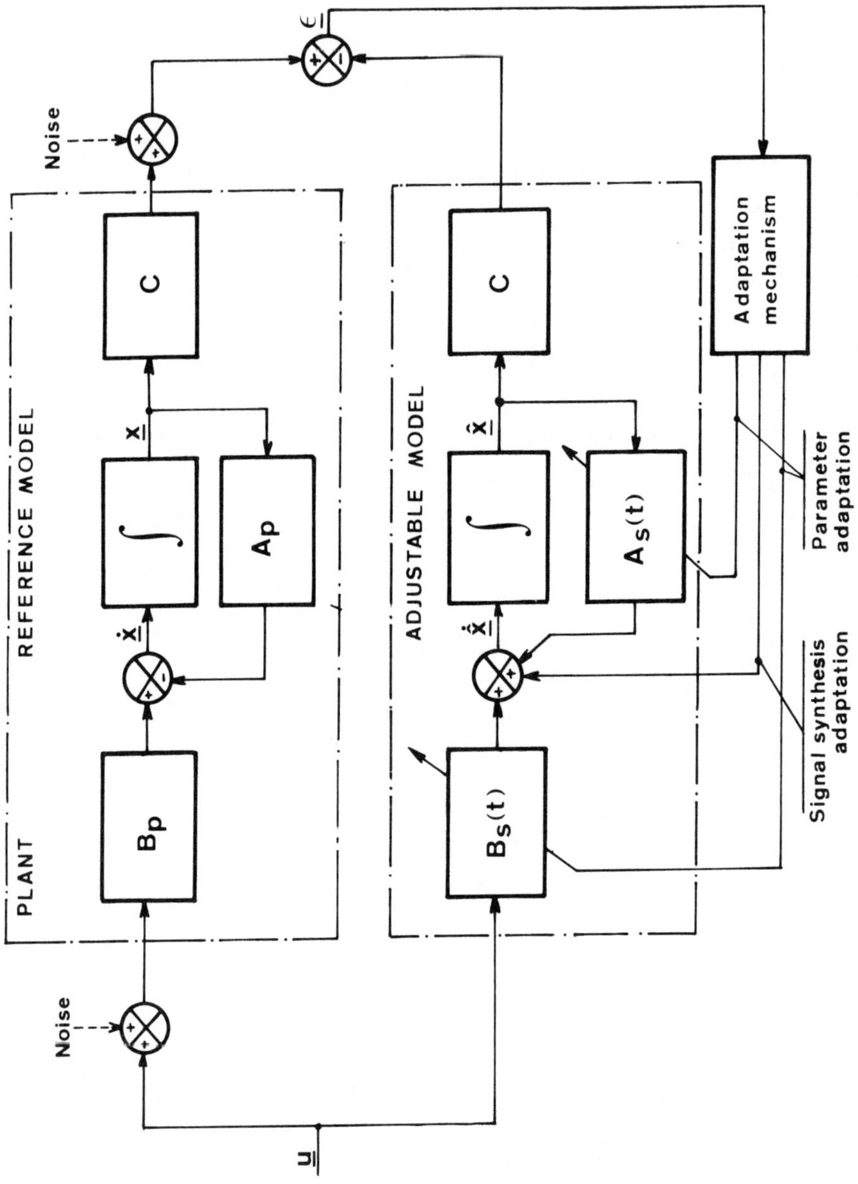

Fig. 1.3-6 Identifier and state observer with an adjustable model.

cannot be obtained. In this case, an adaptive approach must be used. Such an adaptive state observer is represented in Fig. 1.3-6. The observer is now an *adjustable model* having the same mathematical structure as the plant. Under the effect of the adaptation mechanism, the parameters and the states of the adjustable model will track the states and the parameters of the plant. This adaptive state observer performs state observation and parameter identification. In the presence of noise, with a convenient choice of the structure of the adaptation mechanism, an *estimate* of the state of the plant may be obtained. This type of system is called an *identifier and state estimator observer with an adjustable model*, an *identifier with an adjustable model*, a *model reference adaptive identifier and observer*, or an *adaptive state variable observer and identifier*. This type of scheme can be applied for all types of identification problems having a recursive character. As will be shown in Chap. 7, all the recursive approaches to parameter identification lead to model reference adaptive system structures. Some nonlinear identification problems have also been solved successfully using the model reference adaptive systems approach [10].

The configuration represented in Figs. 1.3-3 and 1.3-6 bring up the *dual* character of these two types of configurations; note that one can pass from one configuration to the other by making the following substitutions:

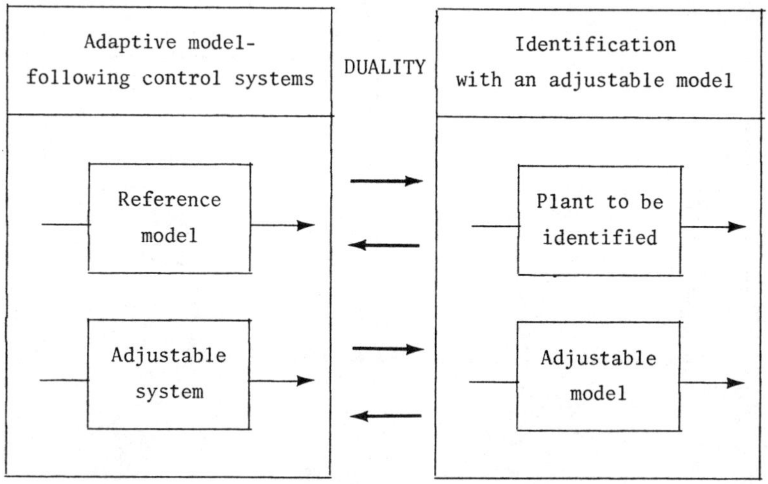

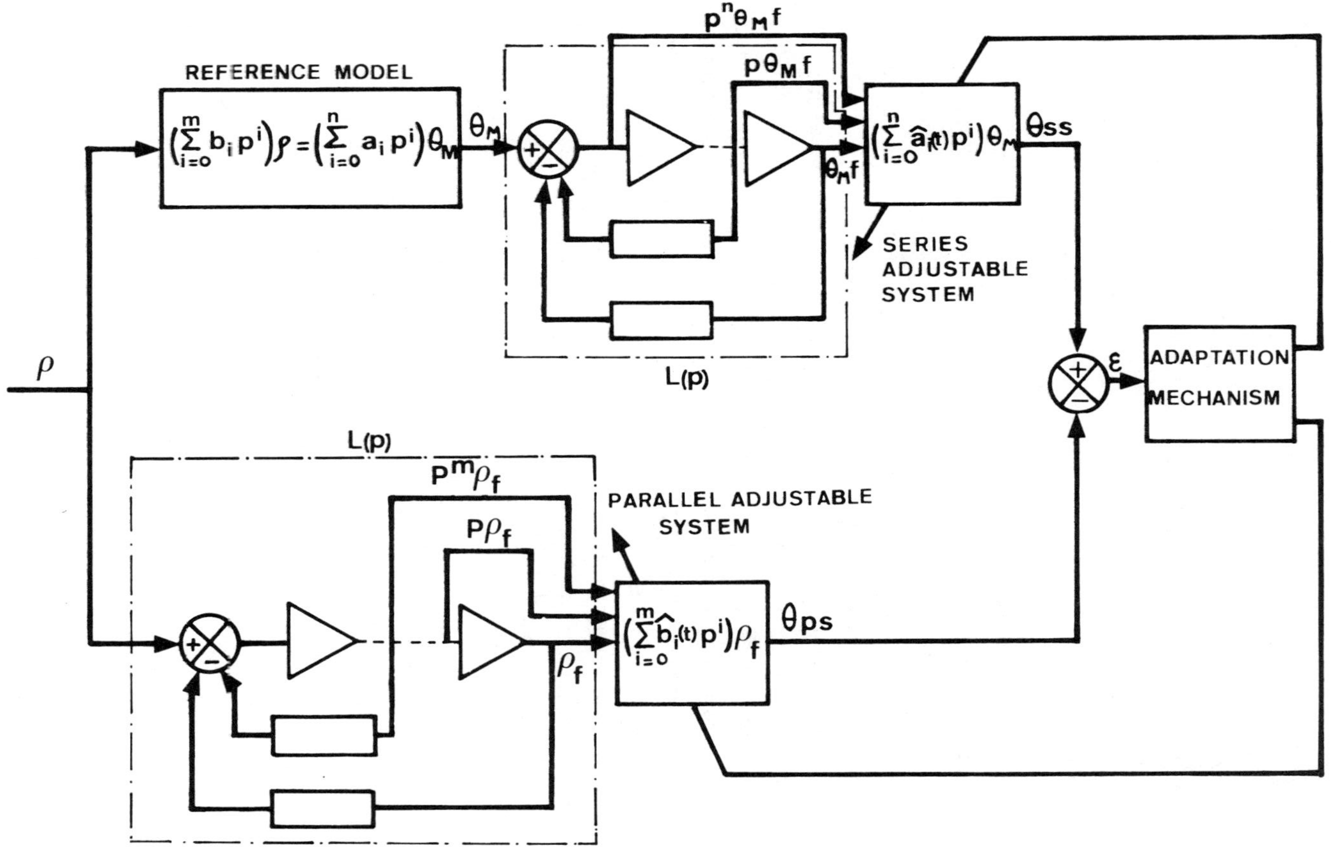

Fig. 1.3-7 Series-parallel MRAS.

Because of the duality existing between AMFC and identification with an adjustable model, the analysis and synthesis of the two types of schemes can be accomplished from the same theoretical bases, with many of the design problems being, to a certain extent, similar. As a consequence, the solutions for one type of MRAS may be extrapolated to the other type of MRAS.

The MRAS schemes presented above are characterized by the fact that the reference model was disposed in *parallel* with the adjustable system. For parameter identification, another typical MRAS configuration characterized by a series-parallel placement of the adjustable model is very popular (Fig. 1.3-7). The low-pass filter L(p), where p = d/dt, is introduced in both paths because it avoids the need to use pure differentiators in the adjustable system. At equilibrium when the parameters of the adjustable system are equal to those of the reference model, the error between the two paths becomes zero. In the field of identification, this configuration, which is a *series-parallel* MRAS, is called the *equation error method* [11, 12] or the *generalized error method* [13], while the previous parallel MRAS (Fig. 1.3-6) is called the *output error method* [4, 13]. The series-parallel MRAS's can also be used in control schemes. In such cases, the reference model is partitioned into two parts: one in series with the adjustable system and one in parallel with the adjustable system.

In Sec. 1.2-1, it was shown that for some types of adaptive systems the identification of dynamic parameters of the plant is used as a measure of the IP of the control system. By using the scheme shown in Fig. 1.3-6, it is possible, for example, to implement adaptive state-variable control or regulation when the parameters of the plant are unknown and the states are inaccessible. Such a scheme is represented in Fig. 1.3-8. This type of system is called an *adaptive state regulator with adjustable plant model* or an *adaptive state-variable controller*. Adaptation is made in this case under the effect of the input which acts on the plant. The quality of the adaptation depends on the nature of this signal [4, 14].

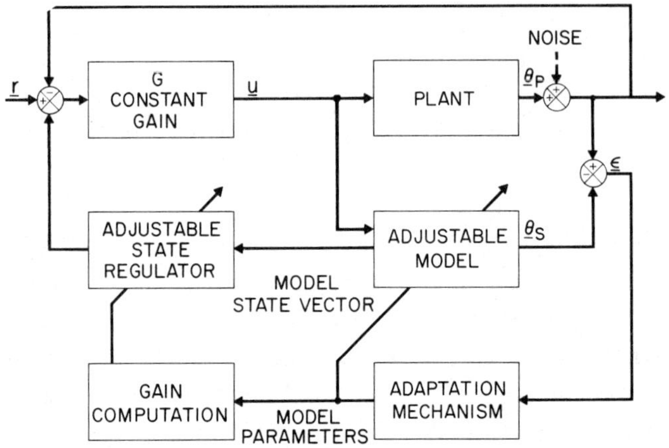

Fig. 1.3-8 Adaptive state regulator with adjustable plant model.

When not enough information exists about the system, in order to obtain a rapid identification sometimes a test signal of small amplitude (preferably random) is superposed on the plant input. A compromise should be found between the identifier performance and the degradation of precision of the control system due to such test signals.

Adaptive state-variable controllers are used when the reference r is almost constant (or slowly varying) and large variations of plant parameters occur. This enables one to assure dynamic regulator performances in the presence of state disturbances in all situations. Adaptive state-variable controllers are also used when the desired dynamic characteristics of the control systems are constrained by the values of the plant parameters.

It is evident that a combination of the adaptive model-following control system and the adaptive state regulator with adjustable plant model is also possible. Such a system (represented in Fig. 1.3-9) may be considered as a *two-level adaptive scheme* in which the adaptive model-following control system superposed on the adaptive loop with state regulator serves to improve the performance of the latter [4].

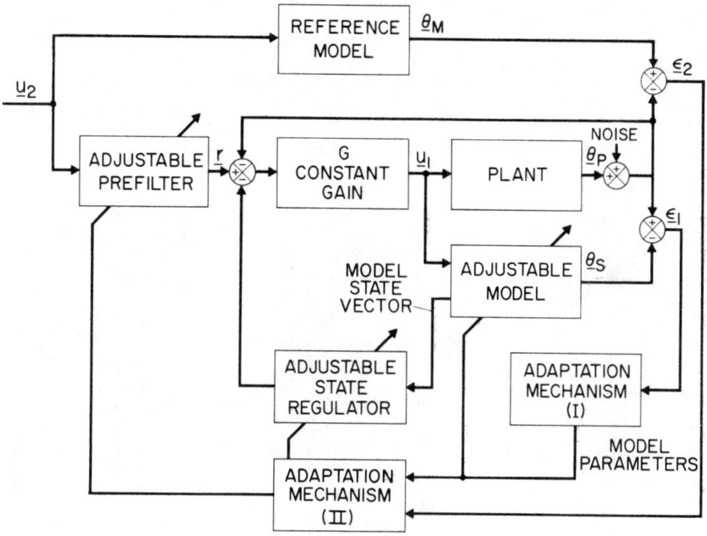

Fig. 1.3-9 Two-level MRAS.

The MRAS's can also be used for synthesizing extremal control systems, as shown in Fig. 1.3-10. An adjustable model is used to identify the parameters of the nonlinear plant characteristics approximated by a parabola and from this information to compute the input which will drive the plant to the extremum [4, 15].

Finally, under the assumption that high speed of adaptation can be assured, the reference model can be a slowly time-varying one whose parameters and inputs are optimized by an optimalizing device (in order to minimize or maximize an index of performance). Such a scheme is represented in Fig. 1.3-11.

Model reference adaptive systems suppose in general the existence of an input signal applied simultaneously to the reference model and the adjustable system. If this input signal is sufficiently rich, both the parametric distance and the state (or output) distance will be minimized. In the absence of input signals only the state distance can, in general, be significantly minimized. Nevertheless, in some cases when the states of both the reference model and the

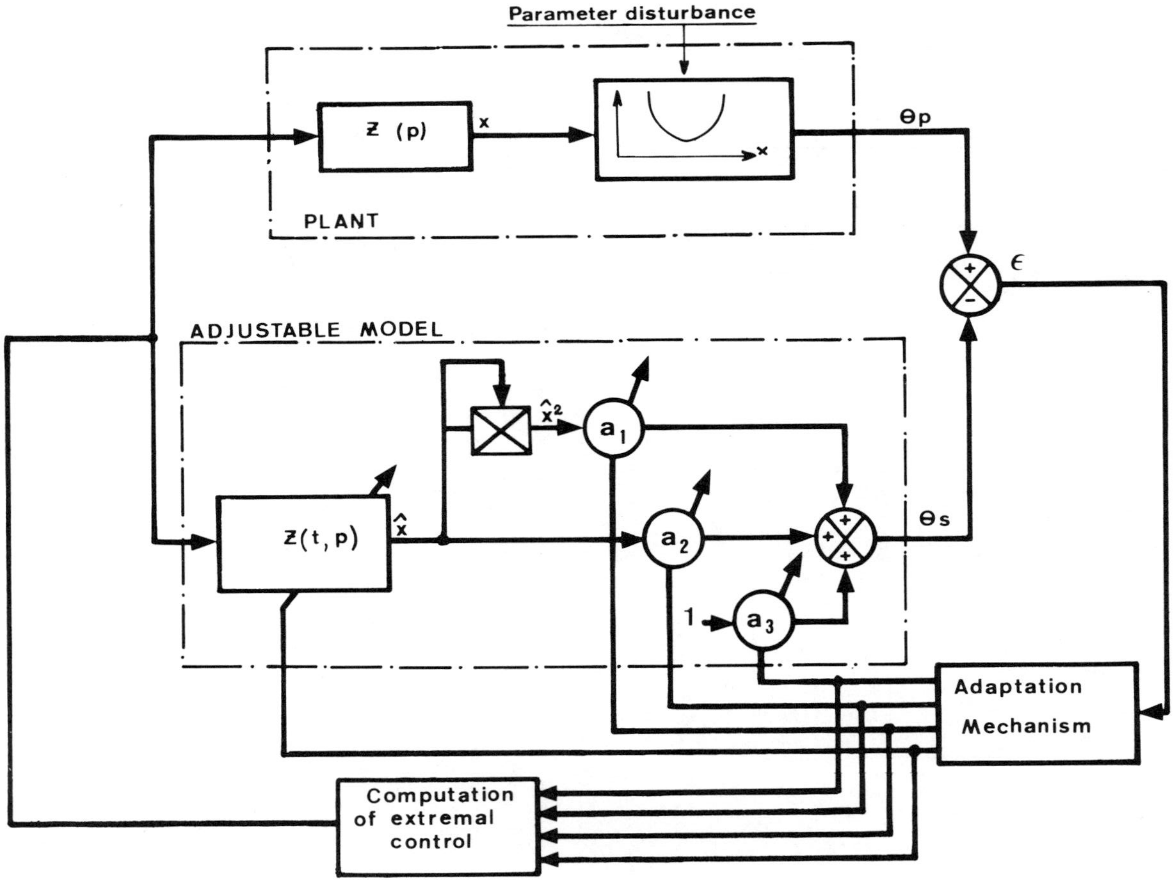

Fig. 1.3-10 Extremal controller using a MRAS.

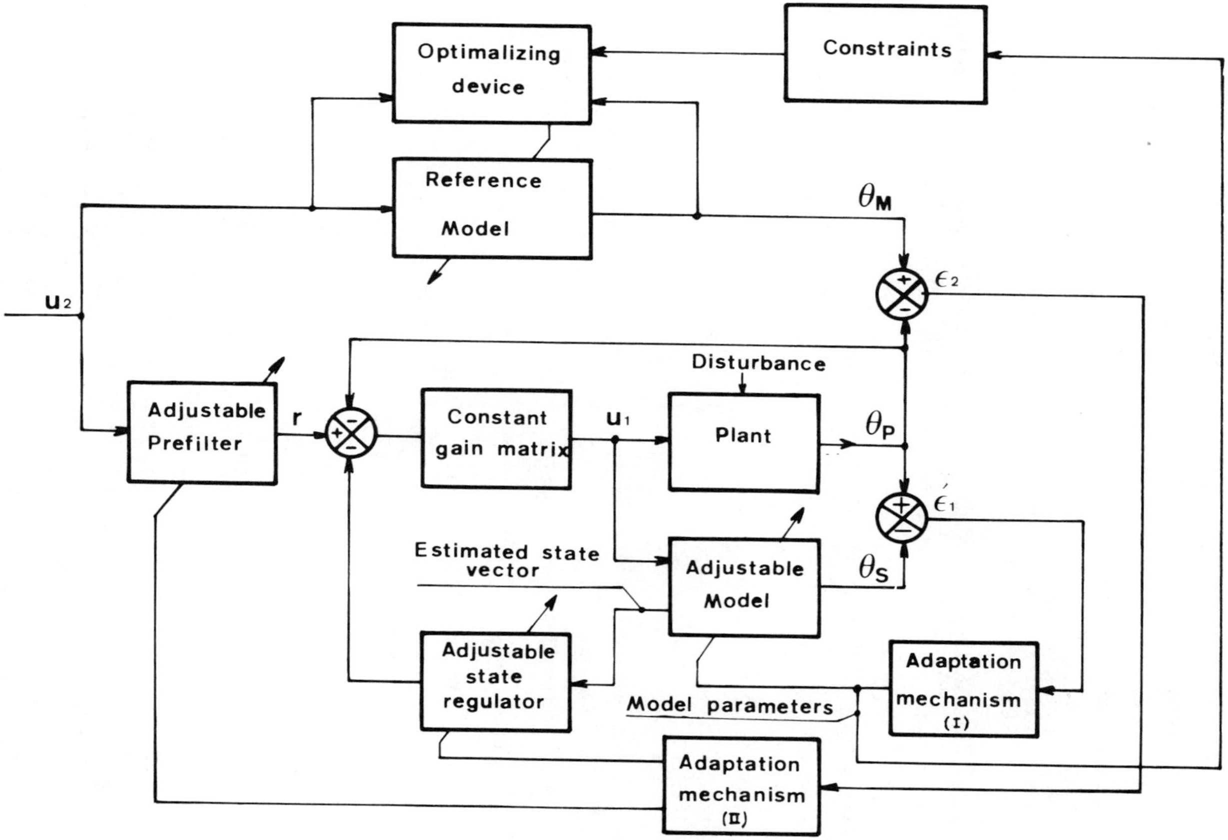

Fig. 1.3-11 Self-optimizing adaptive system using a MRAS.

adjustable system are initialized together almost identically, a minimization of the parametric distance can be obtained.

When the minimization of the parametric distance is required, one is obliged in some cases to introduce *test signals* if the input or the initialization of the states is not enough to ensure good operation of the adaptation mechanism.

1.3-4 Classification of Model Reference Adaptive Systems

There are many types of MRAS's, and for this reason it is impossible to consider only one criterion for the classification of all the typical structures. On the other hand, by considering several criteria for classification and applying these criteria to a given configuration it will be easy to relate the specific configuration to some typical MRAS. In addition to the general classification of adaptive systems given in Sec. 1.2-1, MRAS's also offer the following possibilities for specific classifications [4]:

1. Structure
2. Index of performance
3. Type of application
4. Type of parameter disturbances

Structure

1. Parallel MRAS [Fig. 1.3-12(a)]
2. Series-parallel MRAS [Figs. 1.3-12(b) and (c)]
3. Series MRAS [Fig. 1.3-12(d)]

The most popular structure is the *parallel* configuration (often called the *output error method* when used for identification). The *series-parallel* configuration, often called the *equation error method*, is used in general for identification with an adjustable model. The *series* configuration is also used for identification and is often called the *input error method* [13]. Generally, these three structures are discussed from separate points of view, but a unified approach to their analysis and synthesis is possible [16]. For the implementation

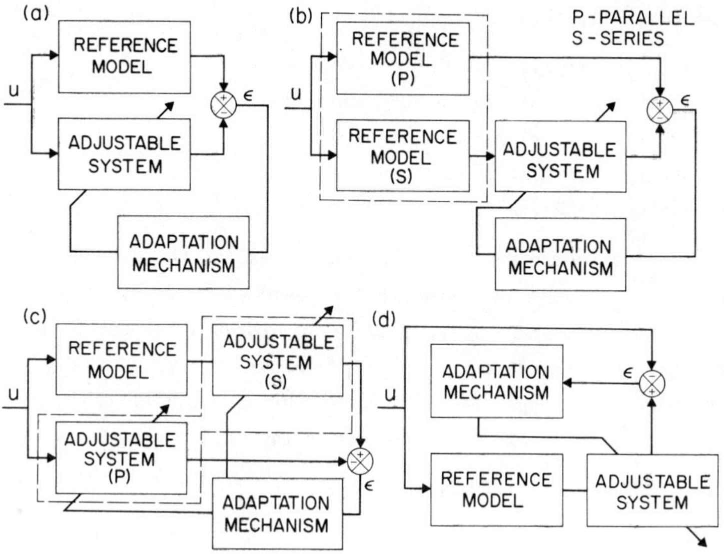

Fig. 1.3-12 Basic structures of MRAS's. (a) Parallel. (b) and (c) Series-parallel. (d) Series.

of series-parallel MRAS's, see Fig. 1.3-7 and the related comments. Similar implementation is used for series MRAS's (for details, see Sec. 2.1). These structures are also used for AMFC.

Index of Performance $(IP)_{RM}$

1. Minimization of a norm of the output generalized error $\underline{\varepsilon}$ and its derivatives
2. Minimization of the state distance
3. Minimization of the parametric distance

The indices of performance specified here must not be confused with the "desired" index of performance specified by the model. The indices of performance specified here serve only to design the adaptation mechanism and express a measure of the difference between the "desired" IP of the reference model and the "real" IP of the adjustable system. The list of indices of performance mentioned above is not exhaustive.

Type of Application

1. Adaptive model-following control system
2. Identification with an adjustable model
3. State observation (estimation)
4. Adaptive regulation
5. Extremal control

The index of performance $(IP)_{RM}$ considered for the design of the adaptation mechanism varies with the field of application. For example, the minimization of the parametric distance is essential for parameter identification, but it is of less importance for adaptive model-following control systems where the minimization of the state (or output) generalized error is the main purpose.

Type of Parameter Disturbances

1. Unknown but constant parameters (either for the reference model or for part of the adjustable system)
2. Frequent nonmeasurable changes of the parameters (either for the reference model or for a part of the adjustable system)

The first case is encountered when MRAS's are used to automatically optimize a control system with an unknown but constant plant parameters[3] or for the identification of time-invariant plants. The second case corresponds to the use of MRAS's for the real-time tracking of parameters of a plant subject to frequent parameter changes or for the control of plants featuring the same characteristics.

The MRAS's may also be classified into continuous, discrete, or hybrid types according to the type of realization of the adjustable system, of the reference model, and of the adaptation mechanism.

---

[3]Control systems with unknown but constant plant parameters which are self-optimizing are also called *self-tuning* control systems [17].

1.3-5 Examples of Application

In this section, we shall briefly present some applications of MRAS techniques to solve specific control problems. The applications reported here have been implemented on real plants. The design aspects for some of these applications will be discussed in detail in the case studies of Chaps. 6 and 7.

Since the application of model reference adaptive systems for the control of real processes is still in its early stages, these presentations cannot fully reflect all the potential of this approach for solving real problems which require adaptive control. They will merely illustrate the various applications (with which we have had experience to date) of the schemes presented in Sec. 1.3-3.

Adaptive Speed Controller for Electrical Drives [18, 19]

The specific problems raised by a high-performance electrical drive in practice are summarized below:

1. The shape of the time response is the main index of performance.
2. It is difficult to determine, in some cases, the values of the dynamic parameters for processes with known structures (for example, the squirrel cage induction motor supplied by variable frequency and voltage).
3. Large variations of plant parameters could occur during normal operation (for example, change of the load moment of inertia, change of the field current, change of the dynamic gain of the thyristor rectifiers, etc.).
4. The necessity to reduce or to eliminate the adjustment of the controllers (dictated by the high cost of the adjustment of the process controller).

To overcome these difficulties in applying optimal quadratic control theory for systems where the shape of the time response is prespecified, one can use a LMFC system of the type represented in Fig. 1.3-2. The model will specify the desired shape of the response for the given class of inputs. It was shown in [20] that for dc and ac drives it is always possible to define a model corresponding to the optimal closed-loop behavior. To avoid the other difficulties related to the uncertainty of the values of the dynamic plant parameters and their variations, an adaptive control system must be implemented.

Two possibilities are therefore to be considered: (1) an adaptive model-following control system [18, 19] and (2) an adaptive state-variable controller [14, 19]. The first approach seems to be the natural one for most of the situations because in the absence of parameter variations an LMFC system can be used.

Adaptive Position Control

In various cases, when an adaptive control must be implemented, one can split the controlled plant into two parts: one which is subject to parameter variations and the other which has almost known and constant dynamic characteristics. Such a situation is encountered, for example, in the positioning of an optical tracking telescope [8] or navigational course control of a ship [9]. In both cases, one has a constant part of the controlled plant which is the integration relating the angular speed to the position (e.g., of the telescope in the vertical plane or the horizontal plane) and a part which relates the input of the actuator to the angular speed, which is subject to large parameter variations. The variations are provoked, in the case of the telescope, by the variations of the moment of inertia in the horizontal plane (the moment of inertia depends on the vertical slope of the telescope) and the variations of the bearing frictions, which depend on the angular speed. In the case of a ship, the dynamic characteristics of the rudder angle/course angular speed transfer function vary with the shipload, speed, and depth of water, and in some cases the ship can become *course unstable* [9].

In such cases, one uses an adaptive model-following control system as in Fig. 1.3-4(b) in which the adaptation signal assures that for the main controller (in this case, the position controller) the variable part of the plant behaves like the model. That is, the

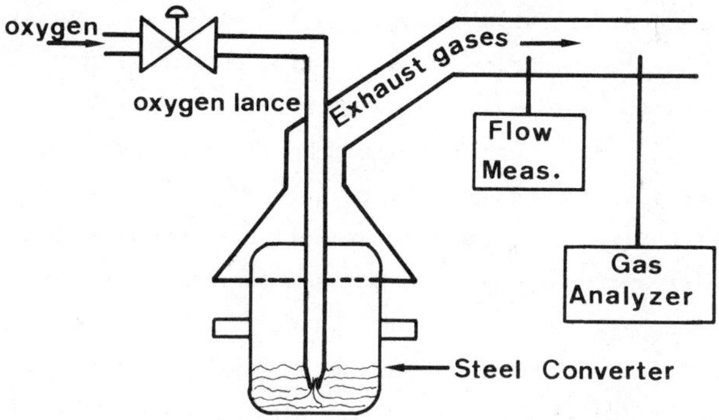

Fig. 1.3-13 Schematic diagram of the steel-making process using a BOF.

controller can be adjusted considering the behavior of the model and the part of the plant with constant and known characteristics.

On-Line Identification and Carbon Percentage Estimation and Prediction in the Steel-Making Process [10]

The basic oxygen furnace (BOF) is actually the method most often used for making steel from molten blast furnace iron. A BOF is represented schematically in Fig. 1.3-13. The reduction of the carbon in the bath is obtained by blowing nearly pure oxygen into the bath through a lance. Carbon reacts with oxygen, producing CO and CO_2. Other impurities, such as phosphorus and silicon, also react with the oxygen. The dynamics of the decarburization process depends on the lance height in the bath and the oxygen flow rate. Actually, the control practice currently includes blowing at the maximum rate at a fixed lance height. The main objective of present controls is to determine the time at which the desired percentage of carbon is attained. For this purpose, metallurgists have established various kinds of models for the last part of the decarburization process, relating the decarburization rate to the carbon content. One successfully tested model is [10].

$$\frac{dC}{dt} = \frac{1}{B + (A'/C^2)} \tag{1.3-11}$$

where dC/dt is the decarburization rate, C is the carbon percentage (content), and B and A' are parameters which are subject to variations from one batch to another (B varies in a range of 1.5/1 and A' in a range of 3/1). dC/dt can be measured (with a delay) using a gas analyzer and flow gas measurements of the exhaust gases which emit from the bath. Of course, the measurements are perturbed by noise. To determine the stopping time, it is necessary first to identify on line the parameters A' and B. Then, given the desired C, one can determine at what value of dC/dt the oxygen blowing must be stopped. But since dC/dt is measured with a delay (15 to 30 sec) and the control of the oxygen also has a delay, one is obliged to have a dynamic model for dC/dt allowing one to predict when dC/dt will reach a desired value. To obtain such a model, one has to take the derivative of Eq. (1.3-11), which becomes

$$\dot{x} = -A[x(1 - Bx)]^{3/2} \tag{1.3-12}$$

where

$$x = \frac{dC}{dt} \quad \text{and} \quad A = \sqrt{\frac{4}{A'}}$$

To identify the nonlinear model of Eq. (1.3-12), an identifier with an adjustable model has been used. Then, once the parameters A and B are identified over a number of measurements, the validity of the model is checked for the next few measurements of dC/dt using the parameters previously obtained. If the results are satisfactory, then the model is run in accelerated time in order to predict at what time dC/dt will reach the value obtained through Eq. (1.3-11) for given C and identified values of A' and B.

Adaptive Voltage Controller for Static dc/ac Converters [19]

The structure of the dc/ac converter considered is represented in Fig. 1.3-14. The system is composed of a thyristorized bridge that synthesizes, from a continuous constant voltage U_0 supplied by the battery, a square waveform having a fundamental component at the desired output frequency (50 Hz for the process being considered).

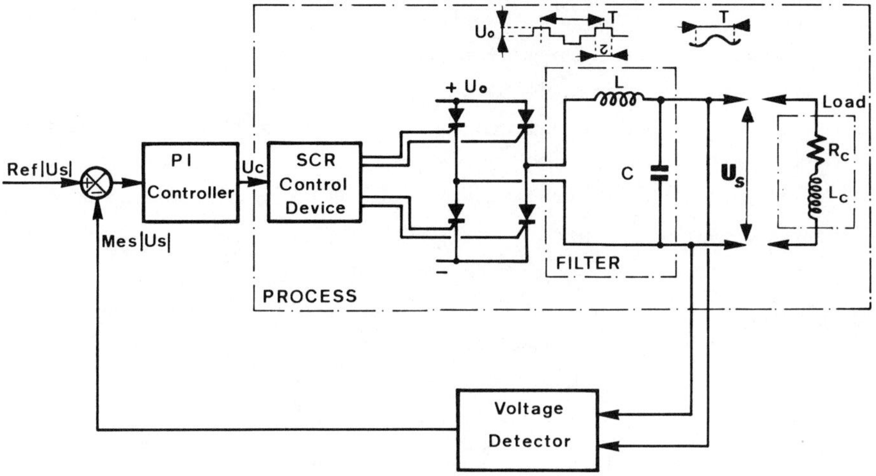

Fig. 1.3-14 Static dc/ac converter.

The SCR control device defines the width τ of the square wave in accordance with its input signal u_C. This control unit is such that the square wave is symmetric with respect to 50-Hz synchronization pulses. A second-order LC filter, which is adjusted in order to be resonant with the fundamental frequency, provides a sinusoidal wave from the square wave. This sinusoidal wave is the output of the converter and supplies a variable load, which is generally an R, L load.

To maintain the output voltage of the converter independent of the load it is necessary to have a control loop. Obviously the dynamic characteristics of the plant are primarily a function of the load impedance, which is rapidly, unpredictably, and widely variable. So, with a normal linear controller, the performance of the control system is often unsatisfactory. One can note that this control problem is not trivial *since it is desired to compensate the influence of process parameter variations on a state variable.*

To obtain good dynamic behavior whatever the working conditions, it is necessary to assure first the stability of the system. So the controller parameters must be adjusted according to the load parameters which should be identified on line. This adaptive device will give a stable system, but it is not clear that it will assure good load perturbation responses as these load perturbations are parameter perturbations. To obtain fast voltage recovery it appeared necessary to adjust the controller parameters and to act simultaneously on the input signal of the plant.

To realize this, a feedforward adaptive control system superposed on the adaptive controller which contains an on-line identification device has been implemented (for details, see Sec. 7.4).

Other applications of MRAS techniques include the following (the list is far from being exhaustive):

- Adaptive model-following control of a heat exchanger [21] and of an electric oven [22]
- Adaptive control of winding machines for the manufacturing of telephone waveguides [2]
- Voltage and frequency control in power systems [23]
- Control of a long-term ventilatory system for lungs [24]
- Dynamic identification and adaptive model-following control of an internal combustion engine [25, 26]
- Adaptive current controller for thyristorized current generators [19]
- Adaptive model-following hydraulic servomechanism [27]
- Recursive identification of a distillation process [28] and of a paper machine [28, 29]

There are also many simulation studies for various applications (many of them concerning aeronautics) which are reported in the literature (for an extensive reference list up to 1974, see Ref. 4).

1.4 CONCLUDING REMARKS

In this chapter, we have reviewed a number of concepts and adaptive structures. We wish to emphasize the following basic ideas:

1. While a *conventional feedback control* is oriented toward the elimination of the effect of state disturbances, an *adaptive control system* is oriented toward the elimination

of the effect of structural disturbances (e.g., parameter variations) upon the performances of the control system.

2. A control system is truly *adaptive* if in addition to a conventional feedback it contains a closed-loop control of its index of performance.
3. The *reference model* concept appears in the theory of linear control systems in two ways: (1) as a means of specifying the desired control performance (linear model-following control) and (2) as the process to be observed in a linear asymptotic observer (the observation model must behave as the process to be observed, which here plays the role of a reference model).
4. Basically, a *model reference adaptive system* is formed (see Fig. 1.3-1) by a *reference model*, which specifies the desired performance; an *adjustable system*, whose performance should be as close as possible to that of the reference model; a substractor, which forms the error between the states or the outputs of the reference model and of the adjustable system (called the *generalized error*); and an *adaptation mechanism*, which processes the generalized error in order to modify accordingly the control or the parameters of the adjustable system.
5. *Model reference adaptive system* structures have a *dual* character. They can be used either for adaptive model-following control or for adaptive state observation and identification. For this reason, the design can be accomplished from the same theoretical bases.
6. There are several structures of model reference adaptive systems, depending on the relative position and structure of the reference model and of the adjustable system as well as on how the generalized error is formed. Three basic structures are relevant: *parallel*, *series-parallel*, and *series* (see Fig. 1.3-12).

REFERENCES

1. J. S. Tyler, The characteristics of model following systems as synthesized by optimal control, IEEE Trans. Autom. Control, *AC-9*, 485-498 (1964).
2. S. Negoesco, B. Courtiol, and C. Françon, Application de la commande adaptive à la fabrication des guides d'ondes hélicoidaux, Automatisme, *23*, 7-14 (January 1978).
3. D. D. Donalson and F. M. Kishi, Review of adaptive control system theories and techniques, in *Modern Control Systems Theory*, Vol. 2 (C. Leondes, ed.), McGraw-Hill, New York, 1965, pp. 228-284.
4. I. D. Landau, A survey of model reference adaptive techniques. Theory and applications, Automatica, *10*, no. 3, 353-379 (July 1974).
5. Ya. Z. Tsypkin, *Adaptation and Learning in Automatic Systems*, Academic Press, New York, 1971.
6. W. D. T. Davies, *System Identification for Self-Adaptive Control*, Wiley-Interscience, New York, 1970.
7. V. W. Eveleigh, *Adaptive Control and Optimization Techniques*, McGraw-Hill, New York, 1967.
8. J. W. Gilbart and G. C. Winston, Adaptive compensation for an optical tracking telescope, Automatica, *10*, 125-131 (1974).
9. J. Van Amerongen and A. J. Udink Ten Cate, Model reference adaptive autopilots for ships, Automatica, *11*, 441-450 (1975).
10. I. D. Landau, L. Muller, G. Dolle, and G. Bianchi, A new method for carbon control in basic oxygen furnace, in *Proceedings of the Second IFAC Symposium on Automation in Mining, Mineral and Metal Processing*, Johannesburg, Sept. 1976.
11. J. Richalet, A. Rault, and R. Pouliquen, *Identification des Processus par la Méthode du Modèle*, Gordon & Breach, New York, 1971.
12. J. M. Mendel, *Discrete Techniques of Parameter Estimation*, Marcel Dekker, New York, 1973.

13. K. J. Åström and P. Eykoff, System identification. A survey, Automatica, *7*, 123-162 (1971).
14. Ed. Sinner, Régulateur adaptatif à variables d'etat pour processus industriels, Rev. RAIRO, *J.1*, 103-123 (1975).
15. I. Dyurki and Y. Kocish, Experimental analog computer investigation of a simple adaptive system, Autom. Remote Control (USSR), *29*, 349-352 (1968)(trans.).
16. I. D. Landau, Design of discrete model reference adaptive systems using the positivity concept, in *Proceedings of the Third IFAC Symposium on Sensitivity, Adaptivity and Optimality*, Ischia, 1973, pp. 307-314.
17. K. J. Åström and B. Wittenmark, On self-tuning regulators, Automatica, *9*, 185-199 (1973).
18. B. Courtiol and I. D. Landau, High speed adaptation system for controlled electrical drives, Automatica, *11*, 119-127 (1975).
19. B. Courtiol, Applying model reference adaptive techniques for the control of electromechanical systems, in *Proceedings of the Sixth IFAC Congress*, Vol. ID, Boston, 1975, pp. 58.2-1-58.2-9.
20. I. D. Landau, A. V. Grossu, and S. Gavat, L'utilisation de la commande adaptative à l'aide d'un modèle dans la régulation des moteurs electriques, Automatisme, *13*, 146-152 (April 1968).
21. G. Bethoux and B. Courtiol, A hyperstable discrete model reference control system, in *Proceedings of the Third IFAC Symposium on Sensitivity Adaptivity and Optimality*, Ischia, 1973, pp. 282-289.
22. R. Benejean, La commande adaptative à modele de référence evolutif (application à la commande sans accés aux variables d'état), Thèse 3me Cycle, Institut National Polytechnique de Grenoble, Grenoble, March 1977.
23. E. Irving, J. P. Barret, C. Charcossey, and J. P. Monville, A method to improve the power network steady state stability and to reduce the unit stress: The multivariable adaptive control of generators, in *Proceedings of the IFAC Symposium on Automatic Control and Protection of Electric Power System*, Melbourne, Feb. 1977.

24. J. Woo and J. Rootengerg, Lyapunov redesign of model reference adaptive control system for long term ventilation of lung, ISA Trans., *14*, no. 1, 89-98 (1975).

25. G. E. Harland, Design of model reference adaptive control for an internal combustion engine, Meas. Control, *6*, 167-173 (April 1973).

26. C. C. Hang, Design studies of model reference adaptive control and identification systems, Ph.D. thesis, University of Warwick, Nov. 1973.

27. B. Porter and L. Tatnall, Performance characteristics of an adaptive hydraulic servo-mechanism, Int. J. Control, *11*, 741-757 (1970).

28. G. Bethoux, Approche unitaire des méthodes d'identification et de commande adaptative des procédés dynamiques, Thèse 3me Cycle, Institut National Polytechnique de Grenoble, Grenoble, July, 1976.

29. S. Gentil, J. P. Sandraz, and C. Foulard, Different methods for dynamic identification of an experimental paper machine, in *Proceedings of the Third IFAC Symposium on Identification and System Parameter Estimation,* Part 1, The Hague, 1973, pp. 473-484.

Chapter Two

THE DESIGN PROBLEM FOR MODEL REFERENCE ADAPTIVE SYSTEMS

The objective of this chapter is to make precise the analytical design problem for model reference adaptive systems as it will be considered throughout the book. However, this supposes the availability of

1. A mathematical description of model reference adaptive systems
2. A set of hypotheses derived from the analysis of the operation conditions

These two aspects will be discussed in Secs. 2.1 and 2.2.

After this, we shall proceed to the examination of the design problem, and we shall emphasize a *two-step* procedure whose first step is directly linked with a *stability* problem.

Then, we shall show that a model reference adaptive system can be equivalently represented (for analytical design purposes) as a nonlinear time-varying feedback system. This will allow us to apply results on the stability of this type of feedback system for the design of the adaptation mechanism.

2.1 MATHEMATICAL DESCRIPTION OF MODEL REFERENCE ADAPTIVE SYSTEMS

2.1-1 *Introduction*

We shall next present the mathematical description of the model reference adaptive systems which will be used throughout the book, using the basic model reference adaptive system configurations of parallel, series-parallel, and series types. Based on the model reference adaptive system duality properties (see Sec. 1.3-3), it will be easy to extend the basic descriptions to specific configurations used for adaptive model-following control or for adaptive state observation and parameter identification.

2.1-2 *Continuous-Time Model Reference Adaptive Systems*

Parallel Model Reference Adaptive Systems

Consider again Fig. 1.3-1 (parallel model reference adaptive system). To describe the reference model, we shall consider the following two formats: (1) state-variable equations and (2) input/output relations in terms of differential operators.

Consider first the state-variable description. One chooses for the *reference model* the linear-type state equation

$$\dot{\underline{x}} = A_M\underline{x} + B_M\underline{u}, \quad \underline{x}(0) = \underline{x}_0 \tag{2.1-1}$$

where $\underline{x}$ is the model state vector (n-dimensional), $\underline{u}$ is the vector input (m-dimensional) belonging to the class of piecewise continuous vector functions, and A_M and B_M are constant matrices [(n × n)- and (n × m)-dimensional, respectively]. The *reference model* is taken to be *stable* and *completely controllable*. The *adjustable system* can either have adjustable parameters or an auxiliary input. In the first case, one uses the following representation:

$$\dot{\underline{y}} = A_S(\underline{e}, t)\underline{y} + B_S(\underline{e}, t)\underline{u} \quad [\underline{y}(0) = \underline{y}_0, A_S(0) = A_{S0}, B_S(0) = B_{S0}] \tag{2.1-2}$$

where $\underline{y}$ is the n-dimensional state vector of the adjustable system and A_S and B_S are time-varying matrices [(n × n)- and (n × m)-dimensional, respectively] whose terms depend (at least) on the *state generalized error vector* $\underline{e}$ through the adaptation law (they can also depend on other variables). In the case of *signal-synthesis adaptation*, one has

$$\dot{\underline{y}} = A_S\underline{y} + B_S\underline{u} + \underline{u}_a(\underline{e}, t) \quad [\underline{y}(0) = \underline{y}_0, \ \underline{u}_a(0) = \underline{u}_{a0}] \tag{2.1-3}$$

where A_S, B_S are now constant matrices and the adaptation signal $\underline{u}_a$ depends (at least) on the state generalized vector $\underline{e}$ through the adaptation law.

The *generalized state error vector*, according to Definition 1.3-1 and Eq. (1.3-5), is given by

$$\underline{e} = \underline{x} - \underline{y} \tag{2.1-4}$$

The design objective in the case of parametric adaptation will be to find an adaptation law such that the parameter matrices $A_S(\underline{e}, t)$ and $B_S(\underline{e}, t)$ are modified so that $\underline{e}$ tends toward 0 for any input $\underline{u}$. Now, if in addition (which is generally the case) we would like the adaptation mechanism to have memory (i.e., it memorizes the good values of the parameters once found), we should consider the inclusion of an integrator in the adaptation mechanism. This will also have the effect that the values of the adjustable parameters at time t will depend not only on $\underline{e}(t)$ but also on its past values: $\underline{e}(\tau)$, $\tau \leq t$. Therefore, the *adaptation law* will be defined in the case of *parametric* adaptation by

$$A_S(\underline{e}, t) = F(\underline{e}, \tau, t) + A_S(0), \quad 0 \leq \tau \leq t \tag{2.1-5}$$

$$B_S(\underline{e}, t) = G(\underline{e}, \tau, t) + B_S(0), \quad 0 \leq \tau \leq t \tag{2.1-6}$$

where F and G denote a functional relation between $A_S(\underline{e}, t)$, $B_S(\underline{e}, t)$, and the values of the vector $\underline{e}$ in the interval $0 \leq \tau \leq t$. In the case of signal-synthesis adaptation, the adaptation law will be defined by

$$\underline{u}_a(\underline{e}, t) = \underline{\mathit{u}}(\underline{e}, \tau, t) + \underline{u}_a(0), \quad 0 \leq \tau \leq t \tag{2.1-7}$$

where $\underline{\mathit{u}}$ also denotes a functional relation between $\underline{u}_a(\underline{e}, t)$ and the values of the vector $\underline{e}$ in the interval $0 \leq \tau \leq t$.

Consider now the description by input/output relations in terms of differential operators. The reference model is described by

$$N(p)\theta_M = M(p)\rho \tag{2.1-8}$$

where $p \triangleq d/dt$,

$$N(p) = \sum_{i=0}^{n} a_i p^i \tag{2.1-9}$$

$$M(p) = \sum_{i=0}^{m} b_i p^i \tag{2.1-10}$$

In Eqs. (2.1-8), (2.1-9), and (2.1-10), ρ is the scalar input, θ_M is the scalar output of the model, and a_i, b_i are the constant coefficients of the differential operators which define the reference model.

The parallel adjustable system is described, in the case of parameter adaptation, by

$$\hat{N}(t, p)\theta_S = \hat{M}(t, p)\rho \tag{2.1-11}$$

where

$$\hat{N}(t, p) = \sum_{i=0}^{n} \hat{a}_i(\varepsilon, t)p^i \tag{2.1-12}$$

$$\hat{M}(t, p) = \sum_{i=0}^{m} b_i(\varepsilon, t)p^i \tag{2.1-13}$$

In Eqs. (2.1-11) through (2.1-13) θ_S is the scalar output of the adjustable system and $\hat{a}_i(\varepsilon, t)$ and $\hat{b}_i(\varepsilon, t)$ are the time-varying coefficients of the differential operators. These coefficients depend on the generalized error ε through the adaptation law. The generalized error is defined as

$$\varepsilon = \theta_M - \theta_S \tag{2.1-14}$$

The *adaptation law* takes the following form:

$$\hat{a}_i(\varepsilon, t) = f_i(\varepsilon, \tau, t) + \hat{a}_i(0), \quad \tau \leq t \tag{2.1-15}$$

$$\hat{b}_i(\varepsilon, t) = g_i(\varepsilon, \tau, t) + \hat{b}_i(0), \quad \tau \leq t \tag{2.1-16}$$

In the case of a signal-synthesis adaptation, the adjustable system is described by

$$\hat{N}(p)\theta_S = \hat{M}(p)[\rho + \mu(\varepsilon, t)] \tag{2.1-17}$$

where

$$\hat{N}(p) = \sum_{i=0}^{n} \hat{a}_i p^i \tag{2.1-18}$$

$$\hat{M}(p) = \sum_{i=0}^{m} \hat{b}_i p^i \tag{2.1-19}$$

$$\mu(\varepsilon, t) = u(\varepsilon, \tau, t) + \mu(0), \quad \tau \leq t \tag{2.1-20}$$

Series-Parallel Model Reference Adaptive Systems

Consider now the series-parallel model reference adaptive systems as presented in Figs. 1.3-7 and 1.3-12(c) in which the adjustable system has two parts: one in series and one in parallel with the reference model. In a state-variable description, the reference model is described by Eq. (2.1-1), and the adjustable system is given by

$$\dot{\underline{y}} = A_S(\underline{e}, t)\underline{x} + B_S(\underline{e}, t)\underline{u} \quad [\underline{y}(0) = \underline{x}_0,\ A_S(0) = A_{S0},\ B_S(0) = B_{S0}] \tag{2.1-21}$$

where the state generalized error vector $\underline{e}$ is given by Eq. (2.1-4). The parallel part of the adjustable system is given by the term $B_S(\underline{e}, t)\underline{u}$ and the series part by $A_S(\underline{e}, t)\underline{x}$ in Eq. (2.1-21).

In the case of a series-parallel reference model [Fig. 1.3-12(b)], one has an adjustable system described by Eq. (2.1-2), and the reference model equation is

$$\dot{\underline{x}} = A_M\underline{y} + B_M\underline{u}, \quad \underline{x}(0) = \underline{y}_0 \tag{2.1-22}$$

In terms of differential operators, one has two types of description of the series-parallel adjustable model. For the first case,

$$\varepsilon = \theta_{Ss} - \theta_{Sp} \tag{2.1-23}$$

where

$$\theta_{Ss} = \left[\sum_{i=0}^{n} \hat{a}_i(\varepsilon, t)p^i\right]\theta_M \tag{2.1-24}$$

and

$$\theta_{Sp} = \sum_{i=0}^{m} \hat{b}_i(\varepsilon, t)p^i\rho \qquad (2.1\text{-}25)$$

In the second case, one has

$$\varepsilon = \theta_M - \theta_S \qquad (2.1\text{-}26)$$

$$\theta_S = -\sum_{i=1}^{n} \hat{a}_i(\varepsilon, t)p^i\theta_M + \sum_{i=0}^{m} \hat{b}_i(\varepsilon, t)p^i\rho \qquad (2.1\text{-}27)$$

The two representations are equivalent if $a_0 = \hat{a}_0 = 1$.

As one can see, the implementation of such a series-parallel model reference adaptive system requires pure derivative operations acting on θ_M and ρ. To avoid this, asymptotically stable, low-pass filters called *state-variable filters* (SVF) are introduced in both paths. The low-pass filter has (n) integrators, allowing one to obtain filtered derivatives up to the order n. (See Fig. 1.3-7.)

Series Model Reference Adaptive Systems

The realization of series model reference adaptive systems is conditioned by the invertibility properties of the reference model. This kind of problem is relatively simple in the case of single-input/single-output (SISO) systems but much more complicated in the case of multivariable systems. Silverman [1] and Singh and Liu [2], among others, have given existence conditions and constructed algorithms for inverse systems. In the following, we shall restrict ourselves to the case of SISO systems. In this case, the reference model is described by Eq. (2.1-8), the series adjustable system by

$$\sum_{i=0}^{m} \hat{b}_i(\varepsilon, t)p^i\theta_S = \sum_{i=0}^{n} \hat{a}_i(\varepsilon, t)p^i\theta_M \qquad (2.1\text{-}28)$$

and the generalized error by

$$\varepsilon = \theta_S - \rho \qquad (2.1\text{-}29)$$

To avoid the necessity of pure derivatives in the implementation of this type of model reference adaptive system, state-variable filters are introduced in both paths similar to the series-parallel case, as shown in Fig. 2.1-1. Equations (2.1-28) and (2.1-29) remain of

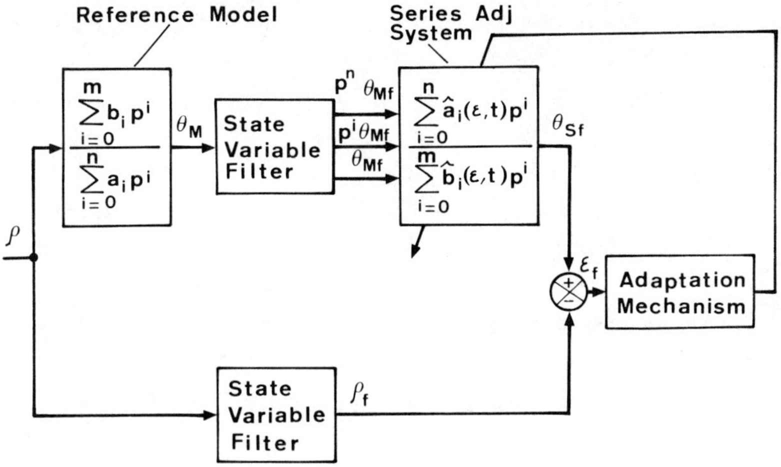

Fig. 2.1-1 SISO series model reference adaptive system.

the same form, but θ_M and ρ are replaced by their filtered values θ_{Mf} and ρ_f, respectively, while the new output of the series adjustable system is θ_{Sf}.

2.2 DESIGN HYPOTHESES

The various representations of model reference adaptive systems given in Sec. 2.1 correspond to the following basic hypotheses:

1. The reference model is a time-invariant linear system.
2. The reference model and the adjustable system are of the same dimension.
3. All the parameters of the adjustable system are accessible for adaptation (in the case of parameter adaptation).
4. During the adaptation process the parameters of the adjustable system depend only on the adaptation mechanism.
5. Except for the input vector $\underline{u}$, there are no other external signals acting on the system (or on part of it).
6. The initial difference between the parameters of the model and those of the adjustable system is unknown.

7. The state generalized error vector and the output generalized error vector are measurable.

This set of hypotheses is called the *ideal* [2] or *basic* case because it allows a straightforward analytical treatment of model reference adaptive system design.

Many real situations closely fit the basic case. Furthermore, the results obtained for this case can be extended to treat the cases where some of the basic hypotheses are violated. The various situations which lead to violation of the basic hypotheses are summarized below:

1. The reference model is a nonlinear time-varying system.
2. The adjustable system also contains nonlinearities.
3. The reference model and the adjustable system do not have the same dimension.
4. Not all of the parameters of the adjustable system are accessible for adaptation.
5. During the adaptation process, the parameters of the adjustable system do not depend only on the adaptation mechanism; i.e., they are subject to external parametric perturbations.
6. Perturbations are applied to various other parts of the system.
7. The measurement of the state generalized error vector or output generalized error vector is obscured by noise.

This set of hypotheses is called the *real* case [3] or the *general* case. The design problems related to the general case are considerably more difficult and, despite some valuable recent results, represent current areas of research.

The basic hypotheses are the same for *adaptive model-following control systems* and for *identifiers with an adjustable model*. One can make the observation that in the first case the parameters of the reference model are known and those of the adjustable system are unknown, while in the second case the situation is reversed but the basic hypothesis 6 is still in force. We also note that the structure of the adjustable system is given, which means that in the case of identification the problem of the process characterization (the determination of the model structure) has already been solved.

In contrast to the *basic* case, in many situations belonging to the *general* case there are noticeable differences between adaptive model-following control and identification with an adjustable model. For example, for adaptive model-following control one often has reference models of lower dimension than the adjustable system, while for the identification with an adjustable model one has adjustable models of lower dimension than the plant. Hypotheses 4 and 5 of the general case apply (practically) only to adaptive model-following control systems. For the most part hypothesis 6 of the general case applies to the adjustable system in the case of adaptive model-following control and to the reference model in the case of identification with an adjustable model.

2.3 DESIGN PROBLEM FORMULATION

The main problem in the synthesis of model reference adaptive systems is the design of the adaptation mechanism for a given configuration of the reference model and of the adjustable system. Consider first the set of basic hypotheses. One can seek an adaptation mechanism which realizes "perfect adaptation" in a finite time for any initial difference between the parameters of the reference model and those of the adjustable system and any initial condition of the model and the adjustable system state vectors. One can also seek to either minimize the adaptation time or, given a desired adaptation time, to minimize a quadratic index of performance.

Now from the engineering point of view it is desirable to design a system which assures the best compromise vis-à-vis performance versus complexity. For example, the necessity of solving, in real time, complicated equations which give the adaptation law should be avoided if possible. To simplify the adaptive system it is very interesting to look for *explicit adaptation laws which do not require real-time solutions of linear or nonlinear differential equations.*

One already knows that the optimal terminal control or minimal time control of linear time-invariant systems requires the knowledge of the initial and final conditions and the real-time resolution of complex equations. It will be a more complicated task to solve such problems in the case of model reference adaptive systems which are nonlinear and time-varying (some of the parameters being functions of the state measurements through the generalized error) and where some initial conditions are unknown.

It seems, then, more natural to look at the design problem of model reference adaptive systems as a "regulator" problem over an infinite time range for the system displaced from its equilibrium (defined by $\underline{x} = \underline{y}$: $A_M = A_S$, $B_M = B_S$). Therefore the design problem can be specified roughly as follows.

Given an unknown initial difference at $t = t_0$ between the parameters of the reference model and those of the adjustable system $[A_M - A_S(t_0), B_M - B_S(t_0)]$ and a known initial state generalized error vector $\underline{e}(t_0) = \underline{x}(t_0) - \underline{y}(t_0)$, find an adaptation law independent of the initial conditions which assures a *perfect asymptotic adaptation* characterized by

$$\lim_{t\to\infty}[\underline{x}(t) - \underline{y}(t)] = \lim_{t\to\infty} \underline{e}(t) = \underline{0} \tag{2.3-1}$$

$$\lim_{t\to\infty} A_S(t) = A_M \tag{2.3-2}$$

$$\lim_{t\to\infty} B_S(t) = B_M \tag{2.3-3}$$

while eventually minimizing a quadratic index of performance given by

$$(IP)_{RM} = \int_{t_0}^{\infty} \{(\underline{x} - \underline{y})^T P(\underline{x} - \underline{y}) + tr[(A_M - A_S)^T Q(A_M - A_S)] + tr[(B_M - B_S)^T \tilde{Q}(B_M - B_S)]\} \, dt \tag{2.3-4}$$

for all inputs $\underline{u}$ belonging to the class of piecewise continuous vector functions. Note that *perfect asymptotic adaptation* implies that $(IP)_{RM} < \infty$, which is a necessary condition for Eq. (2.3-4) to have a minimum.

To directly solve this problem for the model reference adaptive systems is a complicated task because of the time-varying nonlinear character of the system. A *structural* procedure for designing model reference adaptive systems has evolved from the research concerned with this problem. It is a *two-step* design procedure which consists of (1) finding the structures which assure a *perfect asymptotic adaptation* and then (2) selecting the structures and the parameters of the structure in order to minimize Eq. (2.3-4) or to comply with some other index of performance.

The problem of perfect asymptotic adaptation can be interpreted as a *stability* problem. Defining $\underline{e} = \underline{x} - \underline{y}$ as the new state vector of the system, one has a given $\underline{e}(0)$, and the system must be *globally asymptotically stable* in order to have $\lim_{t\to\infty} \underline{e}(t) = \underline{0}$. One must subsequently specify under what conditions we shall also obtain

$$\lim_{t\to\infty} A_S(\underline{e}, t) = A_M, \quad \lim_{t\to\infty} B_S(\underline{e}, t) = B_M$$

The stability approach to model reference adaptive system design therefore appears to be a useful tool. The objective is *to determine the largest family of adaptation laws assuring the overall stability of the adaptive systems among which we shall look for the best one according to a specific* $(IP)_{RM}$.

Various $(IP)_{RM}$'s have been specified in Sec. 1.3-4. Others are concerned, for instance, with the qualitative features of the model reference adaptive systems considered in a specific environment, for example, the necessity of obtaining unbiased parameter estimation in the identification with an adjustable model when the plant output measurements are obscured by noise or to assure a good adaptation when frequent change in parameters occurs (either in the parameters of the model or in the initial values of the parameters of the adjustable system). The choice of weighting matrices in Eq. (2.3-4) will also depend on the type of application. For adaptive model-following control, the main variable is $\underline{e}$; therefore, $(\underline{x} - \underline{y})^T P(\underline{x} - \underline{y})$ needs to be heavily weighted. For identification with an adjustable model the second and third term in Eq. (2.3-4) must be heavily weighted.

2.4 EQUIVALENT REPRESENTATION OF MODEL REFERENCE ADAPTIVE SYSTEMS AS NONLINEAR TIME-VARYING FEEDBACK SYSTEMS

It was previously specified that $\underline{e} = \underline{x} - \underline{y}$ is the main source of information we have upon which differences existing between the adjustable system and the reference model can be based. Thus, we are interested in obtaining a differential equation which characterizes the dynamics of $\underline{e}$. For the case of parallel model reference adaptive systems with accessible state vectors described by Eqs. (2.1-1), (2.1-2), (2.1-4), (2.1-5), and (2.1-6), one obtains, by subtracting Eq. (2.1-2) from Eq. (2.1-1),

$$\dot{\underline{e}} = \dot{\underline{x}} - \dot{\underline{y}} = A_M\underline{x} + [B_M - B_S(\underline{e}, t)]\underline{u} - A_S(\underline{e}, t)\underline{y} \tag{2.4-1}$$

By adding and then subtracting the term $A_M\underline{y}$ on the right-hand side of Eq. (2.4-1) and also using Eqs. (2.1-4), (2.1-5), and (2.1-6), one obtains

$$\dot{\underline{e}} = A_M\underline{e} + [A_M - A_S(0) - F(\underline{e}, \tau, t)]\underline{y} + [B_M - B_S(0) - G(\underline{e}, \tau, t)]\underline{u} \tag{2.4-2}$$

One thus has only one equation for describing the model reference adaptive system, which leads to the equivalent representation of model reference adaptive systems given in Fig. 2.4-1. One can see that the equivalent system has the form of a nonlinear time-varying feedback system.

It was mentioned in the preceding section that perfect asymptotic adaptation implies first that $\lim_{t\to\infty} \underline{e}(t) = \underline{0}$ for any initial conditions. So the design problem becomes the following: Determine $F(\underline{e}, \tau, t)$, $G(\underline{e}, \tau, t)$ *such that the equivalent feedback system defined by* Eq. (2.4-2) will be *globally asymptotically stable for any* $\underline{e}(0)$, $[A_M - A_S(0)]$, $[B_M - B_S(0)]$. Now, classical results on the stability of nonlinear time-varying feedback systems which can be decomposed into a linear time-invariant part and a nonlinear time-varying part already state that the *stability of the feedback system is determined only by the characteristics of the linear part if the nonlinear time-varying part satisfies certain conditions* [4-6].

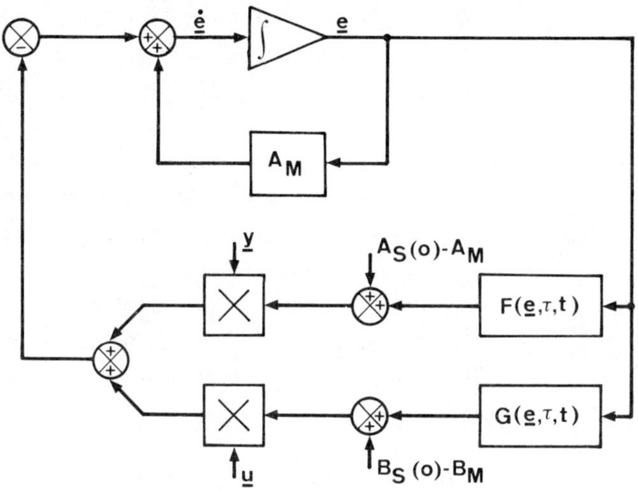

Fig. 2.4-1 Equivalent feedback representation of the generalized state error system equation (2.4-2).

Because A_M is prespecified, it seems reasonable to use for adaptation not the generalized vector $\underline{e}$ directly but another vector $\underline{v}$ obtained from $\underline{e}$ by processing it through a matrix gain (called the *linear compensator*):

$$\underline{v} = D\underline{e} \tag{2.4-3}$$

The matrix D is especially chosen in order to be able to meet the specifications required for the linear part so that the stability of the system is assured. The linear compensator [Eq. (2.4-3)] is therefore the first component of the adaptation mechanism.

If $\underline{u} \not\equiv 0$ and $\underline{y} \not\equiv 0$ for all t, the condition $\lim_{t\to\infty} \underline{e}(t) = \underline{0}$ implies that the adaptation mechanism has memory, which means that once $F(\underline{e}, \tau, t)$ and $G(\underline{e}, \tau, t)$ are found, these values must be conserved when $\underline{e}$ becomes null in order to assure that $A_S(\underline{e}, t) = A_M$, $B_S(\underline{e}, t) = B_M$. This memory can be obtained by using an integrator in the adaptation mechanism. We can therefore write the adaptation law for A_S and B_S in more detail:

$$A_S(\underline{e}, t) = A_S(\underline{v}, t) = \int_0^t \Phi_1(\underline{v}, t, \tau)\, d\tau + \Phi_2(\underline{v}, t) + A_S(0) \tag{2.4-4}$$

$$B_S(\underline{e}, t) = B_S(\underline{v}, t) = \int_0^t \Psi_1(\underline{v}, t, \tau)\, d\tau + \Psi_2(\underline{v}, t) + B_S(0) \tag{2.4-5}$$

where the first terms in Eqs. (2.4-4) and (2.4-5) assure the memory of the adaptation mechanism and the second terms, which are functions only of $\underline{v}(t)$, become null when $\underline{v} = 0$, $\underline{e} = 0$ (they are transient terms which vanish at equilibrium). Φ_1 and Ψ_1 are $n \times n$ and $m \times n$ matrices, respectively, which denote a nonlinear time-varying relation between $A_S(\underline{e}, t)$, $B_S(\underline{e}, t)$, and the values of $\underline{v}$ for $0 \leq \tau \leq t$. Φ_2 and Ψ_2 are matrices of the same dimensions which denote a nonlinear time-varying relation between $A_S(\underline{e}, t)$, $B_S(\underline{e}, t)$, and the values of $\underline{v}(t)$ and are null for $\underline{v} = 0$.

Equations (2.4-3), (2.4-4), and (2.4-5) then define the *adaptation law* (adaptation mechanism), the reference model being specified by Eq. (2.1-1), the adjustable system by Eq. (2.1-2), and the state generalized error vector by Eq. (2.1-4). If we replace $A_S(\underline{e}, t)$ and $B_S(\underline{e}, t)$ in Eq. (2.4-2) by their expressions given by Eqs. (2.4-4) and (2.4-5), we obtain the following equivalent set of equations:

$$\underline{e} = A_M\underline{e} + I\underline{w}_1 \tag{2.4-6}$$

$$\underline{v} = D\underline{e} \tag{2.4-7}$$

$$\underline{w} = -\underline{w}_1 = \left[\int_0^t \Phi_1(\underline{v}, t, \tau)\, d\tau + \Phi_2(\underline{v}, t) + A_S(0) - A_M\right]\underline{y} + \left[\int_0^t \Psi_1(\underline{v}, t, \tau)\, d\tau + \Psi_2(\underline{v}, t) + B_S(0) - B_M\right]\underline{u} \tag{2.4-8}$$

These equations lead to the equivalent representation given in Fig. 2.4-2.

2.5 CONCLUDING REMARKS

This chapter presented the following basic ideas:

1. State variables and input/output relations in terms of differential operators have been considered for describing model reference adaptive systems. However, other formats can also be used.

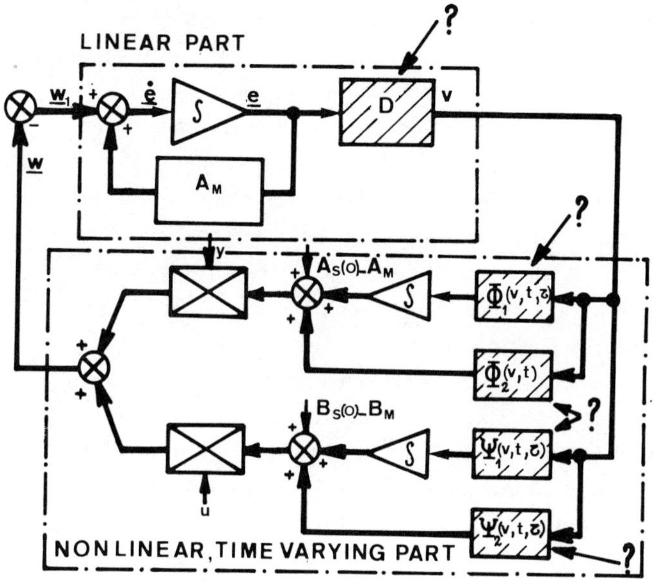

Fig. 2.4-2 Equivalent feedback representation of parallel model reference adaptive systems.

2. For series-parallel model reference adaptive systems there are several possible configurations.
3. For implementing model reference adaptive systems when only input and output measurements are available, *state-variable filters*, allowing one to obtain filtered derivatives, will be used often.
4. Two sets of hypotheses have been considered: (1) the *basic* case, which allows a smooth analytical design and which closely fits many practical situations, and (2) the *general* case, which leads to far more difficult design problems.
5. The concept of *perfect asymptotic adaptation* (the states and the parameters of the adjustable system tend asymptotically to those of the reference model) allows one to link the design of model reference adaptive systems to a *stability* problem.

6. The design problem for model reference adaptive systems can be solved in two steps: (1) Find the structures which can assure perfect asymptotic adaptation and then (2) select the structures and the parameters of these structures in order to comply with an index of performance.
7. A model reference adaptive system can always be represented under the form of an equivalent feedback system which in the case of the basic hypothesis is composed of a linear time-invariant feedforward block and a nonlinear time-varying feedback block.
8. *Perfect asymptotic adaptation* corresponds to the *global asymptotic stability* of the equivalent feedback system plus some additional conditions assuring also the asymptotic convergence of the parameters of the adjustable system to those of the reference model.

PROBLEMS

2.1 Consider a parallel model reference adaptive system with a reference model given by Eq. (2.1-1) and an adjustable system formed by a control system with adjustable feedforward and feedback gains:

$$\dot{\underline{y}} = A_P\underline{y} + B_P\underline{u}_P$$

$$\underline{u}_P = -K_P(\underline{e}, t)\underline{y} + K_U(\underline{e}, t)\underline{u}$$

The matrices A_P and B_P characterizing the controlled plant are supposed to be constant but unknown. Discuss the structural conditions on A_P, B_P, A_M, and B_M that allow one to obtain identical behavior for the reference model and the adjustable system for certain values of $K_P(\underline{e}, t)$ and $K_U(\underline{e}, t)$.

2.2 Describe a single-input/single-output model reference adaptive system with a series-parallel reference model using input/output relations in terms of differential operators in two cases: (a) The series part of the reference model is placed before the adjustable system. (b) The series part of the reference model is placed after the adjustable system.

2.3 Give an equivalent description for the series-parallel model reference adaptive system in state-variable form [Eqs. (2.1-1) and (2.1-21)] that leads to the same expression of the state generalized error but that has a block diagram identical to that of Fig. 1.3-12(c).

2.4 Construct equivalent feedback systems for the two configurations of series-parallel model reference adaptive systems described by state-space equations [Eqs. (2.1-1) and (2.1-21) and Eqs. (2.1-22) and (2.1-2)]. Compare the results with those obtained in Sec. 2.4 for parallel model reference adaptive systems.

2.5 Construct equivalent feedback systems for (single-input/single-output) parallel, series-parallel, and series model reference adaptive systems described by input/output relations in terms of differential operators for the case without state-variable filters.

REFERENCES

1. L. M. Silvermann, Inversion of multivariable linear systems, IEEE Trans. Autom. Control, *AC-14*, no. 3, 270-276 (June 1968).
2. S. P. Singh and R. Liu, A survey of inverse systems, in *Proceedings of the National Electronics Conference (USA)*, 1970, pp. 380-384.
3. I. D. Landau, A survey of model reference adaptive techniques. Theory and applications, Automatica, *10*, no. 4, 353-379 (July 1974).
4. V. M. Popov, The solution of a new stability problem for controlled systems, Autom. Remote Control (USSR), *24*, 1-23 (Jan. 1963) (trans.).
5. V. M. Popov, *Hyperstability of Control Systems*, Springer-Verlag, New York, 1973 (trans.).
6. K. S. Narendra and J. H. Taylor, *Frequency Domain Criteria for Absolute Stability*, Academic Press, New York, 1973.

7. I. D. Landau, Analyse et synthèse des commandes adaptatives à l'aide d'un modèle par des méthodes d'hyperstabilité, Automatisme, (7-8), 301-309 (July-Aug. 1969).

8. I. D. Landau, Sur une méthode de synthèse des systèmes adaptatifs avec modèles utilisés pour la commande et l'identification d'une classe de procédés physiques, No. C.N.R.S.: A.O.8495, Thèse d'Etat ès Sciences, Université de Grenoble, June 1973 (available on microfilm from CNRS Documentation Center, 26, rue Boyer, Paris-75020, France).

Chapter Three

BASIC METHODS FOR SOLVING THE MODEL REFERENCE ADAPTIVE SYSTEM DESIGN PROBLEM

In this chapter, three basic approaches to the design of model reference adaptive systems are briefly presented. These approaches are based on the use of

1. Local parametric optimization theory
2. Lyapunov functions
3. Hyperstability and positivity concepts

The presentation follows the chronological order of their appearance and the logical evolution from one to another. The links existing between them are emphasized.

The first two approaches are presented only for the sake of completeness. They will not be used elsewhere in this book.

Designs based on the use of the third approach will be discussed in detail in Chap. 4. For this reason, a comparison of these various approaches is delayed until the end of Chap. 4 (Sec. 4.3). Anticipating this comparison, one can say that the stability-based approach, and in particular the use of hyperstability and positivity concepts, is the most successful approach to the design of model reference adaptive systems.

3.1 DESIGN METHODS BASED ON LOCAL PARAMETRIC OPTIMIZATION THEORY

3.1-1 *Introduction*

These methods were the first to be used for model reference adaptive system (MRAS) design. The basic idea of these methods is as follows: A quadratic $(IP)_{RM}$ that expresses the structure and/or the state distance between the reference model and the adjustable system is defined (see Sec. 1.3-2, Definition 1.3-7). The surfaces of $(IP)_{RM}$ = constant are defined in the parameter space, in the neighborhood of the nominal point [where $(IP)_{RM} = 0$]. The methods of parameter optimization theory give us algorithms for changing the parameters of the adjustable system in order to go from one surface of constant $(IP)_{RM}$ to another corresponding to a lower $(IP)_{RM}$ (i.e., to close up the adjustable system and the reference model).

These methods require two supplementary assumptions for solving the design problem:

1. $A_M - A_S(t_0)$, $B_M - B_S(t_0)$ are small (i.e., we start in the neighborhood of the reference model parameter value).
2. The speed of adaptation is low.

The second supplementary assumption is necessary in order to be able to easily separate in the measurement of the generalized error the effects of parameter adjustment from those of input signal variations.

Furthermore, these methods have a general drawback: Nothing is said about the stability of the resulting adaptive system. We thus need a stability analysis, which is not always easy to accomplish. However, despite this drawback, ties exist between the results obtained using these methods and those based on a stability point of view, which will be discussed in Secs. 3.2 and 3.3. Furthermore, they provide more insights to the results that will be obtained in Secs. 3.2 and 3.3.

Among the basic optimization methods used to generate adaptation algorithms, one may mention the following: the gradient method, the steepest descent method, and the conjugate gradient method. The implementation of these methods often requires the use of sensitivity functions which must be generated on line. Only the first of these methods will be considered in detail in what follows, along with several other particular techniques.

3.1-2 *The Gradient Method*

Before we discuss the use of the gradient method in the general case, we shall present a simple example.

An Example

Consider a reference model given by

$$(1 + a_1 p + a_2 p^2)\theta_M = b_0 \rho \tag{3.1-1}$$

and a parallel adjustable model given by

$$(1 + a_1 p + a_2 p^2)\theta_S = \hat{b}_0(\varepsilon, t)\rho \tag{3.1-2}$$

The adjustable gain $\hat{b}_0(\varepsilon, t)$ can be, for example, the product of two gains, one of which is subject to variations and the second of which is adjusted by the adaptation mechanism. Or perhaps we face an identification problem where b_0 is unknown and we would like the value of $\hat{b}_0(\varepsilon, t)$ to converge to that of b_0.

The objective of the adaptation mechanism can be the minimization of

$$(IP)_{RM} = \frac{1}{2}\int_{k}^{t_k+\Delta t} \varepsilon^2 \, dt = \frac{1}{2}\int_{k}^{t_k+\Delta t} L(\varepsilon, t) \, dt \tag{3.1-3}$$

where $L(\varepsilon, t)$ is a quadratic form of $\varepsilon = \theta_M - \theta_S$ and depends indirectly on the difference $b_0 - \hat{b}_0(\varepsilon, t)$. Assuming that $b_0(\varepsilon, 0)$ is already close to the exact value b_0, one can use a *gradient* optimization technique which leads to the basic adaptation rule:

$$\Delta\hat{b}_0(\varepsilon, t) = -k \, \text{grad}(IP)_{RM} = -k \frac{\partial[(IP)_{RM}]}{\partial\hat{b}_0}, \quad k > 0 \tag{3.1-4}$$

where $\Delta\hat{b}_0$ signifies a change in $\hat{b}_0(\varepsilon, t)$ from its last computed value and k is an arbitrary positive constant called the *adaptation gain*. The rate of variation of the adjustable parameter $\hat{b}_0(\varepsilon, t)$ will be

$$\dot{\hat{b}}_0(\varepsilon,\ t) = \frac{d\hat{b}_0}{dt} = -k\,\frac{\partial}{\partial t}\left(\frac{\partial[(IP)_{RM}]}{\partial\hat{b}_0}\right) \tag{3.1-5}$$

Assuming that the adaptation is slow, i.e., the variation in $(IP)_{RM}$ caused by the variation in $b_0(\varepsilon,\ t)$ at each moment is small, the order of differentiation on the right-hand side of Eq. (3.1-5) can be interchanged, yielding

$$\frac{d\hat{b}_0}{dt} = -\frac{1}{2}\,k\,\frac{\partial}{\partial\hat{b}_0}\,L(\varepsilon,\ t) = -k\varepsilon\,\frac{\partial\varepsilon}{\partial\hat{b}_0} \tag{3.1-6}$$

This adaptation law is known as the MIT rule [1, 2].

To implement the adaptation mechanism according to Eq. (3.1-6), $\partial\varepsilon/\partial\hat{b}_0$ must be computed. But

$$\frac{\partial\varepsilon}{\partial\hat{b}_0} = \frac{\partial\theta_M}{\partial\hat{b}_0} - \frac{\partial\theta_S}{\partial\hat{b}_0} = -\frac{\partial\theta_S}{\partial\hat{b}_0} \tag{3.1-7}$$

because θ_M is not a function of the adjustable parameter $\hat{b}_0(\varepsilon,\ t)$.

Therefore, the adaptation law (3.1-6) becomes

$$\dot{\hat{b}}_0(\varepsilon,\ t) = k\varepsilon\,\frac{\partial\theta_S}{\partial\hat{b}_0} \tag{3.1-8}$$

$\partial\theta_S/\partial\hat{b}_0$ is simply the sensitivity function of the adjustable model with respect to $\hat{b}_0$.

By taking the partial derivative of both sides of Eq. (3.1-2) with respect to the parameter $\hat{b}_0$, one obtains

$$\frac{\partial\theta_S}{\partial\hat{b}_0} = \rho - a_1\,\frac{\partial\dot{\theta}_S}{\partial\hat{b}_0} - a_2\,\frac{\partial\ddot{\theta}_S}{\partial\hat{b}_0} \tag{3.1-9}$$

Again, assuming slow adaptation, $\hat{b}_0$ is changing with a slow rate, so the order of the differentiation in the right-hand side of Eq. (3.1-9) can be interchanged, yielding

$$\frac{\partial\theta_S}{\partial\hat{b}_0} \simeq \rho - a_1\,\frac{\partial}{\partial t}\left(\frac{\partial\theta_S}{\partial\hat{b}_0}\right) - a_2\,\frac{\partial^2}{\partial t^2}\left(\frac{\partial 0_S}{\partial\hat{b}_0}\right) \tag{3.1-10}$$

or

$$(1 + a_1 p + a_2 p^2)\,\frac{\partial\theta_S}{\partial\hat{b}_0} \simeq \rho \tag{3.1-11}$$

For the particular problem considered, by comparing Eq. (3.1-11) with Eq. (3.1-1), one sees that

$$\frac{\partial \theta_S}{\partial \hat{b}_0} = \frac{\theta_M}{b_0} \tag{3.1-12}$$

and the adaptation law becomes

$$\dot{\hat{b}}_0(\varepsilon, t) = k'\varepsilon\theta_M \quad (k' = k/b_0, \ k > 0) \tag{3.1-13}$$

The General Case (SISO Systems)

Consider now the general case of a SISO *reference model* given by

$$\left(\sum_{i=0}^{n} a_i p^i\right)\theta_M = \left(\sum_{i=0}^{m} b_i p^i\right)\rho \tag{3.1-14}$$

and an *adjustable system* given by

$$\left[\sum_{i=0}^{n} \hat{a}_i(\varepsilon, t)p^i\right]\theta_S = \left[\sum_{i=0}^{m} \hat{b}_i(\varepsilon, t)p^i\right]\rho \tag{3.1-15}$$

and the $(IP)_{RM}$ given by Eq. (3.1-3). Without loss of generality one assumes $a_0 = a_0 = 1$. Applying the same approach, we shall find the adaptation laws

$$\dot{\hat{a}}_i(\varepsilon, t) = k_i^a \varepsilon \frac{\partial \theta_S}{\partial \hat{a}_i}, \quad i = 1, \ldots, n \tag{3.1-16}$$

$$\dot{\hat{b}}_i(\varepsilon, t) = k_i^b \varepsilon \frac{\partial \theta_S}{\partial \hat{b}_i}, \quad i = 0, \ldots, m \tag{3.1-17}$$

where k_i^a and k_i^b are arbitrary positive constants.

To implement the adaptation mechanism, we must generate the sensitivity functions $\partial\theta_S/\partial\hat{a}_i$ and $\partial\theta_S/\partial\hat{b}_i$. From Eq. (3.1-15), at an arbitrary time $t = t_1$ one has

$$\theta_S = -\left[\sum_{i=1}^{n} \hat{a}_i(\varepsilon, t_1)p^i\right]\theta_S + \left[\sum_{i=0}^{m} \hat{b}_i(\varepsilon, t_1)p^i\right]\rho \tag{3.1-18}$$

We again assume that the rate of variation of the adjustable coefficients is small (slow adaptation) compared with their instantaneous value and that the adjustable system is a slowly time-varying system. With this assumption and by taking the partial derivatives of both sides of Eq. (3.1-18) with respect to $\hat{a}_i$ and $\hat{b}_i$, respectively, and interchanging the order of differentiation, one obtains

$$\left.\frac{\partial \theta_S}{\partial \hat{a}_i}\right|_{t=t_1} \simeq -p^i\theta_S - \left[\sum_{j=1}^{n} \hat{a}_j(\varepsilon, t_1)p^j\right]\frac{\partial \theta_S}{\partial \hat{a}_i} \tag{3.1-19}$$

$$\left.\frac{\partial \theta_S}{\partial \hat{b}_i}\right|_{t=t_1} \simeq p^i\rho - \left[\sum_{j=1}^{n} \hat{a}_j(\varepsilon, t_1)p^j\right]\frac{\partial \theta_S}{\partial \hat{b}_i} \tag{3.1-20}$$

One concludes that the sensitivity functions in the neighborhood of t_1, which are required for the adaptation law, are obtained as the output of a *sensitivity filter* having in all cases the transfer function

$$F_S(s) = \frac{1}{1 + \sum_{i=1}^{n} \hat{a}_i(\varepsilon, t_1)s^i} \tag{3.1-21}$$

and the input $-p^i\theta_S$ for obtaining $\partial\theta_S/\partial\hat{a}_i$ and $p^i\rho$ for $\partial\theta_S/\partial\hat{b}_i$.

Furthermore, writing

$$\left.\frac{\partial \theta_S}{\partial \hat{a}_{i-1}}\right|_{t=t_1} = -p^{i-1}\theta_S - \left[\sum_{j=1}^{n} \hat{a}_j(\varepsilon, t_1)p^j\right]\frac{\partial \theta_S}{\partial \hat{a}_{i-1}} \tag{3.1-22}$$

and comparing with Eq. (3.1-19), one obtains, under the hypothesis of an almost *frozen* adjustable system at $t = t_1$,

$$\frac{\partial \theta_S}{\partial \hat{a}_i} \simeq \frac{\partial}{\partial t}\left(\frac{\partial \theta_S}{\partial \hat{a}_{i-1}}\right) \simeq \frac{\partial^{i-1}}{\partial t^{i-1}}\left(\frac{\partial \theta_S}{\partial \hat{a}_1}\right) \tag{3.1-23}$$

and for $\partial\theta_S/\partial\hat{b}_i$ one obtains

$$\frac{\partial \theta_S}{\partial \hat{b}_i} \sim \frac{\partial}{\partial t}\left(\frac{\partial \theta_S}{\partial \hat{b}_{i-1}}\right) \simeq \frac{\partial^{i}}{\partial t^{i}}\left(\frac{\partial \theta_S}{\partial \hat{b}_0}\right) \tag{3.1-24}$$

We conclude that in order to generate the sensitivity functions, one needs only two sensitivity filters, one for generating $\partial\theta_S/\partial\hat{a}_1$ and the other for generating $\partial\theta_S/\partial\hat{b}_0$. The other sensitivity functions are the time derivatives of $\partial\theta_S/\partial\hat{a}_1$ and $\partial\theta_S/\partial\hat{b}_0$, which are directly available from the sensitivity filter used to generate $\partial\theta_S/\partial\hat{a}_1$ and $\partial\theta_S/\partial\hat{b}_0$ [see Eqs. (3.1-19) and (3.1-20)].

The remaining problem is to construct the sensitivity filter, which requires the knowledge of $\hat{a}_i(\varepsilon, t)$ and $\hat{b}_i(\varepsilon, t)$. In the case of identification with an adjustable model, the values of $\hat{a}_i(\varepsilon, t)$ and $\hat{b}_i(\varepsilon, t)$ are available, but this implies the realization of two sensitivity filters with variable coefficients (for which the hardware will be complex). In the case of an adaptive model-following control, the coefficients $\hat{a}_i(\varepsilon, t)$ and $\hat{b}_i(\varepsilon, t)$ are not available, since they are in fact the sum or the product of unknown coefficients of the controlled plant and of the adjustable controller unit. For these two reasons, one takes advantage of the first supplementary hypothesis considered when one applies local parametric optimization techniques [i.e., that $a_i - \hat{a}_i(\varepsilon, 0)$ are already small] to overcome the difficulties in realizing the sensitivity filters. *This allows one to replace the adjustable coefficients* $\hat{a}_i(\varepsilon, t)$ in Eqs. (3.1-19), (3.1-20), and (3.1-21) by a_i or $\hat{a}_i(\varepsilon, 0)$. Therefore, one obtains as the basic diagram of such a model reference adaptive system (MRAS) that given in Fig. 3.1-1(a) and (b).

The sensitivity filter used to implement the MRAS represented in Fig. 3.1-1 appears as a particular *state-variable filter* (SVF) delivering filtered derivatives of the input and output of the adjustable systems, which in this case are equal to the sensitivity functions (state-variable filters are also used for implementing series-parallel and series MRAS's--for details, see Secs. 1.3-4 and 2.1-1).

This scheme can be further simplified if one assumes that the speed of adaptation is low such that

$$\frac{\partial\theta_S}{\partial\hat{a}_i} \gg \frac{\partial^j}{\partial t^j}\left(\frac{\partial\theta_S}{\partial\hat{a}_i}\right) \text{ for } j = 1, \ldots, n;\ i = 1, \ldots, n \tag{3.1-25}$$

$$\frac{\partial\theta_S}{\partial\hat{b}_i} \gg \frac{\partial^j}{\partial t^j}\left(\frac{\partial\theta_S}{\partial\hat{b}_i}\right) \text{ for } j = 1, \ldots, n;\ i = 0, \ldots, m \tag{3.1-26}$$

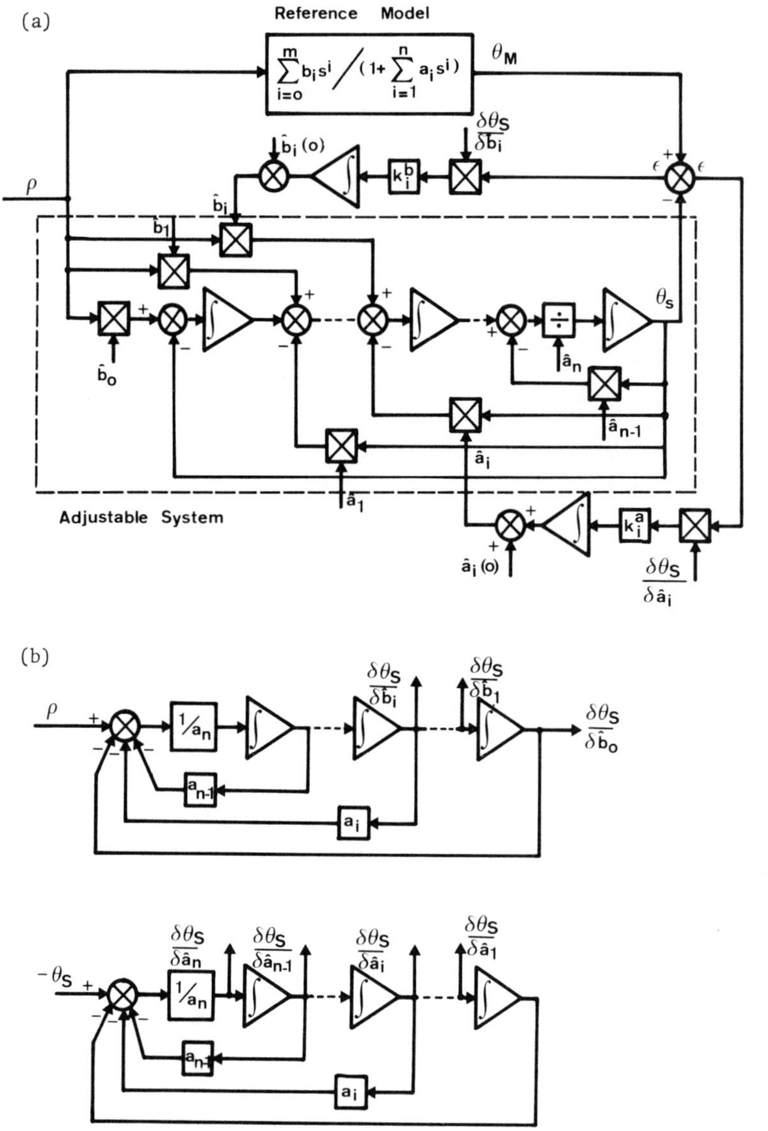

Fig. 3.1-1 MRAS using sensitivity filters. (a) Basic scheme. (b) Sensitivity filters.

In this case, from Eqs. (3.1-19) and (3.1-20) one obtains the following approximation for the sensitivity functions:

$$\frac{\partial \theta_S}{\partial \hat{a}_i} \simeq -p^i \theta_S, \quad i = 1, \ldots, n \tag{3.1-27}$$

$$\frac{\partial \theta_S}{\partial \hat{b}_i} \simeq p^i \rho, \quad i = 0, \ldots, m \tag{3.1-28}$$

For other designs related to this approach, see Refs. 3-12.

3.1-3 Time-Decreasing Criteria

With a somewhat different approach, it will now be shown that adaptation algorithms of the form of Eqs. (3.1-16) and (3.1-17) in which the sensitivity functions are approximated by Eqs. (3.1-27) and (3.1-28) can be directly obtained.

The technique which we shall use to show this (perhaps first introduced by Dressler [13]) states that in order to have

$$\int_{t_2}^{t_3} L(\varepsilon, t)\, dt < \int_{t_1}^{t_2} L(\varepsilon, t)\, dt \tag{3.1-29}$$

for any $t_1 < t_2 < t_3$ with $t_3 - t_2 = t_2 - t_1$ [where $L(\varepsilon, t)$ has been defined in Eq. (3.1-3)], which corresponds to a good adaptive system, it is sufficient to satisfy the following condition:

$$\Delta L(\varepsilon, t) < 0 \quad \text{for all } t > t_1 \tag{3.1-30}$$

or if $t - t_1 \to 0$,

$$\frac{dL(\varepsilon, t)}{dt} < 0 \quad \text{for all } t > t_1 \tag{3.1-31}$$

This means that the integrand of the $(IP)_{RM}$ is decreasing with time. The condition of Eq. (3.1-30) or (3.1-31) is then used to derive the adaptation law.

An Example

Consider the same example as in Sec. 3.1-2, the reference model being characterized by Eq. (3.1-1), the adjustable model by Eq. (3.1-2), and the $(IP)_{RM}$ by Eq. (3.1-3). One requires that for each t

$$\Delta L(\varepsilon, t) = \varepsilon(t)\,\Delta\varepsilon(t) < 0 \tag{3.1-32}$$

$\Delta\varepsilon(t)$ is given by

$$\Delta\varepsilon(t) = \varepsilon(t + \Delta t) - \varepsilon(t) = \theta_M(t + \Delta t) - \theta_S(t + \Delta t) - \varepsilon(t) \tag{3.1-33}$$

Using a Taylor series expansion (with two terms) for $\theta_M(t + \Delta t)$ and $\theta_S(t + \Delta t)$, one obtains

$$\theta_M(t + \Delta t) = \theta_M(t) + \Delta t[b_0\dot{\rho} - a_1\ddot{\theta}_M(t) - a_2\dddot{\theta}_M(t)] \tag{3.1-34}$$

$$\begin{aligned}\theta_S(t + \Delta t) = \theta_S(t) + \Delta t\,\dot{\hat{b}}_0(\varepsilon, t)\rho + \Delta t[\hat{b}_0(\varepsilon, t)\dot{\rho} - a_1\ddot{\theta}_S(t) \\ - a_2\dddot{\theta}_S(t)]\end{aligned} \tag{3.1-35}$$

and

$$\Delta\varepsilon(t) = -\Delta t\,\dot{\hat{b}}_0(\varepsilon, t)\rho + \Delta t\{[b_0 - \hat{b}_0(\varepsilon, t)]\dot{\rho} - a_1\ddot{\varepsilon} - a_2\dddot{\varepsilon}\} \tag{3.1-36}$$

respectively, where $\hat{b}_0(\varepsilon, t)\dot{\rho}$, $\ddot{\varepsilon}$, and $\dddot{\varepsilon}$ are terms estimated at time t and will not depend on the variations in $\hat{b}_0(\varepsilon, t)$ over the interval $(t, t + \Delta t]$. Equation (3.1-36) can also be written in the form

$$\Delta\varepsilon(t) = -\Delta t\,\dot{\hat{b}}_0(\varepsilon, t)\rho + g(t) \tag{3.1-37}$$

where $g(t)$ contains the terms which are not affected by the modification of the adjustable parameters over the interval $(t, t + \Delta t]$.

One assumes, then, that the adaptation law will be such that

$$|\Delta t\,\dot{\hat{b}}_0(\varepsilon, t)\rho| > |g(t)| \tag{3.1-38}$$

and the initial condition of Eq. (3.1-32) is replaced by the condition

$$-\Delta t\,\dot{\hat{b}}_0(\varepsilon, t)\rho\varepsilon < 0 \tag{3.1-39}$$

One immediately obtains the adaptation law

$$\dot{\hat{b}}_0(\varepsilon, t) = k\rho\varepsilon, \quad k > 0 \tag{3.1-40}$$

Note that the adaptation law of Eq. (3.1-40) is the same as the approximate one which is obtained in Sec. 3.1-1 if in the adaptation law of Eq. (3.1-13) the sensitivity function $\partial\theta_S/\partial\hat{b}_0 = \theta_M/b_0$ is replaced by the approximation given by Eq. (3.1-28) for $i = 0$.

The Multivariable Case [7]

Consider the multivariable case where the *reference model* is given by

$$\dot{\underline{x}} = A_M\underline{x} + B_M\underline{u} \qquad (3.1\text{-}41)$$

the *adjustable system* is given by

$$\dot{\underline{y}} = A_S(\underline{e}, t)\underline{y} + B_S(\underline{e}, t)\underline{u} \qquad (3.1\text{-}42)$$

and the generalized error is defined by

$$\underline{e} = \underline{x} - \underline{y} \qquad (3.1\text{-}43)$$

Consider that the integrand of the $(IP)_{RM}$ is given by

$$L(\underline{e}, t) = \frac{1}{2}\underline{e}^T P\underline{e}, \quad P > 0 \qquad (3.1\text{-}44)$$

where P is a symmetric positive definite matrix which weights the components of the state generalized error vector and is chosen by the designer.[1]

In terms of the design rule of Eq. (3.1-30), one has

$$[\Delta\underline{e}(t)]^T P\underline{e}(t) < 0 \quad \text{for all } t > t_1 \qquad (3.1\text{-}45)$$

One defines

$$\Delta\underline{e}(t) = \underline{x}(t + \Delta t) - \underline{y}(t + \Delta t) - \underline{e}(t) \qquad (3.1\text{-}46)$$

where $\underline{x}(t + \Delta t)$ and $\underline{y}(t + \Delta t)$ are computed using a Taylor series expansion (with two terms). Then

$$\Delta\underline{e}(t) = -\frac{\Delta t^2}{2}[\dot{A}_S(\underline{e}, t)\underline{y} + \dot{B}_S(\underline{e}, t)\underline{u}] + g(t)$$
$$= \Delta_S\underline{e} + \underline{g}(t) \qquad (3.1\text{-}47)$$

where g(t) is a vector which contains all the terms evaluated at t which do not depend on the evolution of the adjustable parameters in the interval between t and t + Δt.

---

[1]The notation > 0 in the case of matrices defines a positive definite matrix and the notation ≥ 0 a semidefinite positive (or nonnegative) matrix.

By replacing the condition of Eq. (3.1-45) by the approximate one given by

$$(\Delta_S \underline{e})^T P \underline{e} = -\frac{\Delta t^2}{2}[\dot{A}_S(\underline{e}, t)\underline{y} + \dot{B}_S(\underline{e}, t)\underline{u}]^T P \underline{e} < 0 \tag{3.1-48}$$

in which $\Delta_S \underline{e}$ is obtained from Eq. (3.1-47), one obtains the adaptation laws

$$\dot{A}_S(\underline{e}, t) = F_A \underline{e}\,\underline{y}^T, \quad F_A > 0 \tag{3.1-49}$$

$$\dot{B}_S(\underline{e}, t) = F_B \underline{e}\,\underline{u}^T, \quad F_B > 0 \tag{3.1-50}$$

leading to the satisfaction of the inequality of Eq. (3.1-48). By choosing the terms of the gain matrices F_A and F_B sufficiently large, $\Delta_S \underline{e}$ will dominate in Eq. (3.1-47) so that it is possible to satisfy Eq. (3.1-45). However, large values of the elements of the matrices F_A and F_B lead to instability of the system, as will be shown in Sec. 3.2-1.

An exact solution in deriving the adaptation law in order to comply with the rule of Eq. (3.1-30) can be obtained if one chooses $L(\underline{e}, t)$ as a combination of the square of the state distance and of the structure distance, or

$$\begin{aligned} 2L(\underline{e}, t) &= \underline{e}^T P \underline{e} + \mathrm{tr}\{[A_M - A_S(\underline{e}, t)^T] F_A^{-1} [A_M - A_S(\underline{e}, t)]\} \\ &\quad + \mathrm{tr}\{[B_M - B_S(\underline{e}, t)]^T F_B^{-1} [B_M - B_S(\underline{e}, t)]\}, \quad F_A, F_B > 0 \end{aligned} \tag{3.1-51}$$

But $L(\underline{e})$ as given in Eq. (3.1-51)[2] is in fact (as will be shown shortly) a Lyapunov function for the MRAS considered, and the solution obtained will lead to a globally asymptotically stable MRAS. This, in fact, is the topic of the next section.

---

[2]Recall that given an $(n \times m)$-dimensional matrix H one has

$$\mathrm{tr}\, H^T H = \sum_{i=1}^{n} \sum_{j=1}^{m} h_{ij}^2 = ||H||^2$$

3.2 DESIGN METHODS BASED ON THE USE OF LYAPUNOV FUNCTIONS

3.2-1 Model Reference Adaptive System Stability Problems

Stability problems are inherent in MRAS design due to their time-varying nonlinear character. This becomes clear if one examines the MRAS's designed using local parametric optimization techniques.

Considering the generalized error equation for the example described at the begining of Sec. 3.1.2, one has [by subtracting Eq. (3.1-2) from Eq. (3.1-1)]

$$(1 + a_1p + a_2p^2)\varepsilon = [b_0 - \hat{b}_0(\varepsilon, t)]\rho \tag{3.2-1}$$

and the adaptation law found in Sec. 3.1-3, Eq. (3.1-40), is

$$\dot{\hat{b}}_0(\varepsilon, t) = k\rho\varepsilon, \quad k > 0 \tag{3.2-2}$$

By taking the derivatives of both sides of Eq. (3.2-1) and using Eq. (3.2-2), one obtains (if ρ is a constant signal)

$$(p + a_1p^2 + a_2p^3 + k\rho^2)\varepsilon = 0 \tag{3.2-3}$$

If we restrict ρ to be a constant signal u_0, then Eq. (3.2-3) becomes a third-order, linear differential equation with constant coefficients which can become unstable when either k or u_0 are such that

$$ku_0^2 > a_1/a_2 \tag{3.2-4}$$

(This can be shown using the Routh-Hurwitz criterion.) So the augmentation of the adaptation gain k in order to satisfy the conditions of Eqs. (3.1-38) and (3.1-32) could lead to instabilities even for the case of a particular input equal to a constant. But the instability can also be provoked by the modification of the level of the input signal for a constant adaptation gain.

Similar conclusions are obtained if instead of the adaptation law (3.1-40) one uses the adaptation law (3.1-13) (see [14]). The stability analysis becomes far more complex when one assumes a more realistic situation with a time-varying input.

Thus, to achieve a complete MRAS design one needs to include a stability analysis. Previous design experience has shown that in many cases it is difficult to examine the global stability properties of a MRAS when local parametric optimization is used to design the adaptation mechanism.

It appears that it is definitely much more convenient to formulate the MRAS design problem as a stability problem and then use convenient techniques to solve this nonlinear, time-varying stability problem and come up with a suitable adaptation law that will satisfy the stability requirement. In fact, without doubt, the *first property which a satisfactory MRAS should have is stability of the whole system.* To formulate the design of an MRAS as a stability problem in the *basic case*, one must turn back to the definition of *perfect asymptotic adaptation*. This is given in Sec. 2.3 and, in terms of a stability problem, can be formulated as follows.

DEFINITION 3.2-1 (perfect asymptotic adaptation problem in $\underline{e}$-space). Given $\underline{e}(0) \neq 0$ (known), find an adaptation law (algorithm) such that

$$\lim_{t\to\infty} \underline{e}(t) = 0 \quad \text{for all } \underline{e}(0) \in R^n,\ A_M - A_S(0) \in R^{n\times n},$$
$$B_M - B_S(0) \in R^{n\times m},\ \underline{u}(t) \in C^m,\ \beta(t) \equiv 0$$

where C^m is the input function space composed of the vector piecewise continuous function R^n, $R^{n\times n}$ and $R^{n\times m}$ define the state and the parameter spaces, and the $\beta(t)$ are the perturbations acting on the system which cannot be directly reflected in the adaptation law.

Definition 3.2-1 leads to a global asymptotic stability problem for a dynamic system. Now, in the *general case*, one can formulate a less restrictive design problem as follows.

DEFINITION 3.2-2. Given $\underline{e}(0) \neq 0$ (known), find an adaptation law such that

$$||\underline{e}(t)|| \leq N \quad \text{for all } t > t_0 + T \text{ and for all } \underline{e}(0) \in R^n,$$
$$A_M - A_S(0) \in R^{n\times n},\ B_M - B_S(0) \in R^{n\times m},$$
$$\underline{u}(t) \in C^m,\ \beta(t) \in W$$

where W is the disturbance space.

This second formulation can be related to the *bounded input/bounded output* properties of a dynamic system, where the perturbations $\beta(t)$ are interpreted as an external input and the state generalized error vector as an output. In practical design problems, one can use Definition 3.2-1 to design the structure of the adaptation mechanism and then Definition 3.2-2 for testing the performance of the MRAS's in the presence of perturbations.

As will be shown, the various types of perturbations (either signals or parameter perturbations) can be represented as an external input of an equivalent dynamic system whose output is $\underline{e}(t)$.

3.2-2 Model Reference Adaptive System Design by Lyapunov Functions

We shall now apply the Lyapunov functions to the design of multivariable MRAS's described by state equations.

The *reference model* is given by

$$\dot{\underline{x}} = A_M\underline{x} + B_M\underline{u} \tag{3.2-5}$$

The *adjustable system* is given by

$$\dot{\underline{y}} = A_S(\underline{e}, t)\underline{y} + B_S(\underline{e}, t)\underline{u} \tag{3.2-6}$$

The *generalized error state vector* is given by

$$\underline{e} = \underline{x} - \underline{y} \tag{3.2-7}$$

The equation of the generalized error state vector is therefore

$$\dot{\underline{e}} = A_M\underline{e} + [A_M - A_S(\underline{e}, t)]\underline{y} + [B_M - B_S(\underline{e}, t)]\underline{u} \tag{3.2-8}$$

One defines a Lyapunov function candidate V in the augmented state space $R^nR^{n\times n}R^{n\times m}$, which includes the generalized error state space and the parameter error space. The function chosen is of the form

$$V = \underline{e}^TP\underline{e} + \mathrm{tr}\{[A_M - A_S(\underline{e}, t)]^T F_A^{-1}[A_M - A_S(\underline{e}, t)]\}$$
$$+ \mathrm{tr}\{[B_M - B_S(\underline{e}, t)]^T F_B^{-1}[B_M - B_S(\underline{e}, t)]\} \tag{3.2-9}$$

where P, F_A^{-1}, and F_B^{-1} are positive definite matrices to be defined later. For $\dot{V}$ one obtains the expression

$$\dot{V} = \underline{e}^T(A_M{}^TP + PA)\underline{e} + 2\ \mathrm{tr}\{[A_M - A_S(\underline{e}, t)]^T\ [P\underline{e}\underline{y}^T - F_A{}^{-1}\dot{A}_S(\underline{e}, t)]\}$$
$$+ 2\ \mathrm{tr}\{[B_M - B_S(\underline{e}, t)]^T\ (P\underline{e}\underline{u}^T - F_B{}^{-1}\dot{B}_S(\underline{e}, t)]\} \qquad (3.2\text{-}10)$$

If A_M is a Hurwitz matrix,[3] one has

$$A_M{}^TP + PA_M = -Q \qquad (3.2\text{-}11)$$

where Q is an arbitrary positive definite matrix which therefore allows one to compute a suitable P. The first term in Eq. (3.2-10) will then be negative definite for all $\underline{e} \neq 0$, and the second and third terms will be identically null if one chooses as the adaptation law

$$\dot{A}_S(\underline{e}, t) = F_A(P\underline{e})\underline{y}^T \qquad (3.2\text{-}12)$$

$$\dot{B}_S(\underline{e}, t) = F_B(P\underline{e})\underline{u}^T \qquad (3.2\text{-}13)$$

Integrating Eqs. (3.2-12) and (3.2-13) one obtains:

$$A_S(\underline{e}, t) = \int_0^t F_A(P\underline{e})\underline{y}^T\ d\tau + A_S(0) \qquad (3.2\text{-}14)$$

$$B_S(\underline{e}, t) = \int_0^t F_B(P\underline{e})\underline{y}^T\ d\tau + B_S(0) \qquad (3.2\text{-}15)$$

One should note that with F_A and F_B being arbitrary positive definite matrices *this design assures the global asymptotic stability* of the MRAS *for any magnitude of the adaptation gains* (for which $F_A > 0$, $F_B > 0$) *and for any piecewise continuous input vector function* $\underline{u}$.

This law is exactly equivalent to those which can be obtained in the case of local parametric-optimization techniques requiring $[dL(\underline{e}, t)]/dt < 0$ for all t for the particular choice of $L(\underline{e}, t)$ as a combined sum of the square of the structure distance and state distance [see Eq. (3.1-51)].

The convergence of parameters requires one to test under what conditions $\underline{e}(t) \equiv 0$ implies that $A_S = A_M$, $B_S = B_M$. From Eqs. (3.2-14) and (3.2-15), one obtains

$$\lim_{t\to\infty} \underline{e}(t) = 0 \Rightarrow \begin{cases} \lim\limits_{t\to\infty}[A_M - A_S(\underline{e}, t)] = A \\ \lim\limits_{t\to\infty}[B_M - B_S(\underline{e}, t)] = B \end{cases} \qquad (3.2\text{-}16)$$

[3] A Hurwitz matrix is a matrix which has all eigenvalues with negative real parts and therefore verifies the Hurwitz stability criterion.

where A and B are constant matrices expressing the asymptotic parameter difference. Then, if $\underline{e}(t) \equiv 0$, from Eq. (3.2-8) one obtains

$$A\underline{y} + B\underline{u} \equiv 0 \tag{3.2-17}$$

Identity (3.2-17) can be satisfied for any t if

1. $\underline{y}$ and $\underline{u}$ are linearly dependent and A and B $\neq 0$,
2. $\underline{y}$ and $\underline{u}$ are identically null, or
3. $\underline{y}$ and $\underline{u}$ are linearly independent and A and B = 0.

Therefore, only the third situation leads surely to parameter convergence. Because parameter convergence is concerned mainly with identification, explicit conditions on $\underline{u}$ to assure independence of $\underline{u}$ and $\underline{y}$ will be discussed in Chap. 7.

For a survey of various MRAS Lyapunov designs, see [15, 16].

The approach by Lyapunov functions opens the question of how to choose a class of Lyapunov functions in order to widen the class of adaptation laws which lead to globally stable MRAS's. This is an important problem because to accomplish a complete design we must be able to choose from the class of stable adaptation laws the most suitable for a specific type of application.

An answer to the search for a large class (or classes) of stable adaptation algorithms is found when the hyperstability approach and positivity concepts are used to design globally stable MRAS's.

3.3 HYPERSTABILITY AND POSITIVITY CONCEPTS IN MODEL REFERENCE ADAPTIVE SYSTEM DESIGN

Lyapunov functions can be successfully used for designing stable MRAS's. However, their use is somehow limited because we do not know in general how to widen the class of suitable Lyapunov functions leading to adaptation laws other than those types given in Sec. 3.2. This is an important problem since we are interested in having the largest possible family of adaptation laws assuring the stability of an MRAS from which we can then select the most suitable for a specific application.

This problem can be circumvented to some extent, and a large family of adaptation laws leading to stable MRAS's can be obtained

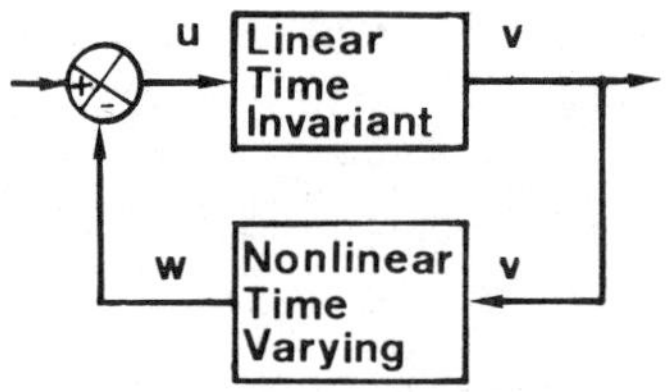

Fig. 3.3-1 Standard nonlinear and/or time-varying feedback system.

through the use of *hyperstability* theory in conjunction with the properties of the *positive dynamic systems*.

We shall next present a pragmatic approach without formal generalized and proven results. The detailed results for the general case and detailed mathematical aspects can be found in Chap. 4 (for the discrete case in Chap. 5) and Appendices B, C, and D. For a reference list concerning the use of the hyperstability and positivity concepts in MRAS design, see Ref. 7.

The *hyperstability* concept mainly concerns the stability properties of a class of feedback systems which can be split into two blocks as shown in Fig. 3.3-1 (which will be referred to as the standard feedback system).[4] The feedback block belongs to the family of linear, nonlinear, and/or time-varying systems satisfying an input/output relation of the form

$$\eta(0, t_1) \triangleq \int_0^{t_1} \underline{v}^T \underline{w}\, dt \geq -\gamma_0^2 \quad \text{for all } t_1 \geq 0 \tag{3.3-1}$$

where $\underline{v}$ is the input vector, $\underline{w}$ is the output vector of the feedback block, and γ_0^2 is a finite positive constant (which does not depend on t_1). For the remainder of the book, this integral inequality will be referred to as the *Popov integral inequality*. If the feedforward block is such that the *feedback system is globally (asymptotically) stable for all feedback blocks satisfying the Popov integral inequality*

---

[4](1) In fact, the position of the two blocks in Fig. 3.3-1 can be interchanged. (2) The external input in Fig. 3.3-1 is null when examining the stability of the feedback system.

of Eq. (3.3-1), one then says that the *feedback system is (asymptotically) hyperstable* and that the *feedforward block is a hyperstable block.*

It turns out that the conditions which allow one to have such hyperstable systems are expressed in terms of the *positivity* properties of the feedforward dynamic block.

For example, in the case of a single-input/single-output linear time-invariant feedforward block, the hyperstability conditions are that the transfer function of the feedforward block be *positive real* for hyperstability and *strictly positive real* for asymptotic hyperstability. For a transfer function h(s) to be *strictly positive real,* it must have the following properties:

1. h(s) is real for real s.
2. The poles of h(s) should lie in $Re[s] < 0$.
3. For all real ω, one has

$$Re[h(j\omega)] > 0, \quad -\infty < \omega < \infty$$

It is obvious that if the transfer function h(s) is reduced to a positive gain, then the above properties are all satisfied.

If we refer to the Nyquist plot of the transfer function h(s), from the definitions of strictly positive real and positive real transfer function's (see also Appendix B) we conclude that their hodograph is entirely situated in the fourth quadrant for $\omega > 0$ (being eventually tangent to the vertical axis $Re[h(j\omega)] = 0$ for some $0 < \omega < \infty$ in the case of real positive transfer functions). This also implies that the input/output phase lag is always less than or equal to 90° (one can roughly say that strictly positive real functions behave like a first-order transfer function).

The concept of *positive dynamic systems* is a cornerstone in hyperstability theory. Furthermore, positive dynamic systems also verify a Popov-type integral inequality, which appears (as shown below) to be very useful in designing MRAS's. The main results concerning positivity and hyperstability are given in Appendices B and C.

To solve a stability problem using the hyperstability approach, one must first be able to cast the original problem as a stability problem related to a feedback system and then be able to isolate one part which is such that it verifies the *Popov integral inequality* of Eq. (3.3-1) while the remainder should verify a corresponding *positivity* condition which will assure the hyperstability of the whole system [17].

It is important to note that the hyperstability approach allows one *to look at a stability problem related to the whole system as a consequence of the properties of its components*. In many cases this can drastically simplify the stability problem. The application of the hyperstability approach to the design of an MRAS is done in several basic steps.

First Step. Transform the MRAS into the form of an equivalent feedback system composed of two blocks, one in the feedforward path and one in the feedback path. (This is always possible, as was shown in Sec. 2.4.)

Second Step. Find solutions for the part of the adaptation laws which appears in the feedback path of the equivalent system such that the Popov integral inequality of Eq. (3.3-1) be satisfied.

Third Step. Find solutions for the remaining part of the adaptation law which appears in the feedforward path of the equivalent system such that the feedforward path be a hyperstable block (which will imply the hyperstability and therefore the global stability of the whole system).

Fourth Step. Return to the original MRAS in order to specify the adaptation law explicitly (i.e., the structure of the adaptation mechanism).

Note. In some cases part of the fourth step is directly connected with first step in order to assure that a *desired* equivalent feedback configuration can be effectively obtained.

An Example

Consider the example (a second-order system) from Sec. 3.1-2.

Subtracting Eq. (3.1-2) from Eq. (3.1-1), one has

$$(1 + a_1p + a_2p^2)\varepsilon = [b_0 - \hat{b}_0(\varepsilon, t)]\rho \qquad (3.3\text{-}2)$$

where $\hat{b}_0(\varepsilon, t)$, according to the general structure of the adaptation laws considered in Sec. 2.4, will be given by an expression having the form

$$\hat{b}_0(\varepsilon, t) = \int_0^t \psi_1{}^0(\nu, t, \tau)\, d\tau + \psi_2{}^0(\nu, t) + \hat{b}_0(0) \qquad (3.3\text{-}3)$$

where ν is the output of a linear block which will process the generalized error:

$$\nu = D(p)\varepsilon \qquad (3.3\text{-}4)$$

The right-hand side of Eq. (3.3-2) can be considered as an input ω_1,

$$\omega_1 = [b_0 - \hat{b}_0(\varepsilon, t)]\rho \qquad (3.3\text{-}5)$$

to a linear time-invariant block defined by the left-hand side of Eq. (3.3-2) and by Eq. (3.3-4). This block is therefore described by the equations

$$(1 + a_1p + a_2p^2)\varepsilon = \omega_1 \qquad (3.3\text{-}6)$$

$$\nu = D(p)\varepsilon \qquad (3.3\text{-}4)$$

If one considers the expression for $\hat{b}_0(\varepsilon, t)$ given by Eq. (3.3-3), the signal ω_1 given by Eq. (3.3-5) appears as the output of a block (nonlinear and time-varying) which has ν, the output of the linear block, as an input. So we have a feedback structure having a *linear feedforward block* described by Eqs. (3.3-6) and (3.3-4) in the feedforward path and a *feedback block* defined by the following expression:

$$\omega = -\omega_1 = [\hat{b}_0(\varepsilon, t) - b_0]\rho$$

$$= \rho\left[\int_0^t \psi_1{}^0(\nu, t, \tau)\, d\tau + \psi_2{}^0(\nu, t) + \hat{b}_0(0) - b_0\right] \qquad (3.3\text{-}7)$$

The equivalent scheme is given in Fig. 3.3-2. And the first step in applying the hyperstability approach has been achieved.

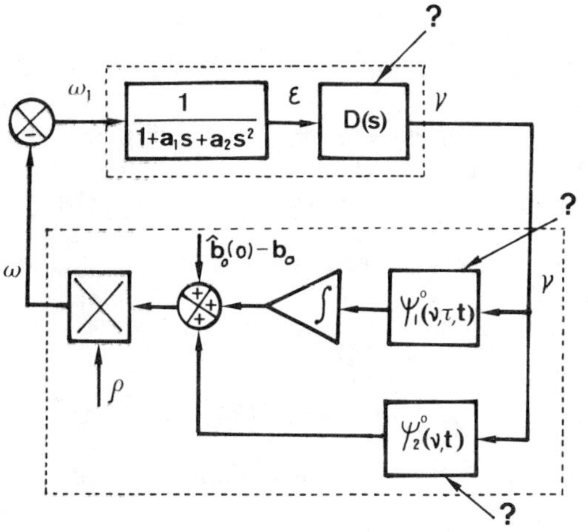

Fig. 3.3-2 Equivalent feedback system for the MRAS defined by Eqs. (3.1-1) to (3.3-4).

Second Step. One must now find solutions for ψ_1^0, ψ_2^0 such that the equivalent feedback block defined by Eq. (3.3-7) verifies the Popov integral inequality of Eq. (3.3-1), which in this case, using Eq. (3.3-7), becomes

$$\int_0^{t_1} \nu\omega\, dt = \int_0^{t_1} \nu\rho\left[\int_0^t \psi_1^0(\nu, t, \tau)\, d\tau + \psi_2^0(\nu, t) + \hat{b}_0(0) - b_0\right] dt \geq -\gamma_0^2 \tag{3.3-8}$$

To solve this problem, one can directly apply the general results given in Appendix D. However, for the learning needs of this particular case, we shall continue to solve it directly in order to emphasize how various solutions can be obtained.

Inequality (3.3-8) can be split into two inequalities, I_1 and I_2, as follows:

$$I_1 \triangleq \int_0^{t_1} \nu\rho\left[\int_0^t \psi_1^{\,0}(\nu, t, \tau)\, d\tau + \hat{b}_0(0) - b_0\right] dt \geq -\gamma_1^{\,2} \qquad (3.3\text{-}9)$$

$$I_2 \triangleq \int_0^{t_1} \nu\rho\psi_2^{\,0}(\nu, t)\, dt \geq -\gamma_2^{\,2} \qquad (3.3\text{-}10)$$

If both these inequalities are satisfied, this will imply that inequality (3.3-8) also will be satisfied because the latter is the sum of Eqs. (3.3-9) and (3.3-10). This is a sufficient condition.

Consider first inequality (3.3-10). If the integrand on the left-hand side is positive, the inequality will be satisfied for $\gamma_2^{\,2} = 0$. To have such an integrand, a first solution is to have

$$\psi_2^{\,0}(\nu, t) = k_2(t)\nu\rho, \quad [k_2(t) \geq 0] \quad \text{for all } t \geq 0 \qquad (3.3\text{-}11)$$

From this form, one can immediately derive the following particular solutions, which will be useful later:

1. *Proportional adaptation:*

$$\psi_2^{\,0}(\nu, t) = k_2\nu\rho, \quad k_2 \geq 0 \qquad (3.3\text{-}12)$$

2. *Relay adaptation*: If one chooses $k_2(t) = k_2/|\nu|$,[5] one gets

$$\psi_2^{\,0}(\nu, t) = k_2(\nu/|\nu|)\rho = k_2(\text{sgn } \nu)\rho, \quad k_2 \geq 0 \qquad (3.3\text{-}13)$$

For $k_2(t) = k_2/|\rho|$ and $k_2(t) = k_2/|\nu\rho|$, one gets

$$\psi_2^{\,0}(\nu, t) = k_2(\text{sgn } \rho)\nu, \quad k_2 \geq 0 \qquad (3.3\text{-}14)$$

and

$$\psi_2^{\,0}(\nu, t) = k_2 \text{ sgn}(\nu\rho), \quad k_2 \geq 0 \qquad (3.3\text{-}15)$$

Consider now inequality (3.3-9). A first solution can be found if one uses the following well-known relation: *Given a function* $f(t)$, *which admits a derivative with respect with time* $\dot{f}(t)$ *and a positive constant* k_1, *one has*

---

[5]The sign $|\cdot|$ defines the modulus of a scalar quantity, and it is always positive.

$$\int_0^{t_1} \dot{f}(t)k_1 f(t)\, dt = \frac{k_1}{2}[f^2(t_1) - f^2(0)] \geq -\frac{1}{2}k_1 f^2(0), \quad k_1 > 0 \tag{3.3-16}$$

Inequality (3.3-16) can be used in Eq. (3.3-9) by identifying

$$\dot{f}(t) = \nu\rho \tag{3.3-17}$$

and

$$k_1 f(t) = \int_0^t \psi_1{}^0(\nu, t, \tau)\, d\tau + \hat{b}_0(0) - b_0 \tag{3.3-18}$$

By differentiating Eq. (3.3-18) with respect to time and using Eq. (3.3-17), one gets

$$\psi_1{}^0(\nu, t, \tau) = k_1\dot{f}(\tau) = k_1\nu\rho, \quad k_1 > 0 \tag{3.3-19}$$

This is a first solution known as *integral adaptation*. It is shown in Appendix D that an inequality of the form of Eq. (3.3-9) is also satisfied if the positive constant k_1 in Eq. (3.3-19) is replaced by a *positive definite kernel* $k_1(t - \tau)$ having as Laplace transform a positive real transfer function with a pole at $s = 0$. To illustrate this kind of solution, consider the following kernel:

$$k_1(t - \tau) = 1 + e^{-(t-\tau)} \tag{3.3-20}$$

Its Laplace transform is

$$\begin{aligned} K(s) &= \int_0^\infty k_1(t)e^{-st}dt = \int_0^\infty (1 + e^{-t})e^{-st}dt = \frac{1}{s} + \frac{1}{s+1} \\ &= \frac{1 + 2s}{s(s+1)} \end{aligned} \tag{3.3-21}$$

One can see that we have a pole at $s = 0$ and that this transfer function verifies the definition of a positive real transfer function (see Appendix B.1); i.e., the transfer function is real for s real, and its poles are in $\mathrm{Re}[s] \leq 0$; there is a simple pole on $\mathrm{Re}[s] = 0$, and its residue (at $s = 0$) is positive. One also has

$$\mathrm{Re}[K(j\omega)] = \frac{1}{1 + \omega^2} > 0, \quad -\infty < \omega < +\infty \tag{3.3-22}$$

So a more general solution that Eq. (3.3-19) for $\psi_1{}^0(\nu, \tau, t)$ will be

$$\psi_1^0(\nu, t, \tau) = k(t - \tau)\nu\rho \tag{3.3-23}$$

where $k(t - \tau)$ is a positive definite kernel with the properties specified above. In fact, $k(t - \tau)$ in Eq. (3.3-23) represents the impulse response of a linear filter characterized by a positive real transfer function with a pole at $s = 0$.

From the implementation point of view, the above result means that in order to verify the Popov integral inequality *one can replace the integrator in the adaptation mechanism which corresponds to the solution for* ψ_1^0 *given by Eq. (3.3-19) by any other filter characterized by a positive real transfer function and having a pole at* $s = 0$. Among such transfer functions, we have

$$K(s) = \frac{K_I}{s} + K_p, \quad K_I, K_p > 0 \tag{3.3-24}$$

which corresponds to *proportional + integral adaptation* (when $\psi_2^0 \equiv 0$),[6] and

$$K(s) = \frac{K_I}{s} + K_p + K_d s, \quad K_I, K_p, K_d > 0 \tag{3.3-25}$$

which corresponds to *proportional + integral + derivative adaptation* (it can be implemented if derivatives or filtered derivatives of the product $\nu\rho$ can be obtained).

We recall that the adaptation law is obtained by summing solutions for ψ_1^0 and ψ_2^0. While ψ_2^0 can be identically null, ψ_1^0 cannot, since it provides the memory of the adaptation mechanism.

Third Step. By finding solutions for $\psi_1^0(\nu, t, \tau)$ and $\psi_2^0(\nu, t)$ allowing one to verify the Popov integral inequality, the problem has been cast into a hyperstability-type problem. One must now find the conditions for the feedforward equivalent path which will assure the asymptotic hyperstability, which, of course, will imply the asymptotic stability of the MRAS. Applying the Popov theorem for asymptotic hyperstability, one must have a *feedforward block characterized by a strictly positive real transfer function*. From

[6] K_p in Eq. (3.3-24) has the same effect as k_2 in Eq. (3.3-12).

Eqs. (3.3-6) and (3.3-4) of the equivalent feedforward block, one concludes that

$$h(s) = \frac{D(s)}{1 + a_1 s + a_2 s^2} \tag{3.3-26}$$

should be a strictly positive real transfer function. To verify the definition of a strictly positive real transfer function given in the first part of this section, one must first have an asymptotically stable reference model, and the coefficients a_1, a_2 and those of D(s) should be real. The third condition will allow us to find an explicit expression for D(s).

Choosing the expression

$$D(s) = d_0 + d_1 s \tag{3.3-27}$$

for D(s), one finds that

$$\text{Re}[h(j\omega)] = \frac{d_0 + (d_1 a_1 - d_0 a_2)\omega^2}{(1 - a_2\omega^2)^2 + \omega^2 a_1^2} \tag{3.3-28}$$

So that $\text{Re}[h(j\omega)] > 0$ for all real ω, we have the condition

$$d_0 > 0, \ \frac{d_1}{d_0} \geq \frac{a_2}{a_1} \quad (\text{or } d_0 = 1, \ d_1 \geq a_2/a_1 \tag{3.3-29}$$

The problem of finding D(s) which allow us to obtain a strictly real positive transfer function can also be solved in another way by using a state realization of h(s) and the formulation of equivalent positivity conditions in terms of the internal description of the system.

From Eqs. (3.3-26) and (3.3-27), one gets the following state realization for h(s):

$$\dot{\underline{e}} = A\underline{e} + \underline{b}\omega_1 \tag{3.3-30}$$

$$\nu = \underline{d}^T\underline{e} \tag{3.3-31}$$

where

$$A = \begin{bmatrix} 0 & 1 \\ -\frac{1}{a_2} & -\frac{a_1}{a_2} \end{bmatrix}, \quad \underline{b} = \begin{bmatrix} 0 \\ \frac{1}{a_2} \end{bmatrix}$$

$$\underline{d}^T = [d_0, d_1], \qquad \underline{e}^T = [\varepsilon, \dot{\varepsilon}]$$

According to the positive real lemma (Appendix B, Lemma B.2-1), the statement "h(s) is a real positive transfer function" is equivalent to the following with respect to the minimal realization $[A, \underline{b}, \underline{d}^T]$ of h(s): There exists a symmetric positive definite (or semidefinite) matrix Q and a symmetric positive definite matrix P such that

$$A^T P + PA = -Q \tag{3.3-32}$$

$$\underline{d} = \underline{b}^T P \tag{3.3-33}$$

One recognizes the first equation as the Lyapunov equation for the reference model. If Q is positive definite, which corresponds to an asymptotically stable reference model, one obtains a characterization of a strictly positive real transfer function (see Appendix B, Lemma B.2-3). For the particular example considered, solving Eqs. (3.3-32) and (3.3-33) for Eqs. (3.3-30) and (3.3-31), one gets

$$\underline{d}^T = [d_0, d_1] = \frac{1}{a_2}[P_{12}, P_{22}] \tag{3.3-34}$$

where

$$P = \begin{bmatrix} P_{11} & P_{12} \\ P_{12} & P_{22} \end{bmatrix}$$

Using, for the example considered, a particular matrix Q,

$$-Q = \begin{bmatrix} -\frac{2}{a_2} & 0 \\ 0 & 2(1 - \alpha) \end{bmatrix}, \quad \alpha > 1 \tag{3.3-35}$$

and solving Eq. (3.3-32), one obtains

$$P_{11} = \frac{a_1}{a_2} + \frac{\alpha}{a_1}, \quad P_{12} = 1, \quad P_{22} = \alpha \frac{a_2}{a_1} \tag{3.3-36}$$

This leads to a condition of the same type as when we used h(jω) directly [Eq. (3.3-29)]; namely,

$$\frac{d_1}{d_0} = \frac{P_{22}}{P_{12}} = \alpha \frac{a_2}{a_1} > \frac{a_2}{a_1} \tag{3.3-37}$$

Fourth Step. The return to the original scheme in this case is obvious because one already explicitly has the terms of the adaptation law, namely, the linear compensator D(s) and the terms which appear in the expression of the adjustable parameter $\hat{b}_0(\varepsilon, t)$.

The full scheme of the adaptive system using a *proportional + integral adaptation* is given in Fig. 3.3-3. One sees that the implementation of the MRAS requires access to the derivatives of ε, or, the same thing, access to the full state vectors of the reference model and of the adjustable system (as also shown in the following chapter). This is a significant difficulty in implementing asymptotically stable MRAS's (the same type of problem appears when Lyapunov design is used) when only some of the states of the model or of the adjustable system are directly available. This problem and various solutions will be discussed in Chaps. 4 and 5.

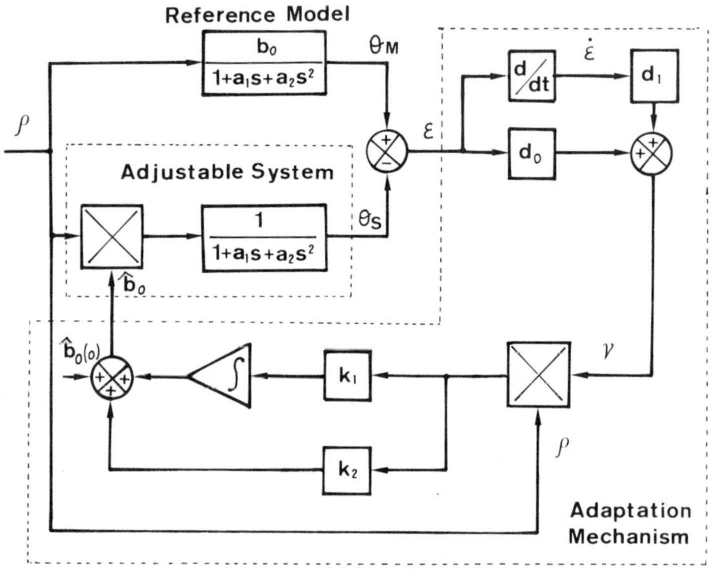

Fig. 3.3-3 Hyperstable MRAS with integral + proportional adaptation.

PROBLEMS

3.1 Examine the positive real and strictly positive real character (if any) of the following transfer functions:

(a) $\dfrac{k(1 + T_1 s)}{1 + T_2 s}$ (b) $\dfrac{K(1 + T_1 s)}{(1 + T_2 s)(1 + T_3 s)}$ (c) $\dfrac{k}{s}$

(d) $\dfrac{k}{s(1 + T_1 s)}$ (e) $\dfrac{k}{s^2 + \omega_0^2}$ (f) $\dfrac{ks}{1 + a_1 s + a_2 s^2}$

Discuss the eventual parameter constraints leading to a positive real or strictly positive real transfer function.

3.2 Consider the reference model given in Eq. (3.1-1) where $a_1 = 1.2$, $a_2 = 1$, and an adjustable system given by $[1 + \hat{a}_1(\varepsilon, t)p + \hat{a}_2(\varepsilon, t)p^2]\theta_S = \hat{b}_0(\varepsilon, t)\rho$, where ε is the output generalized error given by $\varepsilon = \theta_M - \theta_S$.

(a) Synthesize the adaptation laws for $\hat{a}_1(\varepsilon, t)$, $\hat{a}_2(\varepsilon, t)$, and $\hat{b}_0(\varepsilon, t)$ using, successively, the methods given in Sec. 3.1 (MIT rule, time-decreasing criterion), Secs. 3.2 and 3.3.

(b) Give a complete block diagram for each design.

(c) Compare the various schemes from the implementation point of view.

3.3 Consider the reference model $(a_0 + a_1 p + p^2)\theta_M = b_0 \rho$ with $a_0 = 2$ and $a_1 = 1.2$ and the adjustable system $[\hat{a}_0(\varepsilon, t) + \hat{a}_1(\varepsilon, t)p + p^2]\theta_S = \hat{b}_0(\varepsilon, t)\rho$.

(a) Synthesize the adaptation laws using the methods given in Secs. 3.2 and 3.3.

(b) Compare the resulting schemes to that obtained in Problem 3.2 when using the same designs.

3.4 [17] For most electrical drives, the desired response of the speed control loop to a change in the reference input ρ can be expressed in terms of a reference model having the transfer function

$$h_M(s) = \frac{1}{1 + 2sT + 2s^2T^2}$$

which is a second-order transfer function with a damping factor of 0.7. The process to be controlled, composed of the speed controller and the drive, can often be described by $[2Tp(pT + 1)]\theta_S = K_V K_a(\varepsilon, t)(\rho - \theta_S)$, where θ_S is the speed of the motor and ρ is the reference input. The process gain K_V is subject to variations or is unknown, and $K_a(\varepsilon, t)$ is an adjustable gain. Note that for $K_a K_V = 1$, the speed control loop behaves like the reference model.

(a) Design an adaptive model-following speed control system using the methods indicated in Secs. 3.1 to 3.3.

(b) Give a block diagram for each design.

3.5 [14, 17] Consider a reference model given by $(p^2 + K_1\beta p + K_1K_2)\theta_M = K_1K_2\rho$ and an adjustable control system given by $[p^2 + K_1K_VK_a(\varepsilon, t)p]\theta_S = K_1K_2(\rho - \theta_S)$, where the gain K_V is subject to variations and K_a is an adjustable gain. Note that for $K_VK_a = \beta$, both systems behave identically.

(a) Design an adaptive model-following control system using the methods indicated in Secs. 3.1 to 3.3.

(b) Give a block diagram for each design.

REFERENCES

1. P. V. Osborn, H. P. Whitaker, and A. Kezer, New developments in the design of model reference adaptive control systems, IAS Paper No. 61-39, Institute of Aeronautical Sciences, 29th Annual Meeting, New York, Jan. 1961.
2. C. C. Hang and P. C. Parks, Comparative studies of model reference adaptive control systems, IEEE Trans. Autom. Control, *AC-18*, 419-428 (1973).
3. D. D. Donalson and F. M. Kishi, Review of adaptive control system theories and techniques, in *Modern Control Systems Theory*, Vol. 2 (C. Leondes, ed.), McGraw-Hill, New York, 1965, pp. 228-284.

4. V. A. Serdyukov, Synthesis of a generalized adaptation algorithm for a non-search self-adjusting system with reference model, Autom. Remote Control, *31*, 1078-1084 (1970).
5. A. E. Pearson, An adaptive control algorithm for linear systems, IEEE Trans. Autom Control, *AC-14*, 497-503 (1969).
6. A. E. Pearson and V. R. Vanguri, A synthesis procedure for parameter adaptive control systems, IEEE Trans. Autom. Control, *AC-16*, 440-449 (1971).
7. I. D. Landau, A survey of model reference adaptive techniques. Theory and applications, Automatica, *10*, 353-379 (1974).
8. P. M. Lion, Rapid identification of linear and nonlinear systems, AIAA J., , 1835-1842 (1967).
9. J. M. Mendel, *Discrete Techniques of Parameter Estimation. The Equation Error Formulation,* Marcel Dekker, New York, 1973.
10. V. A. Taran, Conditions for the existence and uniqueness of an optimal linear model for dynamic objects, Autom. Remote Control, *31*, 1811-1820 (1970).
11. P. V. Kokotovic and R. S. Rutman, Sensitivity of automatic control systems, Autom. Remote Control, *26*, 730-750 (1965).
12. P. V. Kokotovic, J. V. Medanic, M. J. Vuskovic, and S. P. Bingulac, Sensitivity method in the experimental design of adaptive control systems, in *Proceedings of the Third International Federation of Automatic Control Congress*, London, 1966, pp. 45B.1-12. (Congress preprints.)
13. R. M. Dressler, An approach to model referenced adaptive control systems, IEEE Trans. Autom. Control, *AC-12*, 75-80 (1967).
14. P. C. Parks, Lyapunov redesign of model reference adaptive control systems, IEEE Trans. Autom. Control, *AC-11*, 362-367 (1966).
15. D. P. Lindorff and R. L. Caroll, Survey of adaptive control using Lyapunov design, Int. J. Control, *18*, 897-914 (1973).
16. K. S. Narendra and P. Kudva, Stable adaptive schemes for system identification and control, Parts I and II, IEEE Trans. Syst. Man Cybern., *SMC-4*, 542-560 (1974).
17. I. D. Landau, Analyse et synthèse des commandes adaptatives à l'aide d'un modèle par des méthodes d'hyperstabilité, Automatisme, *14*, 301-309 (1969).

Chapter Four

DESIGN OF CONTINUOUS-TIME MODEL REFERENCE ADAPTIVE SYSTEMS USING THE HYPERSTABILITY AND POSITIVITY APPROACHES

In this chapter, we shall discuss in detail the methodology of designing various basic model reference adaptive system (MRAS) configurations using the hyperstability and positivity approaches. The design will follow the basic four steps emphasized in Sec. 3.3. However, the particularities of each structure and the presentation of the general results for the adaptation laws require a detailed discussion. Many of the most important results will make extensive use of the hyperstability and positivity results given in Appendices B and C as well as of the solution for an integral-type inequality presented in detail in Appendix D.

For the sake of completeness, the design of almost all basic structures of MRAS's, which will appear in the applications chapters (Chaps. 6 to 8), is discussed.

Section 4.1 will present the basic results concerning the design of MRAS's when using the description by state-space equations.

The important problem of designing MRAS's using only input and output measurements is then discussed in Sec. 4.2, where several cases are considered.

In the accompanying table, we indicate in which sections the designs presented in Chap. 4 will be used.

Sections of Chap. 4	*Where the results are used*
4.1-1	4.3, 6.3, 6.6, 6.7, 7.2-2
4.2-2	7.2-3, 7.6
4.2-3	5.5, 6.4, 8 (Problems)
4.2-4	5.5, 6.4, 8 (Problems)
4.2-5	7.2-3, 5.5

Section 4.3 discusses the potential of the various design approaches considered, namely, those based on (1) local parametric optimization theory, (2) Lyapunov functions and (3) hyperstability and positivity, and emphasizes the differences among, the advantages of, and the links which eventually exist among the various approaches.

The various designs presented in this chapter, which will be used in Chaps. 6 to 8, will be evaluated in conjunction with the various types of applications which will be considered.

4.1 DESIGN OF MODEL REFERENCE ADAPTIVE SYSTEMS DESCRIBED BY STATE EQUATIONS

In the following, we shall discuss in detail the design of parallel and series-parallel model reference adaptive systems described by state equations. The design follows the basic steps described in Sec. 3.3, but here we shall explore in detail the various adaptation laws which lead to a stable MRAS.

4.1-1 Parallel Model Reference Adaptive Systems

The basic equations for this type of MRAS in the case of a state-space representation and parameter adaptation have been given in Chap. 2 and are repeated below (for a block diagram, see Fig. 4.1-1):

The *reference model*:

$$\dot{\underline{x}} = A_M\underline{x} + B_M\underline{u}, \quad \underline{x}(0) = \underline{x}_0 \tag{4.1-1}$$

The *parallel adjustable system*:

$$\dot{\underline{y}} = A_S(v, t)\underline{y} + B_S(\underline{v}, t)\underline{u}, \quad [\underline{y}(0) = \underline{y}_0, \; A_S(0) = A_{S0}, \; B_S(0) = B_{S0}] \tag{4.1-2}$$

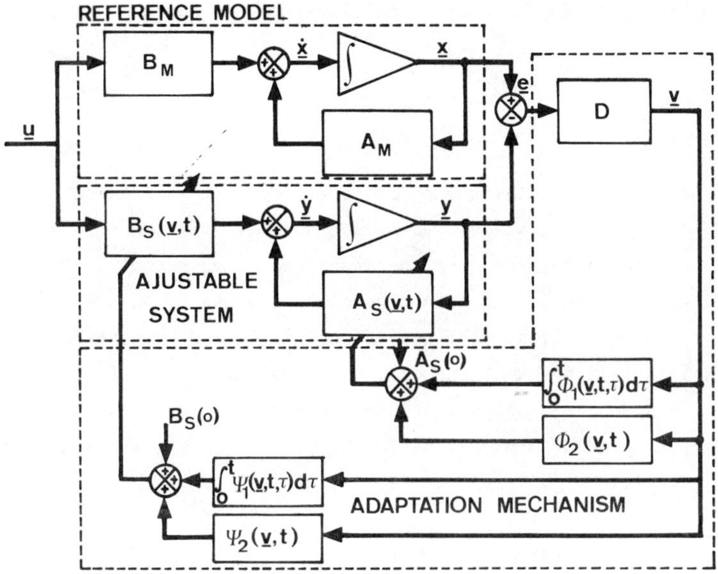

Fig. 4.1-1 Parallel MRAS in state-space representation.

The *state generalized error*:

$$\underline{e} = \underline{x} - \underline{y} \tag{4.1-3}$$

The *adaptation mechanism*:

$$\underline{v} = D\underline{e} \tag{4.1-4}$$

$$A_S(\underline{v}, t) = \int_0^t \Phi_1(\underline{v}, t, \tau)\, d\tau + \Phi_2(\underline{v}, t) + A_S(0) \tag{4.1-5}$$

$$B_S(\underline{v}, t) = \int_0^t \Psi_1(\underline{v}, t, \tau)\, d\tau + \Psi_2(\underline{v}, t) + B_S(0) \tag{4.1-6}$$

The design problem to obtain perfect asymptotic adaptation is as follows:

1. Determine D, Φ_1, Φ_2, Ψ_1, Ψ_2 such that $\lim_{t\to\infty} \underline{e}(t) = \underline{0}$ for any initial conditions $\underline{x}(0)$, $\underline{y}(0)$, $A_M - A_S(0)$, $B_M - B_S(0)$ and for all input vectors $\underline{u}$ belonging to the class of piecewise continuous vector functions

2. Find the supplementary conditions which lead to

$$\lim_{t\to\infty} A_S(\underline{v}, t) = A_M$$

$$\lim_{t\to\infty} B_S(\underline{v}, t) = B_M$$

As will be shown, the *second objective of the design will be attained if the first is achieved by adding some supplementary assumptions on the characteristics of the input vector function* $\underline{u}$. This second aspect of the design mainly concerns the use of MRAS's for parameter identification. The use of positivity and hyperstability concepts in designing MRAS's described by Eqs. (4.1-1) to (4.1-6) follows the same basic steps mentioned in Sec. 3.3, as shown next, and the corresponding expressions for D, Φ_1, Φ_2, Ψ_1, Ψ_2 are summarized in Table 4.1-1.

First Step: Equivalent Representation as a Feedback System

This has already been done in Sec. 2.4 [Eqs. (2.4-1) to (2.4-8)], and the result is repeated here for continuity. The equivalent feedback system for the parallel MRAS specified by Eqs. (4.1-1) to (4.1-6) is described by the following equations:

$$\dot{\underline{e}} = A_M\underline{e} + I\underline{w}_1 \tag{4.1-7}$$

$$\underline{v} = D\underline{e} \tag{4.1-8}$$

$$\underline{w} = -\underline{w}_1 = \left[\int_0^t \Phi_1(\underline{v}, t, \tau)\, d\tau + \Phi_2(\underline{v}, t) + A_S(0) - A_M\right]\underline{y} + \left[\int_0^t \Psi_1(\underline{v}, t, \tau)\, d\tau + \Psi_2(\underline{v}, t) + B_S(0) - B_M\right]\underline{u} \tag{4.1-9}$$

where Eqs. (4.1-7) and (4.1-8) describe the equivalent feedforward linear block and Eq. (4.1-9) describes the equivalent feedback, nonlinear, time-varying block.

Second Step

Search for solutions to Φ_1, Φ_2, Ψ_1, Ψ_2 such that the equivalent feedback block verifies the Popov integral inequality:

$$\eta(0, t_1) \triangleq \int_0^{t_1} \underline{v}^T\underline{w}\, dt \geq -\gamma_0^2 \quad \text{for all } t_1 \geq 0 \tag{4.1-10}$$

Table 4.1-1 Parallel MRAS's Described by State Equations [Eqs. (4.1-1) to (4.1-6)]

	General forms	Particular forms
Adaptation laws	$\Phi_1(\underline{v}, t, \tau) = F_A(t - \tau)\underline{v}(\tau)[G_A\underline{y}(\tau)]^T, \quad \tau \leq t$ $\Phi_2(\underline{v}, t) = F_A'(t)\underline{v}(t)[G_A'(t)\underline{y}(t)]^T$ $\Psi_1(\underline{v}, t, \tau) = F_B(t - \tau)\underline{v}(\tau)[G_B\underline{u}(\tau)]^T, \quad \tau \leq t$ $\Psi_2(\underline{v}, t) = F_B'(t)\underline{v}(t)[G_B'(t)\underline{u}(t)]^T$ $F_A(t - \tau)$, $F_B(t - \tau)$ = positive definite matrix kernels whose Laplace transforms are positive real transfer matrices with a pole at $s = 0$; $G_A, G_B > 0$; $F_A'(t)$, $F_B'(t)$, $G_A'(t)$, $G_B'(t) \geq 0$ for all $t \geq 0$	$F_A(t - \tau) = F_A > 0, \quad F_B(t - \tau) = F_B > 0$ for all $(t - \tau) \geq 0$ (integral adaptation) ——— $F_A'(t) = F_A' \geq 0, \quad G_A'(t) = G_A' \geq 0$ $F_B'(t) = F_B' \geq 0, \quad G_B'(t) = G_B' \geq 0$ (proportional adaptation) ——— $F_A'(t) = F_B'(t) = \mathrm{diag}[\frac{1}{v_1}]$ or $F_A'(t)\underline{v}(t) = F_B'(t)\underline{v}(t) = \mathrm{sgn}\ \underline{v}$ $\mathrm{sgn}\ \underline{v}^T = [\mathrm{sgn}\ v_1, \ldots, \mathrm{sgn}\ \underline{v}_n]$ (relay adaptation)
Stability condition	$H(s) = D(sI - A_M)$ = strictly positive real transfer matrix or: D is a positive definite matrix solution of $DA_M + A_M^T D = -Q \quad (Q > 0)$.	

where γ_0^2 is an arbitrary positive finite constant. By finding these solutions one can change the design problem of an asymptotically stable feedback system described by Eqs. (4.1-7) to (4.1-9) into a hyperstability problem.

By using the expression for $\underline{w}$ given by Eq. (4.1-9), inequality (4.1-10) becomes

$$\eta(0, t_1) = \int_0^{t_1} \underline{v}^T\left[\int_0^t \Phi_1(\underline{v}, t, \tau)\, d\tau + \Phi_2(\underline{v}, t) + A_0\right]\underline{y}\, dt + \int_0^{t_1} \underline{v}^T\left[\int_0^t \Psi_1(\underline{v}, t, \tau)\, d\tau + \Psi_2(\underline{v}, t) + B_0\right]\underline{u}\; dt \geq -\gamma_0^2 \tag{4.1-11}$$

where $A_0 = A_S(0) - A_M$ and $B_0 = B_S(0) - B_M$.

So that the inequality of Eq. (4.1-11) will be satisfied, it is sufficient that each of the two terms on the left-hand side satisfies the same type of inequality. The solutions for this type of inequality are discussed in Appendix D. Using, for example, Lemmas D.1-2 and D.1-5 for Φ_1, Φ_2, Ψ_1, Ψ_2 one gets the following solutions:

$$\Phi_1(\underline{v}, t, \tau) = F_A(t - \tau)\underline{v}(\tau)[G_A\underline{y}(\tau)]^T, \quad \tau \leq t \tag{4.1-12}$$

$$\Phi_2(\underline{v}, t) = F_A'(t)\underline{v}(t)[G_A'(t)\underline{y}(t)]^T \tag{4.1-13}$$

$$\Psi_1(\underline{v}, t, \tau) = F_B(t - \tau)\underline{v}(\tau)[G_B\underline{u}(\tau)]^T, \quad \tau \leq t \tag{4.1-14}$$

$$\Psi_2(\underline{v}, t) = F_B'(t)\underline{v}(t)[G_B'(t)\underline{u}(t)]^T \tag{4.1-15}$$

where $F_A(t - \tau)$ and $F_B(t - \tau)$ are *positive definite matrix kernels* whose Laplace transforms are *positive real transfer matrices with a pole at* $s = 0$. G_A, G_B are *positive definite constant matrices*, $F_A'(t)$, $F_B'(t)$, $G_A'(t)$ and G_B' are *time-varying positive definite (or semidefinite) matrices for all* $t \geq 0$.

Other solutions can be obtained if instead of Lemma D.1-2 one uses any of Lemmas D.1-1, D.1-3, or D.1-4 and instead of Lemma D.1-5 one uses Lemma D.1-6. It is important to note that $F_A(t - \tau)$ *and* $F_B(t - \tau)$ *have the significance of an impulse response of a dynamic system whose outputs are the column vectors of* $\Phi_1(\underline{v}, t, \tau)$ and $\Psi_1(\underline{v}, t, \tau)$, *respectively, and whose inputs are the column vectors of the matrices* $\underline{v}(t)[G_A\underline{y}(t)]^T$ or $\underline{v}(t)[G_B\underline{u}(t)]^T$.

Third Step: Application of the Popov Hyperstability Theorem (Appendix C, Theorem C-2) to the Equivalent Feedback System [1]

Once the equivalent feedback block satisfies the Popov integral inequality it remains to determine what conditions should be satisfied by the equivalent linear block so that the feedback system will be asymptotically hyperstable [which will imply that $\lim_{t\to\infty} \underline{e} = \underline{0}$ for all Φ_1, Φ_2, Ψ_1, Ψ_2, $A_S(0) - A_M$, $B_S(0) - B_M$, $\underline{y}$, $\underline{u}$ which satisfy the integral inequality of Eq. (4.1-11)]. Applying Theorem C-2, one finds that the transfer matrix of the equivalent feedforward block (Eqs. (4.1-7) and (4.1-8)], which is

$$H(s) = D(sI - A_M)^{-1} \tag{4.1-16}$$

must be a strictly positive real transfer matrix. This means, in fact, according to Corollary C-1 and Theorem B.2-1, that the equivalent feedforward block is a *positive* linear time-invariant system. Using Lemma B.2-3, for a given matrix A_M one can compute a suitable matrix D (called the *linear compensator*). One obtains

$$D = P \tag{4.1-17}$$

where P is a positive definite matrix which is the solution of the Lyapunov equation

$$A_M^T P + P A_M = -Q \tag{4.1-18}$$

in which Q is a *positive definite matrix*. Equation (4.1-18) always has a positive definite matrix solution P if and only if the reference model is asymptotically stable, which is generally the case (see Theorem A-2, Appendix A).

The second and third steps in the design can be formulated in a more concise form using the following theorem: Consider the feedback system

$$\dot{\underline{e}} = A\underline{e} + M\underline{w}_1 \tag{4.1-19}$$

$$\underline{v} = D\underline{e} \tag{4.1-20}$$

$$\underline{w} = -\underline{w}_1 = \sum_{q=0}^{p} \left[\int_0^t \Phi_{1q}(\underline{v}, t, \tau)\, d\tau + \Phi_{2q}(\underline{v}, t) + A_{0q}\right]\underline{y}_q. \tag{4.1-21}$$

THEOREM 4.1-1. For the system of Eqs. (4.1-19), (4.1-20), and (4.1-21), $\lim_{t\to\infty} \underline{e} = \underline{0}$ for all $\underline{e}(0) \in R^n$, $A_{0q} \in R^{n\times m_q}$, $\underline{y}_q \in C^{m_q}$ if

1. $\int_0^{t_1} \underline{v}^T[\int_0^t \Phi_{1q}(\underline{v}, t, \tau)\, d\tau + A_{0q}]\underline{y}_q\, dt \geq -\gamma_1^2$
   for all $t_1 \geq 0$, $q = 0, \ldots, p$ (4.1-22)

2. $\int_0^{t_1} \underline{v}^T\Phi_{2q}(\underline{v}, t)\underline{y}_q\, dt \geq -\gamma_2^2$ for all $t_1 \geq 0$,
   $q = 0, \ldots, p$ (4.1-23)
3. $H(s) = D(sI - A)^{-1}M$ (4.1-24)

is a *strictly positive real transfer matrix.*

A feedback system which will satisfy the conditions of Theorem 4.1-1 will also be termed *asymptotically hyperstable.*

Proof: The proof is immediate. Conditions 1 and 2 assure that the equivalent feedback block will satisfy the Popov integral inequality, while condition 3 is a direct consequence of applying Theorem C-2. Explicit solutions for $\Phi_{1q}(\underline{v}, t, \tau)$ and $\Phi_{2q}(\underline{v}, t)$ leading to the verification of conditions 1 and 2 are given in Appendix D (Lemmas D.1-1 to D.1-6).

Fourth Step: Return to the Original Model Reference Adaptive System

This step is implicitly accomplished in this case because the adaptation laws are directly specified by the various solutions for Φ_1, Φ_2, Ψ_1, Ψ_2, and the matrix D [see Eqs. (4.1-4) to (4.1-6)]. Therefore one can summarize the adaptation laws obtained.

THEOREM 4.1-2. The *parallel* MRAS described by Eqs. (4.1-1) to (4.1-6) is *globally asymptotically stable* (in $\underline{c}$-space) if

1. $\Phi_1(\underline{v}, t, \tau)$, $\Phi_1(\underline{v}, t)$, $\Psi_1(\underline{v}, t, \tau)$, $\Psi_2(\underline{v}, t)$ are given by Eqs. (4.1-12) to (4.1-15)
2. The transfer matrix H(s) given by Eq. (4.1-16) is *strictly positive real*

For the particular choice of Φ_1, Φ_2, Ψ_1, Ψ_2 given by Eqs. (4.1-12) to (4.1-15) in which

$$F_A(t - \tau) = F_A > 0, \quad F_B(t - \tau) = F_B > 0 \quad \text{for all } t - \tau \geq 0 \tag{4.1-25a}$$

$$F'_A(t) = F'_A, \quad F'_B(t) = F'_B, \quad G'_A(t) = G'_A, \quad G'_B(t) = G'_B \quad \text{for all } t \geq 0 \tag{4.1-25b}$$

one obtains the so called *integral + proportional* adaptation law. Note that integral + proportional adaptation can also be obtained if [2]

$$\Phi_2(\underline{v}, t) \equiv 0, \quad \Psi_2(\underline{v}, t) \equiv 0 \tag{4.1-26}$$

and

$$F_A(t - \tau) = F'_A\,\delta(t - \tau) + F_A \cdot 1(t - \tau) \quad \text{for all } t - \tau \neq 0 \tag{4.1-27}$$

$$F_B(t - \tau) = F'_B\,\delta(t - \tau) + F_B \cdot 1(t - \tau) \quad \text{for all } t - \tau \neq 0 \tag{4.1-28}$$

where F_A, F'_A, F_B, F'_B are *positive definite matrices and* $\delta(t - \tau)$ is the δ-function and $1(t - \tau)$ is the unit step function (see Appendix B for definitions). Therefore, one concludes that in this case the column vectors of $\Phi_1(\underline{v}, t, \tau)$ are obtained as the output of a linear filter characterized by a positive real transfer matrix,

$$F^A(s) = F'_A + \frac{F_A}{s} \tag{4.1-29}$$

and whose input is the corresponding column of the matrix $\underline{v}[G\underline{y}]^T$. Similar conclusions are also obtained for $\Psi_1(\underline{v}, t, \tau)$.

Integral + relay adaptation is obtained, for example, if in Eq. (4.1-25), $F'_A(t)$, $F'_B(t)$ are given by

$$F'_A(t) = F'_B(t) = \text{diag}\left[\frac{1}{|v_i|}\right] \tag{4.1-30}$$

and therefore

$$\Phi_2(\underline{v}, t) = \text{sgn } \underline{v}[G'_A\underline{y}]^T, \quad \Psi_2(\underline{v}, t) = \text{sgn } \underline{v}[G'_B\underline{u}]^T \tag{4.1-31}$$

where

$$\text{sgn } \underline{v}^T = [\text{sgn } v_1, \text{sgn } v_2, \ldots, \text{sgn } v_n] \tag{4.1-32}$$

The computing blocks for the integral + proportional adaptation law and integral + relay adaptation law are schematically represented in Figs. 4.1-2(a) and (b) for the case where $G_A = G'_A = I$ and F_A and F'_A are diagonal matrices.

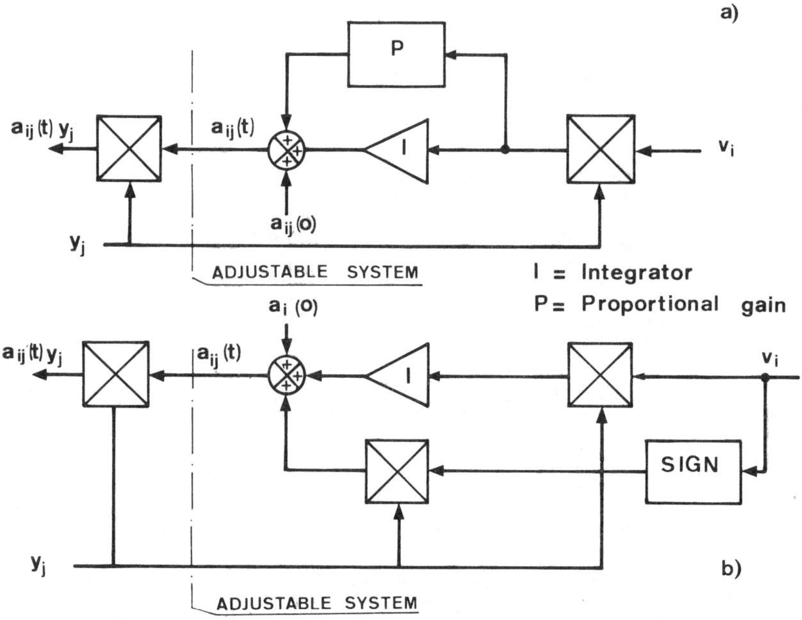

Fig. 4.1-2 Two typical adaptation laws. (a) Integral + proportional adaptation. (b) Integral + relay adaptation.

Note that integral, integral + proportional, and integral + relay adaptations are simple adaptation laws that have proven to yield satisfactory results in practice. However, there are also other possible adaptation laws satisfying conditions 1 and 2 of Theorem 4.1-1. These possibilities have not yet been fully explored.

The Convergence Problem for the Parameters

At this stage the MRAS designed will assure the convergence of the states of the adjustable system to those of the reference model for any initial parameter or state difference between the reference model and the adjustable system. For adaptive control design this is sufficient. However, in identification problems one is also interested in solving the second point of the design problem, namely, finding the conditions under which

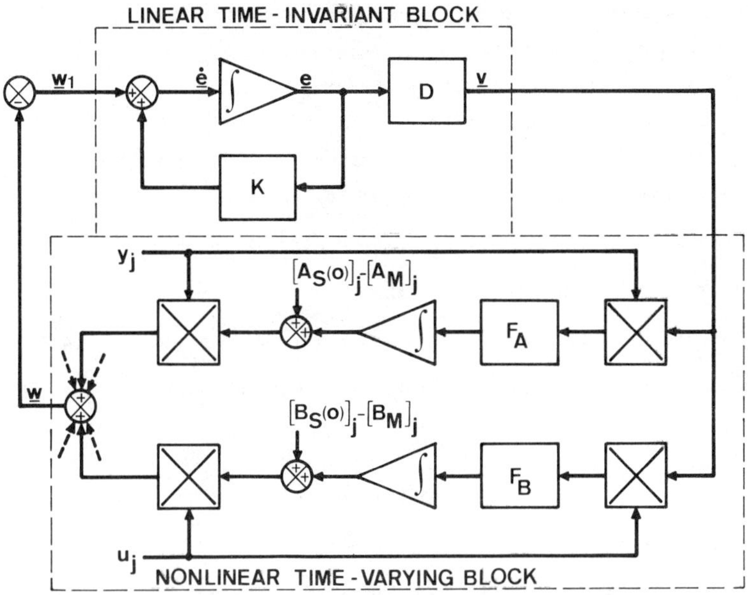

Fig. 4.1-3 Equivalent representation of an MRAS with integral adaptation.

$$\lim_{t\to\infty} A_S(\underline{v}, t) = A_M, \quad \lim_{t\to\infty} B_S(\underline{v}, t) = B_M \tag{4.1-33}$$

Let us first examine the equivalent feedback representation of the MRAS which is shown in Fig. 4.1-3 for the case of integral adaptation. In Fig. 4.1-3, $[A_S(0)^T]_j$, $[A_M]_j$, $[B_S(0)]_j$, and $[B_M]j$ denote the jth column vectors of the matrices $A_S(0)$, A_M, $B_S(0)$, and B_M, respectively. A simple inspection of this scheme allows us to state that a necessary condition for Eq. (4.1-33) to hold is that the components y_j and u_j of the adjustable system state vector $\underline{y}$ and input vector $\underline{u}$ are nonidentically null for $t \geq 0$. This is the case because if, for example, $y_j \equiv 0$ for all $t \geq 0$, one of the feedback links will no longer contribute to the evolution of the complete system, and therefore the difference between the jth column of $A_S(\underline{v}, t)$ and A_M will remain equal to its initial value $[A_S(0)]_j - [A_M]_j$. Since $\underline{y}$ depends on $\underline{u}$, one concludes that a necessary condition for parameter convergence is that $\underline{u}$ has components which are not identically null and have enough broad frequency spectrum and energy content in order to assure that $\underline{y}$ also has components which are not identically null. For a more detailed analysis, let us examine Eq. (4.1-7), rewritten as [obtained directly by subtracting Eq. (4.1-2) from Eq. (4.1-1)]

$$\dot{\underline{e}} = A_M\underline{e} + [A_M - A_S(\underline{v}, t)]\underline{y} + [B_M - B_S(\underline{v}, t)]\underline{u} \qquad (4.1\text{-}34)$$

One first assumes that $\underline{u}$ is such that

$$x_j \not\equiv 0, \quad y_j \not\equiv 0 \quad \text{for all } t \geq t_0, \; j = 1, \ldots, n \qquad (4.1\text{-}35)$$

Because the system is asymptotically stable in $\underline{e}$-space ($\underline{e} = \underline{0}$ is the equilibrium point), we shall have

$$\lim_{t\to\infty}(\underline{x} - \underline{y}) = \lim_{t\to\infty} \underline{e}(t) = \underline{0}, \quad \lim_{t\to\infty} \dot{\underline{e}}(t) = \underline{0} \qquad (4.1\text{-}36)$$

However, this implies [from Eq. (4.1-34)] that

$$[A_M - \lim_{t\to\infty} A_S(\underline{v}, t)]\underline{y} + [B_M - \lim_{t\to\infty} B_S(\underline{v}, t)]\underline{u} = \underline{0} \qquad (4.1\text{-}37)$$

Since we are only interested in the limit behavior of the above expression, taking into account Eq. (4.1-36), Eq. (4.1-37) can be rewritten as

$$[A_M - \lim_{t\to\infty} A_S(\underline{v}, t)]\underline{x} + [B_M - \lim_{t\to\infty} B_S(\underline{v}, t)]\underline{u} = \underline{0} \qquad (4.1\text{-}38)$$

If $\underline{u}$ *and* $\underline{x}$ *are linearly independent vector functions, then* Eq. (4.1-34) *holds only if* $\lim_{t\to\infty} A_S(\underline{v}, t) = A_M$ and $\lim_{t\to\infty} B_S(\underline{v}, t) = B_M$. This means that the parameters of the adjustable system will tend toward the true values of those of the reference model if $\underline{u}$ and $\underline{x}$ are linearly independent. When applying model reference adaptive techniques for parameter identification, this condition will assure unique parameter identification.

It will be shown in Chap. 7 that $\underline{u}$ and $\underline{x}$ will be linearly independent (sufficient conditions) if

1. The reference model is completely controllable.
2. The components of $\underline{u}$ are linearly independent.

3. Each component of $\underline{u}$ contains at least $(n + 1)/2$ distinct frequencies.

One concludes that the conditions for the convergence of the *parameters of the adjustable system to those of the reference model do not depend on the adaptation law found but on the characteristics of the input signal.*

A Specific Design Situation: Reference Model Having a Null Eigenvalue

If A_M has a null eigenvalue, the corresponding transfer matrix H(s) [see Eq. (4.1-16)] will have a pole at s = 0, and one will be able to obtain a matrix H(s) which will be only positive real. This will only assure the *hyperstability* of the equivalent feedback system instead of asymptotic hyperstability (Theorem C-1). As a consequence the MRAS obtained by using the adaptation laws of Eqs. (4.1-12) to (4.1-15) can be made globally stable only in $\underline{e}$-space (i.e., the state generalized state does not necessarily go to zero). To overcome this difficulty, one possibility for obtaining an asymptotic hyperstable equivalent feedback system leading to a *globally asymptotically stable* MRAS is to consider a modified adjustable system given by

$$\dot{\underline{y}} = A_S(\underline{v}, t)\underline{y} + B_S(\underline{v}, t)\underline{u} + K\underline{e} \qquad (4.1\text{-}39)$$

where K is a square matrix (n-dimensional). With this structure of the adjustable system, Eq. (4.1-7) becomes

$$\dot{\underline{e}} = (A_M - K)\underline{e} + I\underline{w}_1 \qquad (4.1\text{-}40)$$

and therefore the equivalent feedforward block will be characterized by the transfer matrix

$$H(s) = D(sI - A_M + K)^{-1} \qquad (4.1\text{-}41)$$

To obtain a strictly positive real transfer matrix one should first choose the matrix K such that the matrix $A_M - K$ only has eigenvalues with negative real parts (the pole at s = 0 will be shifted); then D is computed using Eq. (4.1-18) in which A_M has been replaced by $A_M - K$.

(a)

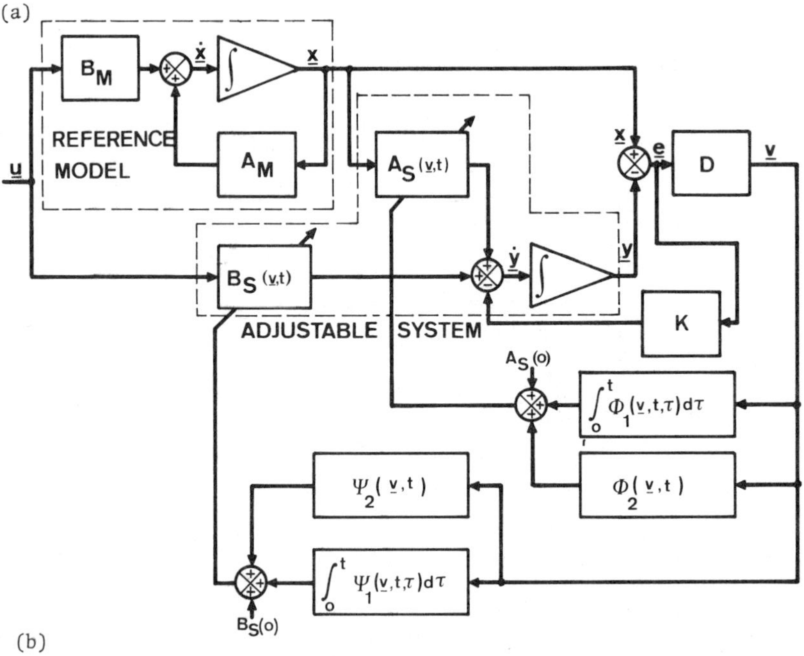

(b)

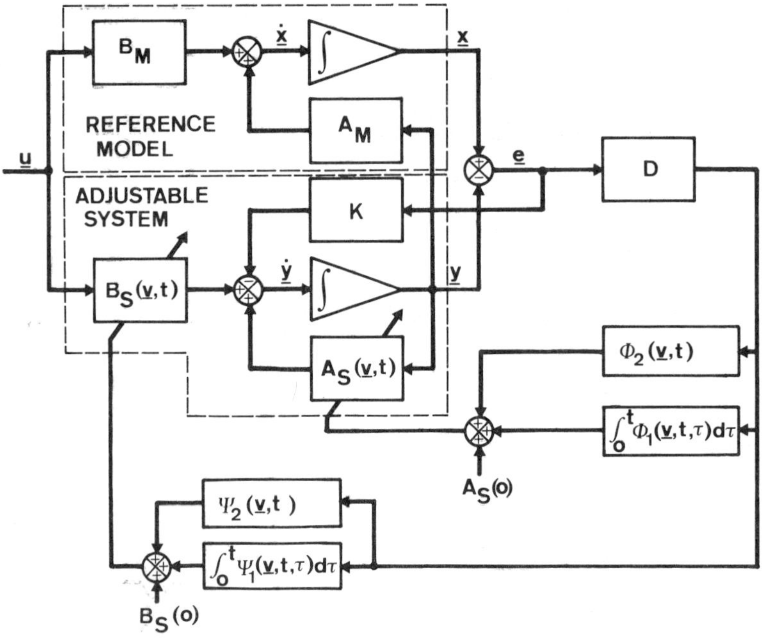

Fig. 4.1-4 Two configurations of series-parallel MRAS's in state-space representation.

4.1-2 *Series-Parallel Model Reference Adaptive Systems*

For the structure of series-parallel MRAS's one can distinguish two basic cases (see also Sec. 2.1) each having a special domain of application (the configuration of case 1 will be used in Sec. 7.2 for parameter identification, and the configuration of case 2 will be used in Sec. 6.4 for adaptive state variable regulators).

Case 1

See Fig. 4.1-4(a). The *reference model* is given by

$$\dot{\underline{x}} = A_M\underline{x} + B_M\underline{u} \tag{4.1-42}$$

The *series-parallel adjustable system* is given by

$$\dot{\underline{y}} = A_S(\underline{v}, t)\underline{x} + B_S(\underline{v}, t)\underline{u} - K\underline{e} \tag{4.1-43}$$

The *generalized state error vector* is

$$\underline{e} = \underline{x} - \underline{y} \tag{4.1-44}$$

The *adaptation mechanism* is given by

$$\underline{v} = D\underline{e} \tag{4.1-45}$$

$$A_S(\underline{v}, t) = \int_0^t \Phi_1(\underline{v}, t, \tau)\, d\tau + \Phi_2(\underline{v}, t) + A_S(0) \tag{4.1-46}$$

$$B_S(\underline{v}, t) = \int_0^t \Psi_1(\underline{v}, t, \tau)\, d\tau + \Psi_2(\underline{v}, t) + B_S(0) \tag{4.1-47}$$

The design follows the same basic steps as for parallel MRAS's.

So that the series-parallel MRAS described by the Eqs. (4.1-42) to (4.1-47) will be *asymptotically hyperstable* the transfer matrix

$$H(s) = D(sI - K)^{-1} \tag{4.1-48}$$

should be a *strictly positive real transfer matrix*. The matrix D, which is positive definite, can be computed explicitly by solving the equation

$$K^TD + DK = -Q, \quad Q > 0 \tag{4.1-49}$$

To obtain D as a positive definite matrix, K should be a Hurwitz matrix (all its eigenvalues must have negative real parts). For Ψ_1 and Ψ_2, one can use Eqs. (4.1-14) and (4.1-15), and Φ_1 and Φ_2 are given by Eqs. (4.1-12) and (4.1-13) in which $\underline{y}$ is replaced by $\underline{x}$, namely, in this case

$$\Phi_1(\underline{v}, t, \tau) = F_A(t - \tau)\underline{v}(\tau)[G_A\underline{x}(\tau)]^T, \quad \tau \leq t \tag{4.1-50}$$

$$\Phi_2(\underline{v}, t) = F_A'(t)\underline{v}(t)[G_A'(t)\underline{x}(t)]^T \tag{4.1-51}$$

where $F_A(t - \tau)$, $F_A'(t)$, G_A, and $G_A'(t)$ have the same meaning as in Eqs. (4.1-12) and (4.1-13).

Note: If K = 0, the equivalent feetforward block is a pure integrator, and its transfer matrix is only positive real. This will assure only the hyperstability of the feedback system and no longer asymptotic hyperstability. (This problem is similar to that encountered in the case of parallel MRAS's when A_M has a null eigenvalue.)

Case 2

In this case [Fig. 4.1-4(b)] the *series-parallel reference model* is described by

$$\dot{\underline{x}} = A_M\underline{y} + B_M\underline{u} \tag{4.1-52}$$

and the *adjustable system* is described by

$$\dot{\underline{y}} = A_S(\underline{v}, t)\underline{y} + B_S(\underline{v}, t)\underline{u} - K\underline{e} \tag{4.1-53}$$

The other equations describing the adaptive system are Eqs. (4.1-44) to (4.1-47) appearing in case 1. Asymptotic hyperstability will be assured if D is computed using Eq. (4.1-49) under the assumption that K is a Hurwitz matrix and if Φ_1, Φ_2, Ψ_1, and Ψ_2 are given, for example, by Eqs. (4.1-12) to (4.1-15).

4.2 DESIGN OF MODEL REFERENCE ADAPTIVE SYSTEMS USING ONLY INPUT AND OUTPUT MEASUREMENTS

4.2-1 Introduction

In this section we shall emphasize some of the basic ideas used in designing MRAS's when either only some of the state vector components are measurable or the derivatives of the outputs of the reference model or of the adjustable system are not available. We shall restrict our presentation in this section to the case of single-input/single-output parallel MRAS's described either by state-variable equations or input/output relations. Extensions and applications of the basic ideas and results that will be presented in this section as well as other approaches to solving this type of design can be found in Chaps. 6, 7, and 8.

We shall first consider a configuration of a single-input/single-output parallel MRAS described by input/output relations in which a state-variable filter is placed before the parallel adjustable system and another one after the reference model in order to provide filtered derivatives of the model output (case 1). The design of this system is a straightforward application of previous results. Then, we shall consider a configuration of a single-input/single-output parallel MRAS described by state-space equations in which state-variable filters are placed on the outputs of both the reference model and the adjustable system (case 2). We shall show how by adding supplementary signals to the adjustable system and using an alternative description by input/output relations one can straightforwardly use the results of the previous design.

After that, we shall consider the design of a single-input/single-output parallel MRAS described by state-space equations in which we shall directly use the output generalized error for implementing the adaptation law (case 3). Despite a different approach to solving the problem compared to the previous case discussed, the results in both cases are very close. In fact, the results obtained by this approach appear to be a particular form of the design obtained in the previous case.

We shall conclude by considering the design of single-input/single-output parallel MRAS's using a nonminimal state realization of the adjustable system which will allow us to directly use the output generalized error for implementing the adaptation law (case 4).

4.2-2 Case 1: State-Variable Filters Placed on the Input of the Adjustable System and on the Output of the Reference Model

In this case, the overall adaptive system structure is shown in Fig. 4.2-1, and it is described by the following:

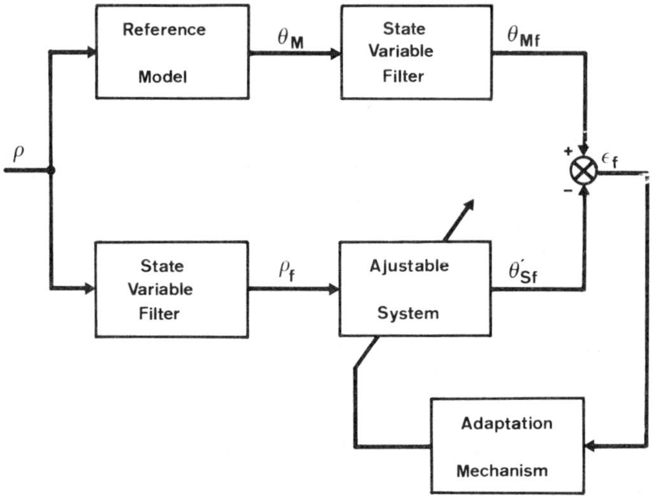

Fig. 4.2-1 Parallel MRAS using state-variable filters placed on the input of the adjustable system and on the output of the reference model.

The reference model:

$$\left(\sum_{i=0}^{n} a_i p^i\right)\theta_M = \left(\sum_{i=0}^{m} b_i p^i\right)\rho, \quad a_n = 1 \tag{4.2-1}$$

The state-variable filter acting on the model output:

$$\left(\sum_{i=0}^{n-1} c_i p^i\right)\theta_{Mf} = \theta_M, \quad c_{n-1} = 1 \tag{4.2-2}$$

The state-variable filter acting on the system input:

$$\left(\sum_{i=0}^{n-1} c_i p^i\right)\rho_f = \rho, \quad c_{n-1} = 1 \tag{4.2-3}$$

The parallel adjustable system:

$$\sum_{i=0}^{n} \hat{a}_i(\nu, t)p^i\theta'_{Sf} = \sum_{i=0}^{m} \hat{b}_i(\nu, t)p^i\rho_f, \quad \hat{a}_n(\nu, t) \equiv 1 \tag{4.2-4}$$

The generalized output error:

$$\varepsilon_f = \theta_{Mf} - \theta'_{Sf} \tag{4.2-5}$$

The adaptation mechanism:

$$\nu = D(p)\varepsilon_f = \left(\sum_{i=0}^{n=1} d_i p^i\right)\varepsilon_f \tag{4.2-6}$$

$$\hat{a}_i(\nu, t) = \int_0^t \phi_{1i}(\nu, t, \tau)d\tau + \phi_{2i}(\nu, t) + \hat{a}_i(0) \tag{4.2-7}$$

$$\hat{b}_i(\nu, t) = \int_0^t \psi_{1i}(\nu, t, \tau)d\tau + \psi_{2i}(\nu, t) + \hat{b}_i(0) \tag{4.2-8}$$

The design objective in this case is to have

$$\lim_{t\to\infty} \varepsilon_f = \lim_{t\to\infty}(\theta_{Mf} - \theta'_{Sf}) = 0 \tag{4.2-9}$$

for any initial error $\varepsilon_f(0)$, $a_i - \hat{a}_i(0)$, $b_i - \hat{b}_i(0)$ and any input ρ belonging to the class of piecewise continuous functions.

The basic adaptation laws for this scheme are summarized in Table 4.2-1, and next we shall show how they can be obtained.

Making the observation that the reference model and the state variable filter are linear and time-invariant, one can obtain the differential equation for θ_{Mf} by considering a zero-state output equivalent configuration for Eqs. (4.2-1) and (4.2-2) in which the state-variable filter is now placed before the reference model. One therefore obtains

$$\left(\sum_{i=0}^{n} a_i p^i\right)\theta_{Mf} = \left(\sum_{i=0}^{m} b_i p^i\right)\rho_f \tag{4.2-10}$$

where ρ_f is the output of the state-variable filter of Eq. (4.2-3). Adding the term $(\Sigma_{i=0}^{n} a_i p^i)\theta'_{Sf}$ to both sides of Eq. (4.2-4) and then subtracting Eq. (4.2-4) from Eq. (4.2-10), one obtains

$$\left(\sum_{i=0}^{n} a_i p^i\right)(\theta_{Mf} - \theta'_{Sf}) - \sum_{i=0}^{n} \hat{a}_i(\nu, t)p^i\theta'_{Sf}$$

$$= -\left(\sum_{i=0}^{n} a_i p^i\right)\theta'_{Sf} + \sum_{i=0}^{m} [b_i - \hat{b}_i(\nu, t)]p^i\rho_f \tag{4.2.11}$$

Table 4.2-1 Parallel MRAS Using State-Variable Filters on the Input of the Adjustable System and the Output of the Reference Model [Eqs. (4.2-1) to (4.2-8)]

	General forms	Particular forms
Adaptation laws	$\phi_{1i} = -k_{ai}(t - \tau)\nu(\tau)p^i\theta'_{Sf}(\tau), \quad \tau \leq t$	$k_{ai}(t - \tau) = k_{ai} > 0$ for all $t - \tau \geq 0$
	$\phi_{2i} = -k'_{ai}(t)\nu(t)p^i\theta'_{Sf}(t)$	$k_{bi}(t - \tau) = k_{bi} > 0$ for all $t - \tau \geq 0$
	$\Psi_{1i} = k_{bi}(t - \tau)\nu(\tau)p^i\rho_f(\tau), \quad \tau \leq t$	(integral adaptation)
	$\Psi_{2i} = k'_{bi}(t)\nu(t)p^i\rho_f(t)$	$k'_{ai}(t) = k_{ai} \geq 0, \quad k'_{bi}(t) = k_{bi} \geq 0$
	$k_{ai}(t - \tau)$, $k_{bi}(t - \tau)$ = positive definite scalar	for all $t \geq 0$ (proportional adaptation)
	kernels whose Laplace transforms are positive	$k'_{ai}(t) = k'_{bi}(t) = \dfrac{1}{\lvert\nu(t)\rvert}$
	real transfer functions with a pole at $s = 0$;	i.e., $k'_{ai}(t)\nu(t) = k'_{bi}(t)\nu(t) = \operatorname{sgn} \nu(t)$
	$k'_{ai}(t) \geq 0$, $k'_{bi}(t) \geq 0$ for all $t \geq 0$	(relay adaptation)
Stability condition	$h(s) = \dfrac{\Sigma_{i=0}^{n-1} d_i s^i}{S^n + \Sigma_{i=0}^{n-1} a_i s^i}$ = strictly positive real transfer function $\left[D(s) = \sum_{i=0}^{n-1} d_i s^i\right]$	

Also, using Eq. (4.2-5), one finally gets

$$\left(\sum_{i=0}^{n} a_i p^i\right)\varepsilon_f = \sum_{i=0}^{n} [\hat{a}_i(\nu, t) - a_i]p^i\theta'_{Sf}$$

$$+ \sum_{i=0}^{m} [b_i - \hat{b}_i(\nu, t)]p^i\rho_f \qquad (4.2\text{-}12)$$

From Eqs. (4.2-6) to (4.2-8) and (4.2-12) one obtains the following equivalent feedback system:

$$\left(\sum_{i=0}^{n} a_i p^i\right)\varepsilon_f = \omega_1 \qquad (4.2\text{-}13)$$

$$\nu = D(p)\varepsilon_f \qquad (4.2\text{-}14)$$

$$\omega = -\omega_1 = -\sum_{l=0}^{n-1} [\textstyle\int_0^t \phi_{1i}(\nu, t, \tau)\, d\tau + \phi_{2i}(\nu, t) + \hat{a}_i(0) - a_i]p^i\theta'_{Sf}$$

$$+ \sum_{i=0}^{m} [\textstyle\int_0^t \psi_{1i}(\nu, t, \tau)\, d\tau + \psi_{2i}(\nu, t) + \hat{b}_i(0) - b_i]p^i\rho_f$$

$$(4.2\text{-}15)$$

where the linear feedforward block (with input ω_1 and output ν) is described by Eqs. (4.2-13) and (4.2-14) and the time-varying nonlinear feedback block (with input ν and output ω) is described by Eq. (4.2-15). Applying Theorem 4.1-1 to the feedback system of Eqs. (4.2-13) to (4.2-15), one should find solutions for ϕ_{1i}, ϕ_{2i}, ψ_{1i}, and ψ_{2i} verifying conditions 1 and 2 of the theorem. By using, for example, Lemmas D.1-4 and D.1-6, one gets

$$\phi_{1i} = -k_{ai}(t - \tau)\nu(\tau)p^i\theta'_{Sf}(\tau), \quad \tau \leq t \qquad (4.2\text{-}16)$$

$$\phi_{2i} = -k'_{ai}(t)\nu(t)p^i\theta'_{Sf}(t) \qquad (4.2\text{-}17)$$

$$\psi_{1i} = k_{bi}(t - \tau)\nu(\tau)p^i\rho_f(\tau), \quad \tau \leq t \qquad (4.2\text{-}18)$$

$$\psi_{2i} = k'_{bi}(t)\nu(t)p^i\rho_f(t) \qquad (4.2\text{-}19)$$

where $k_{ai}(t - \tau)$ and $k_{bi}(t - \tau)$ are *positive definite scalar kernels* whose Laplace transforms are *positive real transfer functions* with a pole at $s = 0$ and $k'_{ai}(t)$, $k'_{bi}(t)$ are *time-varying positive (or nonnegative) gains* for all $t \geq 0$.

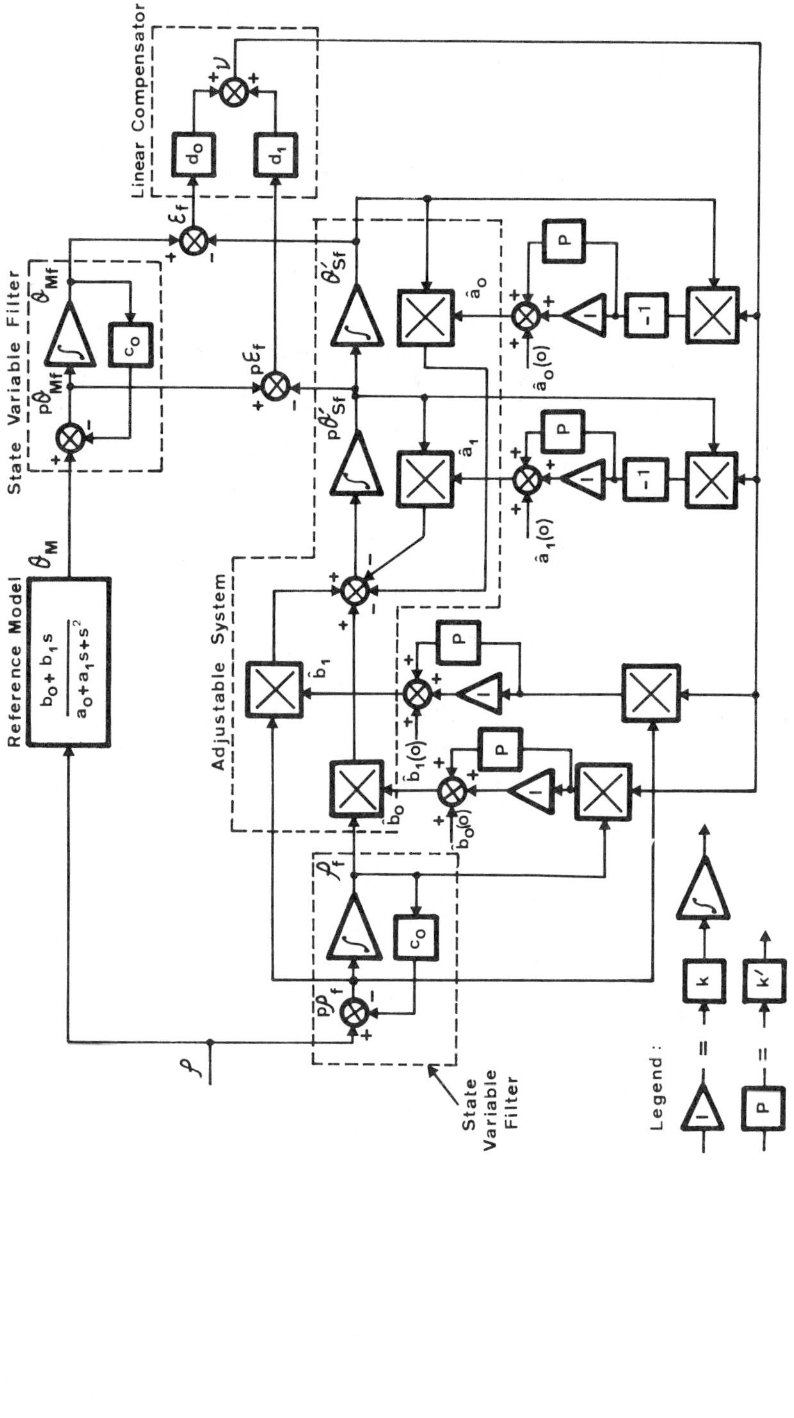

Fig. 4.2-2 Second-order parallel MRAS using state-variable filters on the input of the adjustable system and on the output of the reference model.

To verify condition 3 of Theorem 4.1-1, the transfer function of the equivalent feedforward block,

$$h(s) = \frac{D(s)}{\Sigma_{i=0}^{n} a_i s^i} = \frac{\Sigma_{i=0}^{n-1} d_i s^i}{s^n + \Sigma_{i=0}^{n-1} a_i s^i} \tag{4.2-20}$$

should be strictly positive real. The coefficients d_i can be explicitly computed by considering a state-space realization of h(s) and then applying Lemma B.2-4. An example of implementation for a second-order reference model and integral + proportional adaptation is given in Fig. 4.2-2.

This kind of MRAS is used for parameter identification when only input and output measurements are available (see Sec. 7.2-3). This design can be extended for the case of series-parallel (see Problem 4.2) and series MRAS's. A comparison of various configurations used for parameter identification is presented in Sec. 7.2-3.

4.2-3 Case 2: State-Variable Filters Placed on the Outputs of the Reference Model and of the Adjustable System

In this case, the block diagram of the adaptive system is given in Fig. 4.2-3. In comparison with the scheme given in Fig. 4.2-1, the state-variable filter is now moved from the input of the adjustable system to its output. This kind of configuration is used for building adaptative state observers (see Chap. 8) where the reference model is constituted by a process with unobservable states and unknown (or varying) parameters and the adjustable system will become an observation model. Note that once equilibrium is reached, the outputs of the adjustable system and of the reference model will be identical independently of the type of input, which is not the case for the scheme given in Fig. 4.2-1. The principles of this design are also used for designing adaptive model-following control systems when only input and output measurements are available (see Sec. 6.4). The adaptive system shown in Fig. 4.2-3 is described by the following:

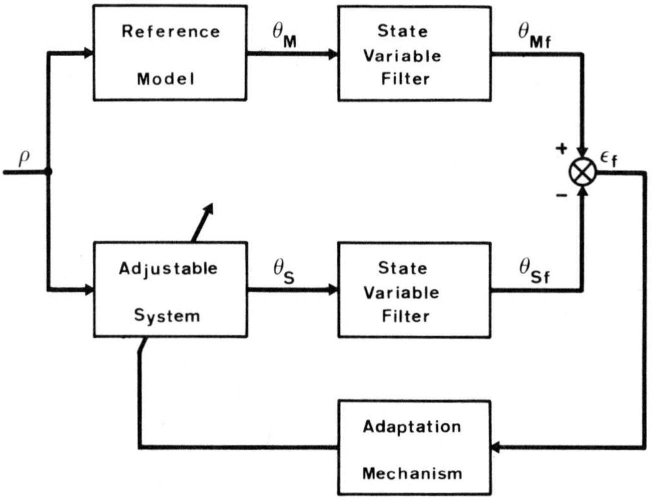

Fig. 4.2-3 Parallel MRAS using state-variable filters placed on the outputs of the reference model and of the adjustable system.

The reference model:

$$\dot{\underline{x}} = \begin{bmatrix} -a_{n-1} & 1 & & & \\ \vdots & & \ddots & & \\ & & & \ddots & \\ -a_1 & 0 & 0 & \cdots & 0 & 1 \\ -a_0 & 0 & 0 & \cdots & 0 & 0 \end{bmatrix} \underline{x} + \begin{bmatrix} 0 \\ \vdots \\ b_m \\ \vdots \\ b_0 \end{bmatrix} = A_M \underline{x} + b_M \rho \qquad (4.2\text{-}21)$$

$$\theta_M = \underline{c}^T \underline{x} = [1\ 0\ 0\ \cdots\ 0] \underline{x} = s_1 \qquad (4.2\text{-}22)$$

Note that this is an observable canonical form [3]; the transfer function of the reference model is expressed by
$h_M(s) = (b_m s^m + b_{m-1} s^{m-1} + \cdots + b_0)/(s^n + a_{n-1} s^{n-1} + \cdots + a_0)$.

The state-variable filter acting on the output of the reference model:

$$\left(\sum_{i=0}^{n-1} c_i p^i \right) \theta_{Mf} = \theta_M, \quad c_{n-1} = 1 \qquad (4.2\text{-}23)$$

The adjustable model:

$$\underline{\dot{y}} = \begin{bmatrix} -\hat{a}_{n-1}(\nu, t) & 1 & \cdots & 0 \\ & & 1 & \\ \vdots & & \ddots & \\ & & 0 & 1 \\ -\hat{a}_0(\nu, t) & 0 & \cdots 0 & 0 \end{bmatrix} \underline{y} + \begin{bmatrix} 0 \\ \vdots \\ b_m(\nu, t) \\ \vdots \\ \hat{b}_0(\nu, t) \end{bmatrix} \rho + \begin{bmatrix} 0 \\ u_{n-2,a} \\ \\ \vdots \\ u_{0,a} \end{bmatrix} + \begin{bmatrix} 0 \\ u_{n-2,b} \\ \\ \vdots \\ u_{0,b} \end{bmatrix}$$

$$= \underline{A}_S(\nu, t)y + \underline{b}_S(\nu, t)\rho + \underline{u}_a(\nu, t) + \underline{u}_b(\nu, t) \tag{4.2-24}$$

$$\theta_S = \underline{c}^T\underline{y} = [1\ 0\ \cdots\ 0]\underline{y} = y_1 \tag{4.2-25}$$

The state-variable filter acting on the output of the adjustable system:

$$\left(\sum_{i=0}^{n-1} c_i p^i\right)\theta_{Sf} = \theta_S, \quad c_{n-1} = 1 \tag{4.2-26}$$

The filtered generalized output error:

$$\varepsilon_f = \theta_{Mf} - \theta_{Sf} \tag{4.2-27}$$

The adaptation mechanism:

$$\nu = D(p)\varepsilon_f = \left(\sum_{i=0}^{n-1} d_i p^i\right)\varepsilon_f \tag{4.2-28}$$

$$\hat{a}_i(\nu, t) = \int_0^t \phi_{1i}(\nu, t, \tau)\, d\tau + \hat{a}_0(0) \tag{4.2-29}$$

$$\hat{b}_i(\nu, t) = \int_0^t \psi_{1i}(\nu, t, \tau)\, d\tau + \hat{b}_i(0) \tag{4.2-30}$$

$$u_{j,a} = \phi_{ja}(\nu, t, \theta_{Sf}), \quad j = 0, 1, \ldots, n-2 \tag{4.2-31}$$

$$u_{j,b} = \psi_{jb}(\nu, t, \rho_f), \quad j = 0, 1, \ldots, n-2 \tag{4.2-32}$$

Our objective is to apply the design results obtained for case 1 (Sec. 4.2-2) to this case, which will assure that $\lim_{t\to\infty} \varepsilon_f = 0$. Note from Eqs. (4.2-23) and (4.2-27) that ε_f is the output of a state-variable filter that has $\varepsilon = \theta_M - \theta_S$ as an input. Therefore, if the state-variable filter is asymptotically stable, $\lim_{t\to\infty} \varepsilon_f = 0$ will also imply that $\lim_{t\to\infty} \varepsilon = 0$. To apply the results from case 1, the present configuration, in which the state-variable filter is now placed on the output of the adjustable system, should have a zero-state output equivalent to the previous configuration of the adjustable system where the state-variable filter is placed on the input [i.e., the differential equations governing θ_{Sf} given by Eq. (4.2-26) and θ'_{Sf} given by Eq. (4.2-4) should be identical]. To obtain this, the adaptation signal vectors $\underline{u}_a$ and $\underline{u}_b$ have been introduced in Eq. (4.2-24), and they will be determined as part of the design. Note that in steady-state situations the two outputs θ_{Sf} and θ'_{Sf} are equivalent if the parameters $\hat{a}_i(\nu, t)$ and $\hat{b}_i(\nu, t)$ in Eqs. (4.2-4) and (4.2-24) are equal and constant. One concludes that the adaptation signal vectors $\underline{u}_a$ and $\underline{u}_b$ should vanish when adaptation is achieved, namely,

$$\underline{u}_a(\nu, t)\Big|_{\varepsilon_f \equiv 0} = \underline{0}, \quad \underline{u}_b(\nu, t)\Big|_{\varepsilon_f \equiv 0} = \underline{0} \tag{4.2-33}$$

and furthermore, since our objective is to have $\lim_{t\to\infty} \varepsilon_f(t) = 0$, we shall also have

$$\lim_{t\to\infty} \underline{u}_a(\nu, t) = \underline{0}, \quad \lim_{t\to\infty} \underline{u}_b(\nu, t) = \underline{0} \tag{4.2-34}$$

The signals $\underline{u}_a$ and $\underline{u}_b$ are therefore transient terms appearing only during the adaptation process. The parameter adaptation laws of Eqs. (4.2-29) and (4.2-30) have been chosen without transient terms since the transients have been transferred to $\underline{u}_a$ and $\underline{u}_b$.

Since we assume that $\underline{u}_a$ and $\underline{u}_b$ will be such that the two outputs θ_{Sf} and θ'_{Sf} are zero-state equivalent (i.e., the differential equations governing θ_{Sf} and θ'_{Sf} are identical), the equivalent feedback system for case 1 [Eqs. (4.2-13) to (4.2-15)] will be applicable to the present case, and the adaptation laws determined for case 1 will be applicable also (θ'_{Sf} being replaced by θ_{Sf}).

Therefore the first step of the design in this case will be the determination of $\underline{u}_a(\nu, t)$ and $\underline{u}_b(\nu, t)$ such that θ'_{Sf} obtained from Eqs. (4.2-3) and (4.2-4) will be zero-state equivalent to θ_{Sf} obtained from Eqs. (4.2-24) to (4.2-26); this will imply that the equivalent feedback system for the MRAS described by Eqs. (4.2-21) to (4.2-32) will be given by Eqs. (4.2-13), (4.2-14), and (4.2-15) with $\phi_{2i} \equiv 0$

and $\psi_{2i} \equiv 0$ [as a consequence of the form of Eqs. (4.2-29) and (4.2-30)].

To illustrate the procedure for the determination of $\underline{u}_a(\nu, t)$ and $\underline{u}_b(\nu, t)$, we shall first consider a second-order example, and then we shall state the results for the general case. Consider a second-order reference model characterized by the transfer function

$$h_M(s) = \frac{b_0 + b_1 s}{a_0 + a_1 s + s^2} \tag{4.2-35}$$

If a state-variable filter is placed on the input of the adjustable system (case 1), the differential equation for θ'_{Sf} is

$$[\hat{a}_0(\nu, t) + \hat{a}_1(\nu, t)p + p^2]\theta'_{Sf} = [\hat{b}_0(\nu, t) + \hat{b}_1(\nu, t)p]\rho_f \tag{4.2-36}$$

where

$$(c_0 + p)\rho_f = \rho \tag{4.2-37}$$

In the case where the state-variable filter is placed on the output of the adjustable system, one has

$$\dot{\underline{y}} = \begin{bmatrix} -\hat{a}_1(\nu, t) & 1 \\ -\hat{a}_0(\nu, t) & 0 \end{bmatrix} \underline{y} + \begin{bmatrix} \hat{b}_1(\nu, t) \\ \hat{b}_0(\nu, t) \end{bmatrix} \rho + \begin{bmatrix} 0 \\ u_{0,a} \end{bmatrix} + \begin{bmatrix} 0 \\ u_{0,b} \end{bmatrix} \tag{4.2-38}$$

$$\theta_S = y_1 \tag{4.2-39}$$

$$(c_0 + p)\theta_{Sf} = \theta_S \tag{4.2-40}$$

By taking sucessive derivatives of Eq. (4.2-39) and using Eq. (4.2-38), one obtains the following differential equation for θ_S:

$$\begin{aligned}[\hat{a}_0(\nu, t) + \hat{a}_1(\nu, t)p + p^2]\theta_S = {} & [\hat{b}_0(\nu, t) + \hat{b}_1(\nu, t)p]\rho \\ & - \dot{\hat{a}}_1(\nu, t)\theta_S + \dot{\hat{b}}_1(\nu, t)\rho \\ & + u_{0,a} + u_{0,b}\end{aligned} \tag{4.2-41}$$

To show that θ_{Sf} and θ'_{Sf} are zero-state equivalent, it is sufficient [from Eq. (4.2-40)] to show that $(c_0 + p)\theta'_{Sf}$ is zero-state equivalent to θ_S. Left-multiplying both sides of Eq. (4.2-36) by $c_0 + p$, one obtains

$$[\hat{a}_0(\nu, t) + \hat{a}_1(\nu, t)p + p^2](c_0 + p)\theta'_{Sf} + \dot{\hat{a}}_0(\nu, t)\theta'_{Sf} + \dot{\hat{a}}_1(\nu, t)p\theta'_{Sf}$$
$$= [\hat{b}_0(\nu, t) + \hat{b}_1(\nu, t)p](c_0 + p)\rho_f + \dot{\hat{b}}_0(\nu, t)\rho_f + \dot{\hat{b}}_1(\nu, t)p\rho_f \tag{4.2-42}$$

Rearranging the terms and also using Eq. (4.2-37), one obtains

$$[\hat{a}_0(\nu, t) + \hat{a}_1(\nu, t)p + p^2](c_0 + p)\theta'_{Sf}$$
$$= [\hat{b}_0(\nu, t) + \hat{b}_1(\nu, t)p]\rho - \dot{\hat{a}}_0(\nu, t)\theta'_{Sf} - \dot{\hat{a}}_1(\nu, t)p\theta'_{Sf}$$
$$+ \dot{\hat{b}}_0(\nu, t)\rho_f + \dot{\hat{b}}_1(\nu, t)p\rho_f \tag{4.2-43}$$

So that $(c_0 + p)\theta'_{Sf}$ will be zero-state equivalent to θ_S (which will imply that θ_{Sf} is zero-state equivalent to θ'_{Sf}), the right-hand sides of Eqs. (4.2-41) and (4.2-43) should be equal. After cancelling the term $[\hat{b}_0(\nu, t) + \hat{b}_1(\nu, t)p]\rho$ which appears in both equations, one obtains

$$-\dot{\hat{a}}_1(\nu, t)\theta_S + \dot{\hat{b}}_1(\nu, t)\rho + u_{0,a} + u_{0,b}$$
$$= -\dot{\hat{a}}_0(\nu, t)\theta'_{Sf} - \dot{\hat{a}}_1(\nu, t)p\theta'_{Sf} + \dot{\hat{b}}_0(t)\rho_f + \dot{\hat{b}}_1(t)p\rho_f \tag{4.2-44}$$

Therefore, θ_{Sf} and θ'_{Sf} will be zero-state equivalent if

$$u_{0,a} = -\dot{\hat{a}}_0(\nu, t)\theta'_{Sf} + \dot{\hat{a}}_1(\nu, t)[\theta'_S - p\theta_{Sf}] \tag{4.2-45}$$
$$u_{0,b} = \dot{\hat{b}}_0(\nu, t)\rho_f + \dot{\hat{b}}_1(\nu, t)[p\rho_f - \rho] \tag{4.2-46}$$

Using Eqs. (4.2-37) and (4.2-40) and replacing θ'_{Sf} by θ_{Sf}, the expressions for $u_{0,a}$ and $u_{0,b}$ can be written as

$$u_{0,a} = -\dot{\hat{a}}_0(\nu, t)\theta_{Sf} + \dot{\hat{a}}_1(\nu, t)c_0\theta_{Sf} \tag{4.2-47}$$
$$u_{0,b} = \dot{\hat{b}}_0(\nu, t)\rho_f - \dot{\hat{b}}_1(\nu, t)c_0\rho_f \tag{4.2-48}$$

With these adaptation signals $u_{0,a}$ and $u_{0,b}$, Eqs. (4.2-38), (4.2-39), and (4.2-40) defining the adjustable system can be replaced by Eqs. (4.2-36) and (4.2-37), and therefore one can apply the design used for the MRAS of the form given by Eqs. (4.2-1) to (4.2-8) in which $\phi_{2i} \equiv 0$ and $\psi_{2i} \equiv 0$. Therefore, the MRAS considered in case 2 will be characterized by the equivalent feedback system of Eqs. (4.2-13) to (4.2-15) with $\phi_{2i} \equiv 0$ and $\psi_{2i} \equiv 0$.

Applying the results for case 1, one finds that the transfer function

$$h(s) = \frac{D(s)}{s^n + \Sigma_{i=0}^{n-1} a_i s^i} = \frac{d_0 + d_1 s}{s^2 + a_1 s + a_0} \tag{4.2-49}$$

should be strictly positive real and $\hat{a}_0(\nu, t)$, $\hat{a}_1(\nu, t)$, $\hat{b}_0(\nu, t)$, $\hat{b}_1(\nu, t)$ are given in the case of an integral adaptation by

$$\hat{a}_0(t) = -\int_0^t k_{a0}\nu(\tau)\theta_{Sf}(\tau)\, d\tau + \hat{a}_0(0) \tag{4.2-50}$$

$$\hat{a}_1(t) = -\int_0^t k_{a1}\nu(\tau)[p\theta_{Sf}(\tau)]\, d\tau + \hat{a}_1(0) \tag{4.2-51}$$

$$\hat{b}_0(t) = \int_0^t k_{b0}\nu(\tau)\rho_f(\tau)\, d\tau + \hat{b}_0(0) \tag{4.2-52}$$

$$\hat{b}_1(t) = \int_0^t k_{b1}\nu(\tau)[p\rho_f(\tau)]\, d\tau + \hat{b}_1(0) \tag{4.2-53}$$

From Eqs. (4.2-47) and (4.2-48) and Eqs. (4.2-50) to (4.2-53), for $u_{0,a}(\nu, t)$ and $u_{0,b}(\nu, t)$ one obtains the expressions

$$u_{0,a} = k_{a0}\nu(t)\theta_{Sf}^2 - k_{a1}\nu(t)(p\theta_{Sf})c_0\theta_{Sf} \tag{4.2-54}$$

$$u_{0,b} = k_{b0}\nu(t)\rho_f^2 - k_{a1}\nu(t)(p\rho_f)c_0\rho_f \tag{4.2-55}$$

The implementation of this MRAS is indicated in Fig. 4.2-4.

Now, comparing the expression of ν, which is

$$\nu = (d_0 + d_1 p)\varepsilon_f \tag{4.2-56}$$

with the equation of ε obtained by subtracting Eq. (4.2-40) from Eq. (4.2-2), which is

$$\varepsilon = (c_0 + p)\varepsilon_f \tag{4.2-57}$$

one can see that choosing $c_0 = d_0$ *and* $d_1 = 1$ *we shall have*

$$\varepsilon = \nu \quad \text{for } c_0 = d_0,\ d_1 = 1 \tag{4.2-58}$$

and we have solved another design problem, which is the direct use of the true generalized output error for implementing the adaptation law. The implementation of the MRAS for this choice of state-variable filter, for the example considered, is represented in Fig. 4.2-5. Note that it is always possible to have h(s) given by Eq. (4.2-20) as a strictly positive real transfer function with $d_{n-1} = 1$ either by a particular choice of the matrix Q in Lemma B.2-3 or simply by dividing the coefficients of D(p) obtained for a given Q by d_{n-1}.

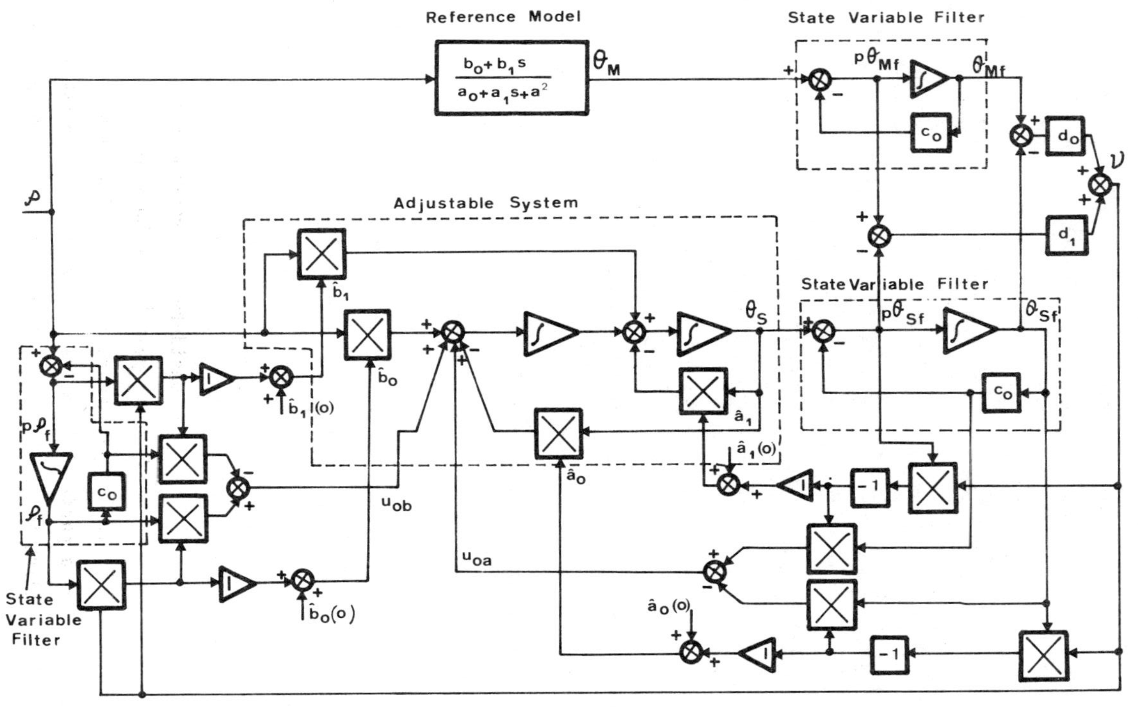

Fig. 4.2-4 Second-order parallel MRAS with integral adaptation using state-variable filters placed on the outputs of the reference model and of the adjustable system.

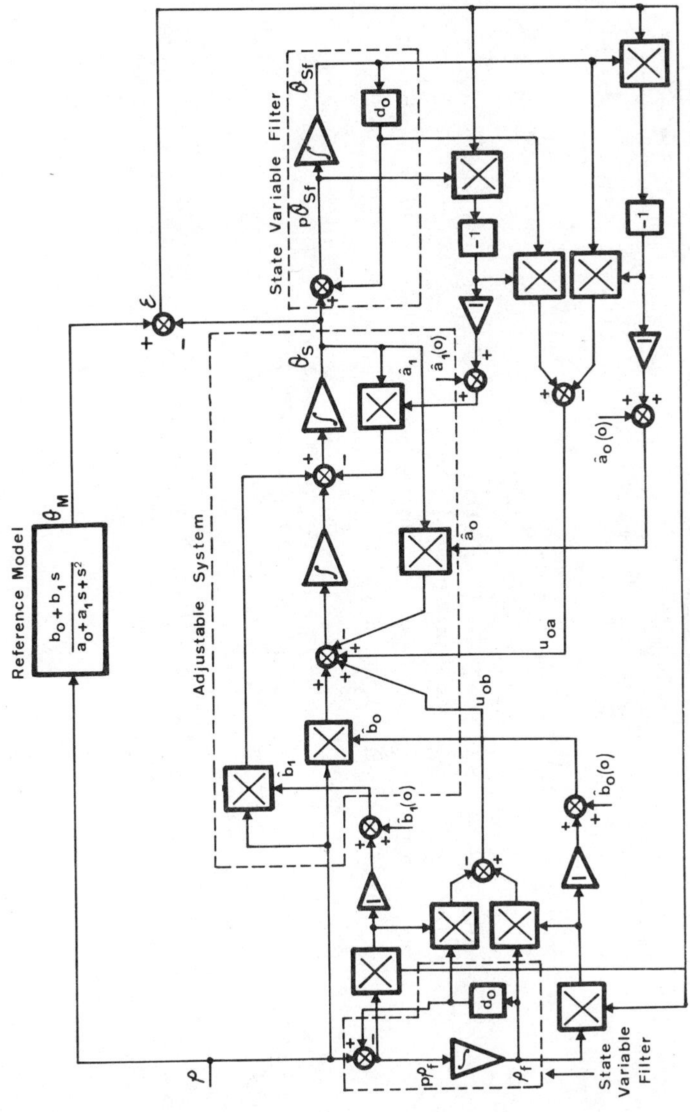

Fig. 4.2-5 Implementation of the parallel MRAS represented in Fig. 4.2-4 for the case $c_0 = d_0$, $d_1 = 1$.

Table 4.2-2 Parallel MRAS Using State-Variable Filters on the Outputs of the Reference Model and of the Adjustable System [Eqs. (4.2-21) to (4.2-32)].

	General forms	Particular forms
Parametric adaptation laws	$\hat{a}_i(\nu, t) = \int_0^t -k_{ai}(t - \tau)\nu(\tau)p^i\theta_{Sf}(\tau)\,d\tau$ $\hat{b}_i(\nu, t) = \int_0^t k_{bi}(t - \tau)\nu(\tau)p^i\rho_f(\tau)\,d\tau$ $k_{ai}(t - \tau)$, $k_{bi}(t - \tau)$ = positive definite scalar kernels whose Laplace transforms are positive real transfer functions with a pole at $s = 0$, and the degree of the numerator of the Laplace transforms is equal to the degree of the denominator minus 1	$k_{ai}(t - \tau) = k_{ai} > 0$ for all $t - \tau \geq 0$ $k_{bi}(t - \tau) = k_{bi} > 0$ for all $t - \tau \geq 0$ i.e., $\dot{\hat{a}}_i(\nu, t) = -k_{ai}\nu(t)p^i\theta_{Sf}(t)$ $\dot{\hat{b}}_i(\nu, t) = k_{bi}\nu(t)p^i\rho_f(t)$ (integral adaptation)
Adaptation signals	$u_{j,a}(\nu, t) = -\sum_{i=0}^{j} \hat{a}_i(\nu, t)\left(p^{n-j-2} + \sum_{r=0}^{n-3-j} c_{r+1+j}p^r\right)p^i\theta_{Sf}$ $+ \sum_{i=j+1}^{n-1} \hat{a}_i(\nu, t)\left(\sum_{r=0}^{j} c_r p^r\right)p^{i-j-1}\theta_{Sf}$, $j = 0, 1, \ldots, n - 2$ $u_{j,b}(\nu, t) = \sum_{i=0}^{j} \hat{b}_i(\nu, t)\left(p^{n-j-2} + \sum_{r=0}^{n-3-j} c_{r+1+j}p^r\right)p^i\rho_f$	

$$- \sum_{i=j+1}^{m} \hat{b}_i(\nu, t)\left(\sum_{r=0}^{j} c_r p^r\right) p^{i-j-1}\rho_f, \quad j = 0, 1, \ldots, n-2$$

Stability condition	$\nu = D(p)\varepsilon$ $\dfrac{\Sigma_{i=0}^{n-1} d_i s^i}{s^n + \Sigma_{i=0}^{n-1} a_i s^i}$ = strictly positive real transfer function $[D(s) = \Sigma_{i=0}^{n-1} d_i s^i]$	$\nu = \varepsilon$ if $\dfrac{s^{n-1} + \Sigma_{i=0}^{n-2} c_i s^i}{s^n + \Sigma_{i=0}^{n-1} a_i s}$ = strictly positive real transfer function [where $(s^{n-1} + \Sigma_{i=0}^{n-2} c_i s^i)^{-1}$ is the transfer function of the state-variable filters)]

Note also that the band pass of a state-variable filter with $c_i = d_i$ will be comparable with the band pass of the reference model. For example, if n = 2, h(s) given by Eq. (4.2-49) will be strictly real positive if $d_0 \leq a_1$, which implies that the pole of the state-variable filter is smaller than or equal to the sum of the poles of the reference model, i.e., the band pass of the state-variable filter will be lower than that of a filter having a pole equal to the sum of the poles of the reference model.

For the general case described by Eqs. (4.2-21) to (4.2-32) the corresponding adaptation laws are summarized in Table 4.2-2. This design can be extended to the case of series-parallel MRAS's (see Problem 4.3).

4.2-4 Case 3: Adaptation Laws Directly Using the Generalized Output Error

We have seen in the previous section that it is possible with a particular choice of state-variable filters to obtain adaptation laws which depend only on the output generalized error. We shall show next how, given an MRAS in state-variable form, it is possible to directly construct adaptation laws which do not use either the state generalized error vector or the derivatives of the output generalized error but only the output generalized error.

Consider the following single-input/single-output reference model:

$$\dot{\underline{x}} = A_M \underline{x} + b_M \rho \tag{4.2-59}$$

$$\theta_M = \underline{c}^T \underline{x} \tag{4.2-60}$$

and a parallel adjustable model:

$$\dot{\underline{y}} = A_S(\varepsilon, t)\underline{y} + b_S(\varepsilon, t)\rho \tag{4.2-61}$$

$$\theta_S = \underline{c}^T \underline{y} \tag{4.2-62}$$

The generalized state error vector is

$$\underline{e} = \underline{x} - \underline{y} \tag{4.2-63}$$

and the generalized output error is

$$\varepsilon = \theta_M - \theta_S = \underline{c}^T \underline{e} \tag{4.2-64}$$

We recall that our objective is to construct an adaptation law which depends on ε and not on $\underline{e}$ and which does not involve derivatives of ε. The adaptation laws desired should be of the form

$$A_S(\varepsilon, t) = \int_0^t \Phi_1(\varepsilon, t, \tau)\, d\tau + A_S(0) \tag{4.2-65}$$

$$\underline{b}_S(\varepsilon, t) = \int_0^t \underline{\Psi}_1(\varepsilon, t, \tau)\, d\tau + \underline{b}_S(0) \tag{4.2-66}$$

and they must assure that

$$\lim_{t\to\infty} \varepsilon = 0 \tag{4.2-67}$$

for all initial conditions $\varepsilon(0)$, $A_S(0) - A_M$, $\underline{b}_S(0) - \underline{b}_M$ and all inputs ρ belonging to the class of piecewise continuous functions.

To be able to design an MRAS which directly uses the generalized output error for the implementation of the adaptation laws, one must build an MRAS whose equivalent feedback system is of the form

$$\dot{\underline{e}}' = A_M\underline{e}' + \underline{d}\omega_1 \tag{4.2-68}$$

$$\varepsilon = \underline{c}^T\underline{e}' \tag{4.2-69}$$

$$\omega = -\omega_1 = \underline{f}^T(t, \varepsilon)\underline{z} + \underline{g}^T(t, \varepsilon)\underline{q} \tag{4.2-70}$$

where ω_1 is a scalar, $\underline{f}^T(t, \varepsilon)$ and $\underline{g}^T(t, \varepsilon)$ are vector functions which depend on ε, and $\underline{z}$ and $\underline{q}$ are auxiliary vector functions. The dimension of $\underline{f}(t, \varepsilon)$ and $\underline{z}$ depend on the number of adjustable parameters of $A_S(\varepsilon, t)$, and the dimension of $\underline{g}(t, \varepsilon)$ and $\underline{q}$ depend on the number of adjustable parameters of $\underline{b}_S(\varepsilon, t)$. The initial objective of the design will be fulfilled if (according to Theorem 4.1-1)

$$\int_0^{t_1} \varepsilon\omega\, dt = \int_0^{t_1} \varepsilon\underline{f}^T(t, \varepsilon)\underline{z}\, dt + \int_0^{t_1} \varepsilon\underline{g}^T(t, \varepsilon)\underline{q}\, dt \geq -\gamma_0^2 \quad \text{for all } t_1 \geq 0 \tag{4.2-71}$$

and the transfer function

$$h'(s) = \underline{c}^T(sI - A_M)^{-1}\, \underline{d} \tag{4.2-72}$$

is *strictly positive* real. Given $\underline{c}^T$ and A_M, $\underline{d}$ can be computed using Lemma B.2-3 (Appendix B), and one has

$$\underline{d} = P^{-1}\underline{c} \tag{4.2-73}$$

where P is the solution of the Lyapunov equation

$$A_M^T P + PA_M = -Q, \quad Q > 0 \tag{4.2-74}$$

The inequality of Eq. (4.2-71) will be verified if, for example, the components of $\underline{f}(t, \varepsilon)$ and $\underline{g}(t, \varepsilon)$ are given by

$$f_i(t, \varepsilon) = \int_0^t k_{fi}(t - \tau) z_i(\tau)\varepsilon(\tau) + f_i(0) \tag{4.2-75}$$

$$g_i(t, \varepsilon) = \int_0^t k_{gi}(t - \tau) q_i(\tau)\varepsilon(\tau) + g_i(0) \tag{4.2-76}$$

where $k_{fi}(t - \tau)$ and $k_{gi}(t - \tau)$ are *positive definite scalar kernels* whose Laplace transforms are *positive real transfer functions* with a pole at s = 0 and f_i and g_i are the components of the vectors $\underline{f}$ and $\underline{g}$, respectively.

However, the use of the equivalent feedback system of Eqs. (4.2-68) to (4.2-70) for the design of the MRAS of the form given by Eqs. (4.2-59) to (4.2-66) implies that ε given by Eqs. (4.2-68) to (4.2-70) is zero-state equivalent to ε obtained from Eqs. (4.2-59) to (4.2-64); i.e., they are governed by the same differential equations. But this cannot be obtained for the MRAS described by Eqs. (4.2-59) to (4.2-64). To obtain an equivalent feedback system which is zero-state output equivalent to that given by Eqs. (4.2-68), (4.2-69), and (4.2-70), one must modify the structure of the adjustable system by including two transient adaptation vector signals $\underline{u}_a(\varepsilon, t)$ and $\underline{u}_b(\varepsilon, t)$. These adaptation signals are such that

$$\underline{u}_a(\varepsilon, t)\Big|_{\varepsilon \equiv 0} = \underline{0}, \quad \underline{u}_b(\varepsilon, t)\Big|_{\varepsilon \equiv 0} = \underline{0} \tag{4.2-77}$$

and as the objective of the adaptation is achieved, namely, $\lim_{t\to\infty} \varepsilon(t) = 0$, we shall also have

$$\lim_{t\to\infty} \underline{u}_a(\varepsilon, t) = \underline{0}, \quad \lim_{t\to\infty} \underline{u}_b(\varepsilon, t) = \underline{0} \tag{4.2-78}$$

The new adjustable system will therefore have the structure

$$\dot{\underline{y}} = A_S(\varepsilon, t)\underline{y} + \underline{b}_S(\varepsilon, t)\rho + \underline{u}_a(\varepsilon, t) + \underline{u}_b(\varepsilon, t) \tag{4.2-79}$$

$$\theta_S = \underline{c}^T \underline{y} \tag{4.2-80}$$

and the adaptation signals $\underline{u}_a(\varepsilon, t)$ and $\underline{u}_b(\varepsilon, t)$ will be determined as part of the design.

Therefore the first step of the design in this case is to determine $\underline{u}_a(\varepsilon, t)$ and $\underline{u}_b(\varepsilon, t)$ such that the equivalent feedback system for the MRAS described by Eqs. (4.2-59), (4.2-60), (4.2-79), and (4.2-80) and Eqs. (4.2-63) to (4.2-66) be given by Eqs. (4.2-68) to (4.2-70).

Consider now the design in the case of an nth-order single-input/single-output parallel MRAS described by the following:

The reference model:

$$\dot{\underline{x}} = \begin{bmatrix} -a_{n-1} & 1 & 0 & & & 0 \\ -a_{n-2} & 0 & 1 & & & \\ & & & \ddots & & \\ \vdots & & & & 1 & 0 \\ & & & & 0 & 1 \\ -a_0 & 0 & 0 & & 0 & 0 \end{bmatrix} \underline{x} + \begin{bmatrix} 0 \\ 0 \\ b_m \\ \vdots \\ b_0 \end{bmatrix} \rho = A_M \underline{x} + \underline{b}_M \rho \tag{4.2-81}$$

$$\theta_M = [1\ 0\ \cdots\ 0]\underline{x} = \underline{c}^T \underline{x} = x_1 \tag{4.2-82}$$

The parallel adjustable system:

$$\dot{\underline{y}} = \begin{bmatrix} -\hat{a}_{n-1}(\varepsilon, t) & 1 & 0 & \cdots & 0 \\ -\hat{a}_{n-2}(\varepsilon, t) & 0 & 1 & & \\ & & & \ddots & \\ \vdots & & & 1 & 0 \\ & & & 0 & 1 \\ -\hat{a}_0(\varepsilon, t) & 0 & \cdots & 0 & 1 \end{bmatrix} \underline{y} + \begin{bmatrix} 0 \\ 0 \\ \hat{b}_m(\varepsilon, t) \\ \vdots \\ \hat{b}_0(\varepsilon, t) \end{bmatrix} \rho + \begin{bmatrix} 0 \\ u_{n-2,a}(\varepsilon, t) \\ \vdots \\ u_{0,a}(\varepsilon, t) \end{bmatrix} + \begin{bmatrix} 0 \\ u_{n-2,b}(\varepsilon, t) \\ \vdots \\ u_{0,b}(\varepsilon, t) \end{bmatrix} \tag{4.2-83}$$

$$\theta_S = [1\ 0\ \cdots\ 0]\underline{y} = \underline{c}^T \underline{y} = y_1 \tag{4.2-84}$$

Table 4.2-3 Parallel MRAS Directly Using the Generalized Output Error [Eqs. (4.2-81) to (4.2-84)

	General forms	Particular forms
Auxiliary variables (z_i, q_i)	$\left(p^{n-1} + \sum_{r=0}^{n-2} d_r p^r\right) z_1 = \theta_S \quad (z_1 = \theta_{Sf},\ z_{i+1} = p^i z_1),$ $i = 0, 1, \ldots, n-1$ $\left(p^{n-1} + \sum_{r=0}^{n-2} d_r p^r\right) q_1 = \rho \quad (q_1 = \rho_f,\ q_{i+1} = p^i q_1),$ $i = 0, 1, \ldots, n-1$	
Parametric adaptation laws	$\hat{a}_i(\varepsilon, t) = -\int_0^t k_{fi+1}(t - \tau) z_{i+1}(\tau)\varepsilon(\tau)\, d\tau + \hat{a}_i(0),$ $i = 0, 1, \ldots, n-1$ $\hat{b}_i(\varepsilon, t) = \int_0^t k_{gi+1}(t - \tau) q_{i+1}(\tau)\varepsilon(\tau)\, d\tau + b_i(0),$ $i = 0, 1, \ldots, m$ $k_{fi+1}(t - \tau),\ k_{gi+1}(t - \tau)$ = positive definite scalar kernels whose Laplace transforms are positive real transfer functions with a pole at $s = 0$, and the degree of the numerator of the Laplace transforms is equal to the degree of the denominator minus 1	$k_{fi+1}(t - \tau) = k_{fi+1} > 0$ for all $t - \tau \geq 0$ $k_{gi+1}(t - \tau) = k_{gi+1} > 0$ for all $t - \tau \geq 0$ i.e., $\dot{\hat{a}}_i(\varepsilon, t) = -k_{fi+1} z_{i+1}(t)\varepsilon(t)$ $\dot{\hat{b}}_i(\varepsilon, t) = k_{gi+1} q_{i+1}(t)\varepsilon(t)$ (integral adaptation)

Adaptation signals	$$u_{j,a}(\varepsilon, t) = -\sum_{i=0}^{j} \dot{\hat{a}}_i(\varepsilon, t)\left(p^{n-2-j} + \sum_{r=0}^{n-3-j} d_{r+1+j}p^r\right)p^i z_1 + \sum_{i=j+1}^{n-1} \dot{\hat{a}}_i(\varepsilon, t)\left(\sum_{r=0}^{j} d_r p^r\right)p^{i-j-1} z_1,$$ $j = 0, 1, \ldots, n-2$ $$u_{j,b}(\varepsilon, t) = \sum_{i=0}^{j} \dot{\hat{b}}_i(\varepsilon, t)\left(p^{n-2-j} + \sum_{r=0}^{n-3-j} d_{r+1+j}p^r\right)p^i q_1 - \sum_{i=j+1}^{m} \dot{\hat{b}}_i(\varepsilon, t)\left(\sum_{r=0}^{j} d_r p^r\right)p^{i-j-1} \rho_f,$$ $j = 0, 1, \ldots, n-2$
Stability condition	$$h'(s) = \frac{s^{n-1} + \Sigma_{i=0}^{n-2} d_i s^i}{s^n + \Sigma_{i=0}^{n-1} a_i s^i} = \text{strictly positive real}$$ transfer function

The adaptation laws are summarized in Table 4.2-3. The results represent a particular case of the design presented in Sec. 4.2-3 when the state-variable filters are chosen such that $c_i = d_i$, $i = 0, 1, \ldots, n - 2$ (see the end of Sec. 4.2-3).

Using the design presented above, we overcome the need for the derivatives of ε, but the state-variable filters acting on the input and output of the adjustable system have a band pass often comparable to that of the model (because of the positivity conditions). This will effect a slowdown of the adaptation process. Therefore, when the level of measurement noise is not too high and its spectrum is outside the band pass of the model, we are interested in using the design given in Sec. 4.2-3 with state-variable filters having a higher band pass. In this case, for implementing the adaptation laws we shall use a linear combination of the filtered output generalized error and its derivatives (which will be available). However, the design presented in Table 4.2-3 allows us to eliminate one state-variable filter (which acts in the design given in Sec. 4.2-3 on the output of the reference model), which means n - 1 integrators less than in the design given in Sec. 4.2-3 (see Fig. 4.2-5). For the case of series-parallel MRAS's, results of the same form have been obtained by Narendra and Kudva [4, 5] (see also Problem 4.3).

4.2-5 Case 4: Nonminimal State-Space Realization of the Adjustable System

In the previous approaches, state-variable filters were introduced in order to implement asymptotically stable MRAS's when only input and output measurements were used. The filters were used for generating auxiliary variables necessary for the implementation of the adaptation laws. In fact, the introduction of the state-variable filters can be interpretated as an augmentation of the state vector of the adjustable system. One, therefore, can ask if there exists a suitable nonminimal state-space realization of the adjustable system such that the need for auxiliary variables can be overcome. The answer to this question is yes, and such solutions for the case of series-parallel MRAS's have been proposed by Lüders and Narendra [6] and Anderson [7]. We shall next consider an extension of the approach proposed by Lüders and Narendra to the case of parallel MRAS's which allows us to point out some problems which do not appear in the series-parallel case (for the series-parallel case, see Problem 4.6).

The basic idea of the present design approach is that any completely controllable and observable n-dimensional, single-input/single-output system can be equivalently described by a nonminimal representation of dimension 2n - 1 having the form

$$\dot{\underline{x}} = \begin{bmatrix} \alpha_1 & \underline{a}^T & \underline{b}^T \\ \underline{r} & F & 0 \\ \underline{0} & 0 & F \end{bmatrix} \underline{x} + \begin{bmatrix} \beta_1 \\ \underline{0} \\ \underline{r} \end{bmatrix} \rho, \quad x(0) = \begin{bmatrix} x_1(0) \\ 0 \\ \underline{x}^3(0) \end{bmatrix} \qquad (4.2\text{-}85)$$

$$\theta_M = \underline{c}^T \underline{x} = x_1 \qquad (4.2\text{-}86)$$

[See Fig. 4.2-6 for an example.]

In Eq. (4.2-85), the state vector $\underline{x}$ is a (2n - 1)-dimensional vector composed of

$$\underline{x}^T = [x_1, \underline{x}^{2T}, \underline{x}^{3T}] \qquad (4.2\text{-}87)$$

where x_1 is a scalar and $\underline{x}^2$ and $\underline{x}^3$ are (n - 1)-dimensional vectors, α_1, β_1 are scalar parameters, and

$$\underline{a}^T = [\alpha_2 \cdots \alpha_n], \quad \underline{b}^T = [\beta_2 \cdots \beta_n] \qquad (4.2\text{-}88)$$

are parameter vectors, (n - 1)-dimensional. The pair ($\underline{r}$, F) can be any completely observable pair. However, a canonical representation or the transfer function characterizing the system can easily be derived if one chooses

$$\underline{r}^T = [1 \cdots 1]$$

$$F = \mathrm{diag}[-\lambda_i], \quad i = 2, 3, \ldots, n, \ \lambda_i \neq \lambda_j, \ i \neq j \qquad (4.2\text{-}89)$$

where all $\lambda_i > 0$, i = 2, 3, ..., n, and $\underline{r}$ is an (n - 1)-dimensional vector.

Table 4.2-4 Parallel MRAS Using a Nonminimal Space Realization of the Adjustable System [Eqs. (4.2-85) to (4.2-89) and Eqs. (4.2-91) to (4.2-94)]

	General forms	Particular forms
Adaptation laws	$\hat{\alpha}_1(\varepsilon, t) = \int_0^t k_{1a}{}^1(t - \tau) y_1(\tau)\varepsilon(\tau)\, d\tau + k_{2a}{}^1(t) y_1(t)\varepsilon(t) + \hat{\alpha}_1(0)$	$k_{1a}{}^1(t - \tau) = k_{1a} > 0$, $k_{1b}{}^1(t - \tau) = k_{1b} > 0$, $K_1{}^a(t - \tau) = K_1{}^a > 0$, $K_1{}^b(t - \tau) = K_1{}^b > 0$ for all $t - \tau \geq 0$
	$\hat{\beta}_1(\varepsilon, t) = \int_0^t k_{1b}{}^1(t - \tau)\rho(\tau)\, d\tau + k_{2b}{}^1(t)\rho(t)\varepsilon(t) + \hat{\beta}_1(0)$	(integral adaptation)
	$\underline{\hat{a}}(\varepsilon, t) = \int_0^t K_1{}^a(t - \tau)\underline{y}^2(\tau)\varepsilon(\tau)\, d\tau + K_2{}^a(t) y^2(t)\varepsilon(t) + \hat{a}(0)$	$k_{2a}{}^1(t) = k_{2a}{}^1 \geq 0$, $k_{2b}{}^1(t) = k_{2b}{}^1 \geq 0$, $K_2{}^a(t) = K_2{}^a \geq 0$, $K_2{}^b(t) = K_2{}^b \geq 0$ for all $t \geq 0$
	$\underline{\hat{b}}(\varepsilon, t) = \int_0^t K_1{}^b(t - \tau)\underline{y}^3(\tau)\varepsilon(\tau)\, d\tau + K_2{}^b y^3(t)\varepsilon(t) + \hat{b}(0)$	(proportional adaptation)
	$k_{1a}{}^1(t - \tau)$, $k_{1b}{}^1(t - \tau)$ and $K_1{}^a(t - \tau)$,	

	$K_1^b(t - \tau)$ are positive definite kernels (scalars and matrices, respectively) whose Laplace transforms are positive real transfer functions and matrices, respectively, with a pole at $s = 0$; $k_{2a}^1(t) \geq 0$, $k_{2b}^1(t) \geq 0$, $K_2^a(t) \geq 0$, $K_2^b(t) \geq 0$ for all $t \geq 0$	
Stability conditions	1. F must be an asymptotically stable matrix. 2. $h(s) = \underline{c}^T(sI - A)\underline{b}^*$ must be a strictly positive real transfer function, where $\underline{c}^T = [1\ 0 \cdots 0]$, $\underline{b}^{*T} = [1\ 0 \cdots 0]$ $A = \begin{bmatrix} \alpha_1 & \underline{a}^T \\ \underline{r} & F \end{bmatrix}$ and $(\underline{r}, F)$ must be completely observable.	1. $F = \text{diag}[-\lambda_i]$, $i = 2, 3, \ldots, n$, $\lambda_i \neq \lambda_j$, $i \neq j$ $\lambda_i > 0$ $\underline{r}^T = [1\ 1 \cdots 1]$ 2. $h(s) = \dfrac{\Pi_{i=2}^n (s + \lambda_i)}{s^n + \Sigma_{i=0}^{n-1} a_i s^i}$ must be a strictly positive real transfer function.

With the above choice for $\underline{r}$ and F in Eq. (4.2-85), the transfer function of the system described by Eqs. (4.2-85) and (4.2-86) becomes

$$h_M(s) = \frac{\Sigma_{i=0}^{n-1} b_i s^i}{s^n + \Sigma_{i=0}^{n-1} a_i s^i}$$

$$= \frac{\beta_1 + \beta_2[1/(s + \lambda_2)] + \cdots + \beta_n[1/(s + \lambda_n)]}{s - \alpha_1 - \alpha_2[1/(s + \lambda_2)] - \cdots - \alpha_n[1/(s + \lambda_n)]} \quad (4.2\text{-}90)$$

One can see that the coefficients of the transfer function (a_i and b_i) depend on the parameters α_i and β_i as well as on the values of λ_i.

Since the reference model characterized by the transfer function of Eq. (4.2-90) can be represented by Eqs. (4.2-85) and (4.2-86), one defines a parallel adjustable system in the form

$$\dot{\underline{y}} = \begin{bmatrix} \hat{\alpha}_1(\varepsilon, t) & \hat{\underline{a}}^T(\varepsilon, t) & \hat{\underline{b}}^T(\varepsilon, t) \\ \underline{r} & F & 0 \\ \underline{0} & 0 & F \end{bmatrix} y + \begin{bmatrix} \hat{\beta}_1(\varepsilon, t) \\ 0 \\ \underline{r} \end{bmatrix} \rho \quad (4.2\text{-}91)$$

$$\theta_S = \underline{c}^T \underline{y} = y_1 \quad (4.2\text{-}92)$$

where the state vector $\underline{y}$ is composed of

$$y^T = [y_1, \underline{y}^{2T}, \underline{y}^{3T}] \quad (4.2\text{-}93)$$

where y_1 is a scalar and $\underline{y}^2$ and $\underline{y}^3$ are (n - 1)-dimensional vectors. Note that $\underline{y}^2$ and $\underline{y}^3$ do not directly depend on the adjustable parameters. The generalized output error is defined by

$$\varepsilon = \theta_M - \theta_S = x_1 - y_1 \quad (4.2\text{-}94)$$

The design follows the basic steps emphasized in Sec. 3.3; however, in this case the equivalent linear time-invariant feedforward block will not be completely controllable. In this case, one should apply Theorem C-5 instead of Theorem C-2 (Appendix C). The corresponding adaptation laws are summarized in Table 4.2-4.

The first stability condition indicated in Table 4.2-4 assures that the uncontrollable states of the equivalent linear feedforward block are asymptotically stable. The second condition assures that

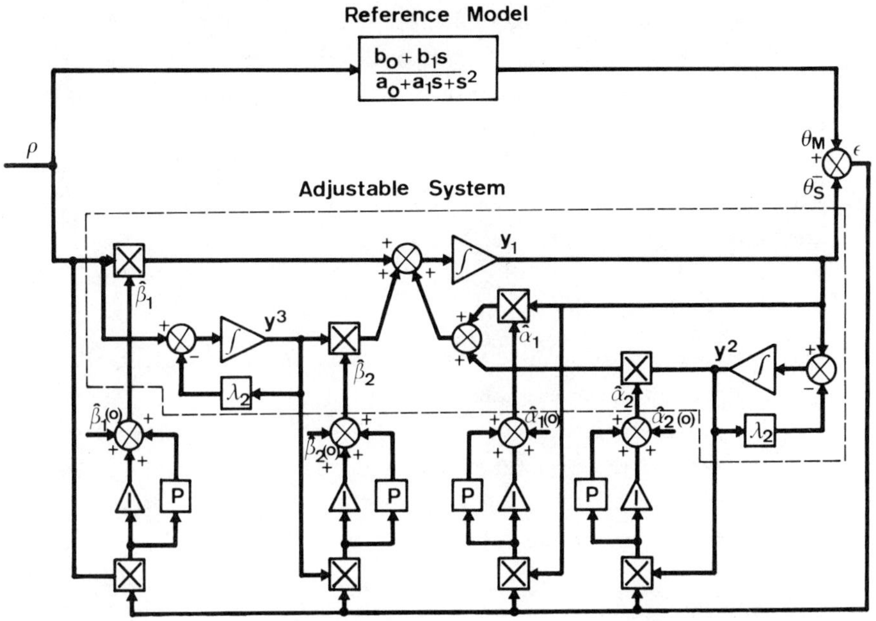

Fig. 4.2-6 Second-order parallel MRAS with integral + proportional adaptation using a nonminimal realization of the adjustable system.

the remaining completely observable and controllable part is strictly positive real, as required by Theorem C-5.

For the case of *integral + proportional adaptation* and n = 2, the complete scheme of the adaptive system is represented in Fig. 4.2-6. Note that compared to cases 1, 2, and 3 (see, for example, Figs. 4.2-2, 4.2-4, and 4.2-5) the number of integrators is lower. In case 2 (Fig. 4.2-5), with the particular choice $c_1 = d_0$, which corresponds to case 3, one needs (without counting the adaptation loop) four integrators, while in Fig. 4.2-6, only three integrators are used. In general, instead of n + 2(n - 1) integrators, for this design we shall need only 1 + 2(n - 1) integrators (n - 1 fewer integrators). However, this approach has several limitations. If we are interested in tracking the parameters of the model transfer function or of a canonical representation, these parameters are not directly obtained (they appear to be combinations of various α_i, λ_i and β_i, λ_i, respectively). Furthermore, this scheme cannot be used for directly observing inaccessible states of the model, since we are always interested in observing the states of a minimal realization. In addition, the number of the adjustable parameters is, in general, higher than in the previous designs—which implies a higher complexity of the adaptive mechanism.

4.2-6 Concluding Comments

Four design techniques for MRAS's using only input and output measurements have been examined.

The designs presented in Secs. 4.2-2 and 4.2-5 can be used when the structure of the adjustable system can be chosen by the designer without any constraint. This situation is encountered when MRAS's are used for parameter identification [5-10]. The comparison of the two approaches has been discussed in Sec. 4.2-5. We shall emphasize here the basic conclusions:

1. The structure of the adjustable system in the design given in Sec. 4.2-5 is simpler than for the design given in Sec. 4.2-2 [$1 + 2(n - 1)$ integrators instead of $n + 2(n - 1)$ integrators are needed].
2. The number of adjustable parameters (and of course the number of adaptation chains which contain two multipliers and a proportional + integral amplifier) for the design given in Sec. 4.2-5 is higher than in the case of the design given in Sec. 4.2-2 if $m \le n - 1$ ($2n$ adjustable parameters instead of $n + m$).
3. The design given in Sec. 4.2-2 allows us to directly track the parameters of the transfer function of the reference model, while in the case of the design given in Sec. 4.2-5 the parameters of the adjustable system must be combined in order to obtain an estimation of the parameters of the

transfer function of the reference model. This can cause additional errors, for example, in the presence of measurement noise.

The designs presented in Secs. 4.2-3 and 4.2-4 have a wider field of application. They can be used for designing adaptive state observers and identifiers as well as (with some modifications) adaptive model-following control systems. Despite their different derivation, the designs are close in the sense that the design given in Sec. 4.2-4 appears as a particular case of the design given in Sec. 4.2-3. Both designs can be summarized as follows.

For the design given in Sec. 4.2-3, we define an MRAS with parametric adaptation and signal adaptation in order to obtain an equivalent feedback system described by

$$\left(p^n + \sum_{i=0}^{n-1} a_i p^i\right)\varepsilon_f = \omega_1 \tag{4.2-95}$$

$$\nu = \left(p^{n-1} + \sum_{i=0}^{n-2} d_i p^i\right)\varepsilon_f \tag{4.2-96}$$

$$\omega = -\omega_1 = \sum_{i=1}^{N} f_i(\nu, t) z_i \tag{4.2-97}$$

where the filtered generalized error ε_f is defined as the output of an asymptotically stable state-variable filter:

$$\left(p^{n-1} + \sum_{i=0}^{n-1} c_i p^i\right)\varepsilon_f = \varepsilon \tag{4.2-98}$$

To obtain an asymptotically hyperstable equivalent feedback system (which will imply that $\lim_{t\to\infty} \varepsilon_f = 0$ and $\lim_{t\to\infty} \varepsilon = 0$) one must have

$$\int_0^t \omega\nu \, d\tau \geq -\gamma_0^2 \quad \text{for all } t \geq 0 \tag{4.2-99}$$

and the transfer function

$$H(s) = \frac{p^{n-1} + \Sigma_{i=0}^{n-2} d_i s^i}{p^n + \Sigma_{i=0}^{n-1} a_i s^i} \tag{4.2-100}$$

must be strictly positive real.

Equations (4.2-99) and (4.2-100) allow us to design the coefficients d_i of the linear compensator and the functions $f_i(\nu, t)$. The parameters of the state-variable filter are free insofar as it remains asymptotically stable. The band pass of the state-variable filter is chosen in accordance with the level and the frequency spectrum of the noise.

For the design given in Sec. 4.2-4, one directly chooses a state-variable filter with $c_i = d_i$ such that H(s) given by Eq. (4.2-100) is strictly positive real. Then we define an MRAS with parametric adaptation and signal adaptation in order to obtain an equivalent system described by

$$\left(p^n + \sum_{i=0}^{n-1} a_i p^i\right)\varepsilon = \left(p^{n-1} + \sum_{i=0}^{n-2} d_i p^i\right)\omega_1 \tag{4.2-101}$$

$$\omega = -\omega_1 = \sum_{i=1}^{n} f_i(\varepsilon, t) z_i \tag{4.2-102}$$

The equivalent feedback system of Eqs. (4.2-101) and (4.2-102) will be asymptotically hyperstable if, in addition to the condition that H(s) given by (4.2-100) be strictly positive real, ω given by Eq. (4.2-102) satisfies the Popov integral inequality of Eq. (4.2-99).

Comparing both designs, one notes the following:

1. The design given in Sec. 4.2-4 is a particular case of the design given in Sec. 4.2-3 for $c_i = d_i$.
2. The design given in Sec. 4.2-4 requires only two state-variable filters instead of the three required by the design given in Sec. 4.2-3.
3. In the design given in Sec. 4.2-4, the band pass of the state-variable filters is comparable to that of the reference model (because of the positivity conditions). For this reason, the speed of adaptation will be, in general, lower than that which can be obtained using the design given in Sec. 4.2-3.

4.3 COMPARISON OF THE VARIOUS MODEL REFERENCE ADAPTIVE SYSTEM DESIGN APPROACHES

In this section, we shall attempt to compare the various approaches to the design of MRAS's which have been presented previously (Chap. 3 and this chapter) and to emphasize the links which can be recognized among them.

As mentioned in Sec. 3.1, the design methods based on local parametric optimization techniques have several drawbacks, some of which are recalled here:

1. The initial difference between the parameters of the adjustable system and of the reference model is assumed to be small.
2. The adaptation process is supposed to be slow.
3. No indication is given regarding the selection of the adaptation gains in order to assure the convergence of the adaptation process (stability of the overall MRAS).

In Sec. 3.2, it was shown how MRAS's designed by local parametric optimization methods can easily become unstable. This is one of the main reasons why the stability approach has been considered a basic tool for designing MRAS's. The success of this approach as the first stage in the design comes from the fact that MRAS's are time-varying, nonlinear systems, and the asymptotic stability of such a system is a fundamental problem which should be solved first.

In using stability considerations as the basis for the design of MRAS's, results have been obtained by either using suitable Lyapunov functions or hyperstability and positivity concepts. The use of these two approaches is based on the fact that an MRAS can always be represented as an equivalent feedback system having (under the hypothesis of the basic case; see Sec. 2.4) a linear, time-invariant feedforward block and a time-varying, nonlinear feedback block. This structure appears explicitly when using the hyperstability and positivity approach and indirectly when using Lyapunov functions. In the latter case, the Lyapunov functions used are always composed of two terms: a quadratic form in terms of the state variables of the linear feedforward block and a second term depending on the form of the feedback block. It is known that such types of Lyapunov functions are suitable for designing asymptotically stable, nonlinear, time-varying feedback systems.

Furthermore, the use of an equivalent feedback system with a particular structure is a crucial step in designing MRAS's which use only input and output measurements. In this case, one defines explicitly the equivalent feedback system, and therefore one finds the adequate structure of the adaptive system which leads to this equivalent feedback system (see Sec. 4.2-4).

In trying to point out the eventual connections between the design of MRAS's using local parametric optimization methods and those using stability approaches, one concludes the following:

1. A Lyapunov function of the form

$$V = \underline{e}^T P \underline{e} + \mathrm{tr}[A_M - A_S(\underline{e}, t)]^T F_A^{-1} [A_M - A_S(\underline{e}, t)] + \mathrm{tr}[B_M - B_S(\underline{e}, t)]^T F_B^{-1} [B_M - B_S(\underline{e}, t)] \qquad (4.3\text{-}1)$$

   as used in Sec. 3.2-2 appears to be at the same time the sum of the square of the state distance and of the structural distance between the model and the adjustable system. Therefore, the design method presented in Sec. 3.1-3 based on the requirement that the derivative of the state and/or structure distance should always be negative leads in this case to a globally asymptotically stable MRAS.
2. If an equivalent feedback system is derived for some of the resulting MRAS's designed using the local parametric optimization method, one concludes that the equivalent linear time-invariant feedforward block is not characterized by a strictly positive real transfer matrix (compare the design of Sec. 3.1-3 with the design of Sec. 4.1-1 for integral adaptation, for example).

3. Comparing the design based on local parametric optimization methods (gradient-like technique) which uses sensitivity filters (Sec. 3.1-2) and the design of MRAS's with only input and output measurements via the hyperstability and positivity approach (Sec. 4.2), one sees that both types of design lead to the use of filters to generate a set of auxiliary variables (in one case we use sensitivity filters, and in the other case we use state-variable filters).

To some extent the latter approach allows us to give an answer to the question, When is an MRAS using sensitivity filters stable? The answer is that these sensitivity filters do not necessarily need to have the same dynamics as the denominator of the reference model transfer function (or as the denominator of the frozen adjustable system) but should be such that the transfer function defined as the ratio between the denominator of the sensitivity filter transfer function and the denominator of the reference model transfer function is a positive real transfer function. In addition to this, to assure the global asymptotic stability, supplementary adaptation signals must be added.

One should next discuss which approach is better when an MRAS is designed from stability considerations. The potential of Lyapunov functions or hyperstability and positivity concepts in solving the design of an MRAS is theoretically the same. However, more general adaptation laws (and algorithms in the discrete case) have been obtained using the hyperstability and positivity concepts. The reason for obtaining these more general results seems to be the fact that the search for a suitable Lyapunov function for the equivalent feedback system is replaced by two positivity problems to be solved independently. The results which were obtained in solving the positivity problem associated with the equivalent nonlinear, time-varying feedback block are, in fact, the basis for the more general results obtained via the hyperstability and positivity approach.

Since the hyperstability approach leads to a globally asymptotically stable MRAS and since for a global asymptotically stable system a Lyapunov function exists, one can ask, What is the Lyapunov function which corresponds to the MRAS's designed via the hyperstability and positivity approach? For the case of continuous-time MRAS's, one has the following result [11].

THEOREM 4.3-1. The expression

$$V(\underline{e}, t) = \underline{e}^T P \underline{e} + 2 \int_{-\infty}^{t} (P\underline{e})^T [\int_{-\infty}^{t'} F_A(t' - \tau) P\underline{e}(G_A \underline{y})^T \, d\tau] \underline{y} \, dt'$$

$$+ 2 \int_{-\infty}^{t} (P\underline{e})^T F_A'(t') P\underline{e}\underline{y}^T G_A'(t) \underline{y} \, dt'$$

$$+ 2 \int_{-\infty}^{t} (P\underline{e})^T [\int_{-\infty}^{t'} F_B(t' - \tau) P\underline{e}(G_B \underline{u})^T \, d\tau] \underline{u} \, dt'$$

$$+ 2 \int_{-\infty}^{L} (P\underline{e})^T F_B'(t') P\underline{e}\underline{u}^T G_A'(t) \underline{u} \, dt' \qquad (4.3\text{-}2)$$

is a Lyapunov function for the *parallel* MRAS described by Eqs. (4.1-1) to (4.1-6) with A_M, a Hurwitz matrix, if the adaptation laws of Eqs. (4.1-12), (4.1-13), (4.1-17), and (4.1-18) are used.

Similar results can be obtained for series-parallel MRAS's. This method of constructing Lyapunov functions can also be extended to the discrete case.

All the design methods discussed above have a common characteristic: *They are of the deterministic type.* As pointed out in [10], it is possible to find in some areas of application of MRAS's (as, for example, identification) that design methods specific for these types of applications can furnish new ideas for developing other design methods for MRAS's. Two directions of investigation seem interesting: One is the use of some techniques derived from estimation theory (for example, the use of instrumental variable techniques in the presence of noise); the second, and the most promising, is the developing of design methods based on *stochastic stability concepts.* Some of these developments, which represent current areas of research, will be discussed in Chap. 7.

PROBLEMS

4.1 Given a reference model described by $(p^n + \Sigma_{i=0}^{n-1} a_i p^i)\theta_M = b_0\rho$ and a parallel adjustable system described by $[p^n + \Sigma_{i=0}^{n-1} \hat{a}_i(\varepsilon, t)p^i]\theta_S = \hat{b}_0(\varepsilon, t)\rho$, where ρ is the input, θ_M is the output of the reference model, θ_S is the output of the adjustable system, and ε is the output generalized error $\varepsilon = \theta_M - \theta_S$, determine adaptation laws for $\hat{a}_i(\varepsilon, t)$ and $\hat{b}_0(\varepsilon, t)$ leading to an asymptotically hyperstable MRAS. (Hint: Use a state-space representation in phase-variable form and apply the results of Sec. 4.1 or extend the method indicated in Sec. 3.3.)

4.2 Determine the adaptation laws for series-parallel MRAS's where the reference model is described by $(\Sigma_{i=0}^{n} a_i p^i)\theta_M = (\Sigma_{i=0}^{m} b_i p^i)\rho$, $a_0 = 1$, and for the two following structures of the adjustable system: (a) The series adjustable system is described by $\theta_{Ss} = \Sigma_{i=0}^{n} \hat{a}_i(\varepsilon_f, t)p^i\theta_M$. The parallel adjustable system is described by $\theta_{Sp} = \Sigma_{i=0}^{m} \hat{b}_i(\varepsilon_f, t)p^i\rho$, and the output generalized error is $\varepsilon = \theta_{Ss} - \theta_{Sp}$. (b) The series-parallel adjustable system is described by $\theta_S = -\Sigma_{i=0}^{n} \hat{a}_i(\varepsilon_f, t)p^i\theta_M + \Sigma_{i=0}^{m} \hat{b}_i(\varepsilon_f, t)p^i\rho$, and the output generalized error is $\varepsilon = \theta_M - \theta_S$ (ε_f is the filtered output generalized error). In the above equations, ρ is the input, θ_M is the output of the reference model, and θ_S is the output of the adjustable system. The use of pure derivatives for implementing the adjustable system and the adaptation laws must be avoided by introducing state-variable filters. (Hint: See Secs. 2.1-2 and 4.2-2.)

4.3 Consider a reference model described by

$$\begin{bmatrix} \dot{x}_1 \\ \dot{x}_2 \end{bmatrix} = \begin{bmatrix} -a_1 & 1 \\ -a_0 & 0 \end{bmatrix} \begin{bmatrix} x_1 \\ x_2 \end{bmatrix} + \begin{bmatrix} b_1 \\ b_0 \end{bmatrix} \rho$$

$$\theta_M = \underline{c}^T \underline{x} = x_1$$

and a series-parallel adjustable system described by

$$\begin{bmatrix} \dot{y}_1 \\ \dot{y}_2 \end{bmatrix} = \begin{bmatrix} -\hat{a}_1(\varepsilon, t) & 1 \\ -\hat{a}_0(\varepsilon, t) & 0 \end{bmatrix} \begin{bmatrix} x_1 \\ y_2 \end{bmatrix} + \begin{bmatrix} \hat{b}_1(\varepsilon, t) \\ \hat{b}_0(\varepsilon, t) \end{bmatrix} \rho + \begin{bmatrix} k_1 \\ k_2 \end{bmatrix} \varepsilon$$

$$+ \underline{u}_a(\varepsilon, t) + \underline{u}_b(\varepsilon, t)$$

$$\theta_S = \underline{c}^T \underline{y} = y_1$$

The output generalized error is given by $\varepsilon = \theta_M - \theta_S = x_1 - y_1$. The gains k_1, k_2 are constant gains to be chosen by the designer. Design a MRAS using (a) the output generalized error for implementing the adaptation laws and (b) a linear combination of the filtered output generalized error and of its first derivative. (c) Compare the adaptation laws obtained with those obtained in Secs. 4.2-3 and 4.2-4 in the case of a parallel adjustable system.

4.4 Extend Problem 4.3 for an n-dimensional reference model characterized by a transfer function having a numerator of degree $m \leq n - 1$.

4.5 For the design given in Sec. 4.2-4, one uses state-variable filters having the transfer function $h_F(s) = (s^{n-1} + \Sigma_{i=0}^{n-2} c_i s^i)^{-1}$, where the coefficients c_i are chosen such that the transfer function

$$h(s) = \frac{s^{n-1} + \Sigma_{i=0}^{n-2} c_i s^i}{s^n + \Sigma_{i=0}^{n-1} a_i s^i}$$

is strictly positive real (the a_i being the coefficients of the denominator of the transfer function of the reference model).

(a) For $n = 2$, find the expression for c_0 allowing us to satisfy the positivity condition and assuring the maximum band pass for the corresponding state-variable filter $(c_0 + s)^{-1}$.

(b) Using a state-space realization for $h(s)$ $(n = 2)$, how should the matrix Q be chosen as a diagonal positive matrix in Lemma B.2-3 in order to maximize the band pass of the state-variable filter.

(c) For $a_0 = 1$, $a_1 = 1.2$, find c_0 which maximizes the band pass of the state-variable filter, and compute the ratio between the band pass of the state-variable filter and that of a reference model having the transfer function $h_M(s) = (1 + 1.2s + s^2)^{-1}$.

(d) Repeat part (c) for $a_0 = 0.4$, $a_1 = 1.2$.

4.6 [5] Design an MRAS where the n-order transfer function of the reference model is given by Eq. (4.2-90) and the series-parallel adjustable system is described by

$$\dot{\underline{y}} = \begin{bmatrix} \hat{\alpha}_1(\varepsilon, t) & \hat{\underline{a}}^T(\varepsilon, t) & \hat{b}^T(\varepsilon, t) \\ \underline{r} & F & 0 \\ \underline{0} & \underline{0} & F \end{bmatrix} \begin{bmatrix} \theta_M \\ \underline{y}^2 \\ \underline{y}^3 \end{bmatrix} + \begin{bmatrix} \hat{\beta}_1(\varepsilon, t) \\ \underline{0} \\ \underline{r} \end{bmatrix} \rho + \begin{bmatrix} \lambda_1 \\ \underline{0} \\ \underline{0} \end{bmatrix} \varepsilon$$

$$\theta_S = y_1$$

where θ_M is the output of the reference model, $\underline{r}^T = [1 \cdots 1]$, $\underline{y}^T = [y_1, \underline{y}^{2T}, \underline{y}^{3T}]$, dim $\underline{r}$ = dim $\underline{y}^2$ = dim $\underline{y}^3$ = n - 1, dim $\underline{y}$ = 2n - 1, $F = \text{diag}[-\lambda_i]$, i = 2, 3, ..., n ($\lambda_i \neq \lambda_j$ for $i \neq j$), and $\lambda_1 > 0$. (Hint: Extend the design given in Sec. 4.2-5.)

4.7 Consider a bilinear reference model described by

$$\dot{\underline{x}} = A_M\underline{x} + (\Sigma_{i=1}^m C_M{}^i u_i)\underline{x} + B_M\underline{u}, \quad \underline{u}^T = [u_1 \cdots u_m]$$

and a series-parallel adjustable system described by

$$\dot{\underline{y}} = A_S(\underline{e}, t)\underline{x} + [\Sigma^m_{i=1} C_S{}^i(\underline{e}, t)u_i]\underline{x} + B_S(\underline{e}, t)\underline{u} - K\underline{e}$$

where $\underline{e} = \underline{x} - \underline{y}$.

(a) Find adaptation laws assuring $\lim_{t\to\infty} \underline{e} = \underline{0}$ for any initial generalized state error and parameter error.

(b) Extend the design for the case when the adjustable system is described by $\dot{\underline{y}} = A_S(\underline{e}, t)\underline{y} + [\Sigma_{i=1}^m C_S{}^i(\underline{e}, t)u_i]\underline{x} + B_S(\underline{e}, t)\underline{u}$.

4.8 Consider a reference model with polynomial nonlinearities on the input, $\dot{\underline{x}} = A_M\underline{x} + \Sigma_{i=1}^{\ell}\underline{b}_i\rho^i$, and an adjustable system described by $\dot{\underline{y}} = A_S(\underline{e}, t)\underline{y} + \Sigma_{i=1}^{\ell}\hat{\underline{b}}_i(\underline{e}, t)\rho^i$, where $\underline{e} = \underline{x} - \underline{y}$. Find adaptation laws assuring $\lim_{t\to\infty} \underline{e} = \underline{0}$ for any initial generalized state error and parameter error.

REFERENCES

1. I. D. Landau, A hyperstability criterion for model reference adaptive control systems, IEEE Trans. Autom. Control, *AC-15*, no. 5, 552-555 (1969).
2. I. D. Landau, A generalization of the hyperstability conditions for model reference adaptive systems, IEEE Trans. Autom. Control, *AC-17*, no. 2, 246-247 (1972).
3. R. L. Caroll and D. P. Lindorff, An adaptive observer for single-input, single-output linear systems, IEEE Trans Autom. Control, *AC-18*, no. 5, 428-435 (1973).
4. P. Kudva and K. S. Narendra, Synthesis of an adaptive observer using Lyapunov's direct method, Int. Control, *18*, 1201-1210 (1973).
5. K. S. Narendra and P. Kudva, Stable adaptive schemes for system identification and control, Parts I and II, IEEE Trans. Syst. Man Cybern., *SMC-4*, 542-560 (1974).
6. G. Lüders and K. S. Narendra, A new canonical form for an adaptive observer, IEEE Trans. Autom. Control, *AC-19*, 117-119 (1974).
7. B. D. O. Anderson, Adaptive identification of multiple-input, multiple-output plants, in *Proceedings of the 1974 IEEE Conference on Decision and Control,* Phoenix, IEEE, New York, 1974, 273-281.
8. C. C. Hang, The design of model reference parameter estimation systems using hyperstability theories, in *Proceedings of the Third IFAC Symposium on System Parameter Estimation,* North Holland/American Elsevier, The Hague, 1973, pp. 741-744.
9. I. D. Landau, Sur une méthode de synthèse des systèmes adaptatifs avec modèle utilisé pour la commande et l'identification d'une classe de procédés physiques, No. C.N.R.S.: A.0.8495, Thèse d'Etat ès Sciences, Université de Grenoble, June 1973.
10. I. D. Landau, A survey of model reference adaptive techniques. Theory and applications, Automatica, *10*, 353-379 (1974).

Chapter Five

DESIGN OF DISCRETE-TIME MODEL REFERENCE ADAPTIVE SYSTEMS

5.1 INTRODUCTION

The practical implementation of model reference adaptive systems (MRAS's) using digital computers requires the derivation of discrete-time adaptation laws. In general, the discretization of continuous systems does not present great difficulties if one is concerned with linear time-invariant systems. But in the case of MRAS's, discretization must be approached with much more caution. There are two reasons for this:

1. The MRAS's are time-varying nonlinear systems.
2. The qualitative features of the adaption process are changed *since an inherent delay of one sample appears in the adaptation loop.*

It is therefore necessary to directly develop adaptation algorithms for discrete-time MRAS's.

To derive adaptation algorithms for discrete-time MRAS's, hyperstability and positivity concepts are again very useful, and the methodology follows the basic steps encountered in the design of continuous-time MRAS's. As for continuous-time MRAS's, the first step of the design is to transform the initial system into an equivalent feedback system. But the equivalent time-varying, nonlinear part also contains a delay caused by the discrete nature of the adaptation process. The presence of this delay introduces a supplementary condition which must be satisfied by the adaptation algorithm and which does not appear in the continuous case.

Another aspect in designing discrete-time MRAS's is the fact that the digital computers used for their implementation have much more flexibility than the analog devices which are used in the continuous case. For this reason, the development of adaptation algorithms with time-varying adaptation gains (or in general using time-varying filters) instead of only constant adaptation gains (as is usual in the continuous case) becomes attractive. These kinds of algorithms feature various properties which allow one to solve diverse adaptation problems which cannot be solved using constant adaptation gains. Their implementation on a digital computer is very convenient because the adaptation gains are computed recursively at each step without requiring real-time solutions of a set of linear or nonlinear equations. Two main classes of problems are encountered for which specific adaptation algorithms should be derived:

> *Class A:* Algorithms used when frequent changes in the values of the parameters could occur either in the reference model or in the adjustable system (e.g., real-time identification or adaptive model-following control)
>
> *Class B:* Algorithms used when only an unknown initial difference between the parameters of the reference model and those of the adjustable system occurs (e.g., identification of linear time-invariant systems or adaptive model-following control of linear time-invariant systems with unknown parameters)

In most cases, the algorithms belonging to class A utilize constant adaptation gains (or in general linear time-invariant positive filters), while the algorithms belonging to class B utilize time-varying adaptation gains (decreasing gain). One should note for the algorithms of class B that once equilibrium is reached, adaptation is no longer necessary. This is the reason why these algorithms have, in general, adaptation gains which decrease in time. Suitable adaptation algorithms with time-varying gains can also be developed for a class of situations between A and B (slowly time-varying linear systems). However, this is still an area of current research.

In what follows, we shall first present an example of designing a second-order single-input/single-output system in order to emphasize the specifics of designing discrete-time MRAS's. We shall then present the derivation of algorithms for class A and B problems involving single-input/single-output MRAS's described by difference equations and multivariable MRAS's described by state-space equations. We shall conclude by discussing the design of discrete-time MRAS's when only input and output measurements are used for the implementation of the adaptation algorithms. We should mention that algorithms for discrete-time MRAS's can also be obtained using gradient techniques combined with the Lyapunov function approach [1, 2].

The designs presented in this chapter will be extensively used in Chaps. 7 and 8 as well as in Sec. 6.3-4 (they will also be experimentally evaluated in conjunction with the types of application).

5.2 DESIGN OF DISCRETE-TIME MODEL REFERENCE ADAPTIVE SYSTEMS. AN EXAMPLE

Consider the following single-input/single-output parallel MRAS:

The *reference model*:

$$\theta_M(k) = a_1\theta_M(k - 1) + a_2\theta_M(k - 2) + b_1\rho(k - 1) \tag{5.2-1}$$

where $\rho(k)$ is the input sequence, $\theta_M(k)$ is the output of the reference model, k is the sample number, and a_1, a_2, and b_1 are the parameters of the reference model.

The *parallel adjustable system*:

$$\theta_S^0(k) = \hat{a}_1(k - 1)\theta_S(k - 1) + \hat{a}_2(k - 1)\theta_S(k - 2) + \hat{b}_1(k - 1)\rho(k - 1) \tag{5.2-2}$$

$$\theta_S(k) = \hat{a}_1(k)\theta_S(k - 1) + \hat{a}_2(k)\theta_S(k - 2) + \hat{b}_1(k)\rho(k - 1) \tag{5.2-3}$$

where $\theta_S^0(k)$ is the a priori output of the adjustable system computed using the value of the adjustable parameters at the instant k - 1 and $\theta_S(k)$ is the a posteriori output of the adjustable system computed using the value of the adjustable parameters at the instant k (i.e., after the adaptation has acted). The expressions $\theta_S^0(k)$ and $\theta_S(k)$ illustrate the adaptation delay of one sample which appears when discrete adaptation is used. (As will be shown next, the introduction of a priori variables is necessary in order to circumvent the one-step delay which inherently appears in the adaptation loop.)

The *generalized output error*:

$$\varepsilon^0(k) = \theta_M(k) - \theta_S^0(k) \tag{5.2-4}$$

$$\varepsilon(k) = \theta_M(k) - \theta_S(k) \tag{5.2-5}$$

The adaptation mechanism will contain, by analogy to the continuous case, a linear compensator generating a signal ν_k:

$$\text{a priori:} \quad \nu_k^0 = \varepsilon_k^0 + \sum_{i=1}^{r} d_i \varepsilon_{k-1} \tag{5.2-6}$$

$$\text{a posteriori:} \quad \nu_k = \varepsilon_k + \sum_{i=1}^{r} d_i \varepsilon_{k-1} \tag{5.2-7}$$

The degree r and the coefficients d_i will be determined as part of the design. The signal ν_k^0 will be used to implement the adaptation algorithms, which for this example will be chosen in the form

$$\hat{a}_i(k) = \hat{a}_i(k - 1) + \phi_i(\nu_k^0) = \sum_{\ell=0}^{k} \phi_i(\nu_\ell^0) + \hat{a}_i(-1), \quad i = 1, 2 \tag{5.2-8}$$

$$\hat{b}_1(k) = \hat{b}_1(k - 1) + \psi_1(\nu_k^0) = \sum_{\ell=0}^{k} \psi_1(\nu_\ell^0) + \hat{b}_1(-1) \tag{5.2-9}$$

However, in developing the design we shall use modified adaptation algorithms of the form

$$\hat{a}_i(k) = \hat{a}_i(k - 1) + \phi_i'(\nu_k), \quad i = 1, 2 \tag{5.2-10}$$

$$\hat{b}_1(k) = \hat{b}_1(k - 1) + \psi_1'(\nu_k) \tag{5.2-11}$$

Further, in the last stage we shall determine the relation between $\phi_i'(\nu_k)$, $\phi_i(\nu_k^0)$ and $\psi_1'(\nu_k)$, and $\psi_1(\nu_k^0)$, respectively, or, more precisely, between ν_k^0 and ν_k. To derive the adaptation algorithms we shall follow the basic steps described in Sec. 3.3 and already used in the continuous-time case.

First Step

Subtracting Eq. (5.2-3) from Eq. (5.2-1) and also using Eq. (5.2-5), we obtain

$$\begin{aligned} \varepsilon_k = a_1\varepsilon_{k-1} + a_2\varepsilon_{k-2} + [a_1 - \hat{a}_1(k)]\theta_S(k - 1) \\ + [a_2 - \hat{a}_2(k)]\theta_S(k - 2) + [b_1 - \hat{b}_1(k)]\rho(k - 1) \end{aligned} \tag{5.2-12}$$

Also using Eqs. (5.2-7), (5.2-10), and (5.2-11), one obtains the following equivalent feedback system:

$$\varepsilon_k = a_1\varepsilon_{k-1} + a_2\varepsilon_{k-2} + \omega_1(k) \qquad (5.2\text{-}13)$$

$$\nu_k = \varepsilon_k + \sum_{i=1}^{r} d_i\varepsilon_{k-i} \qquad (5.2\text{-}14)$$

$$\omega(k) = -\omega_1(k) = \sum_{i=1}^{2}\left[\sum_{\ell=0}^{k}\phi_i'(\nu_\ell) + \hat{a}_i(-1) - a_i\right]\theta_S(k - i)$$
$$+ \left[\sum_{\ell=0}^{k}\psi_1'(\nu_\ell) + \hat{b}_1(-1) - b_1\right]\rho(k - 1) \qquad (5.2\text{-}15)$$

where Eqs. (5.2-13) and (5.2-14) define a linear time-invariant feed-forward block and Eq. (5.2-15) defines a nonlinear time-varying feedback block.

Second Step

To be able to apply the hyperstability Theorem C-4 (Appendix C) to the equivalent feedback system of Eqs. (5.2-13) to (5.2-15) one should first determine $\phi_i'(\nu_\ell)$ and $\psi_1'(\nu_\ell)$ such that the equivalent feedback block defined by Eq. (5.2-15) satisfies an inequality of the form of Eq. (C-8) (Appendix C), which in this case becomes

$$\eta(0, k_1) = \sum_{k=0}^{k_1}\omega(k)\nu_k \geq -\gamma_0^2 \quad \text{for all } k_1 \geq 0 \qquad (5.2\text{-}16)$$

By using Eq. (5.2-15) the inequality of Eq. (5.2-16) becomes

$$\eta(0, k_1) = \sum_{i=1}^{2}\sum_{k=0}^{k_1}\nu_k\theta_S(k - i)\left[\sum_{\ell=0}^{k}\phi_i'(\nu_\ell) \quad \hat{a}_i(-1) - a_i\right]$$
$$+ \sum_{k=0}^{k_1}\nu_k\rho(k - 1)\left[\sum_{\ell=0}^{k}\psi_1'(\nu_\ell) + \hat{b}_1(-1) - b_1\right] \geq -\gamma_0^2$$
$$(5.2\text{-}17)$$

where k = 0 corresponds to the beginning of the adaptation. Note that the adjustable system and the reference model can be excited by the input ρ before the adaptation starts (i.e., for $k < 0$). One observes that inequality (5.2-17) will be satisfied if each of the three terms in the left-hand side satisfy an inequality of the same type. To find solutions for ϕ_i' and ψ_1' leading to the satisfaction of such inequalities, one can either use the results given in Appendix D, Lemmas D.2-1 to D.2-5, or one can find particular solutions by the use of the following relation [3]:

$$\sum_{k=0}^{k_1} x_k\left(\sum_{\ell=0}^{k} x_\ell + c\right) = \frac{1}{2}\left(\sum_{k=0}^{k_1} x_k + c\right)^2 + \frac{1}{2}\sum_{k=0}^{k_1} x_k^2 - \frac{c^2}{2} \geq -\frac{c^2}{2} \tag{5.2-18}$$

Using the relation given by Eq. (5.2-18), one obtains the following particular solutions for ϕ_i' and ψ_1':

$$\phi_i'(\nu_k) = \alpha_i \nu_k \theta_S(k - i), \quad \alpha_i > 0, \ i = 1, 2 \tag{5.2-19}$$

$$\psi_1'(\nu_k) = \beta_1 \nu_k \rho(k - 1), \quad \beta_1 > 0 \tag{5.2-20}$$

Equation (5.2-18) and the form of the left-hand side of inequality (5.2-17) explain why one obtains adaptation laws at this stage which depend on ν_k and not on ν_k^0. The reason is that ν_k which appears in the expression of $\eta(0, k_1)$ should appear in ϕ_i' and ψ_1' in order to satisfy inequality (5.2-17).

Third Step

To have $\lim_{k\to\infty} \varepsilon_k = 0$ for all ε_0, $\hat{a}_i(-1) - a_i$, $\hat{b}_1(-1) - b_1$, and all bounded input sequences ρ_k, the equivalent feedback system defined by Eqs. (5.2-13), (5.2-14), and (5.2-15) should be asymptotically hyperstable. With the feedback block of Eq. (5.2-15) satisfying inequality (5.2-16) one can then apply the Popov hyperstability theorem for discrete systems (Appendix C, Theorem C-4). Therefore, the discrete transfer function of the equivalent feedforward block defined by Eqs. (5.2-13) and (5.2-14), which is

$$h(z) = \frac{1 + \sum_{i=1}^{r} d_i z^{-1}}{1 - a_1 z^{-1} - a_2 z^{-2}} \tag{5.2-21}$$

should be strictly positive real.

So that the transfer function of Eq. (5.2-21) will be strictly positive real, the poles of h(z) should lie in $|z| < 1$. Considering now the parameter plane a_1-a_2, we note that this condition is satisfied if a_1 and a_2 lie within the triangle shown in Fig. 5.2-1. This stability domain is defined by the inequalities

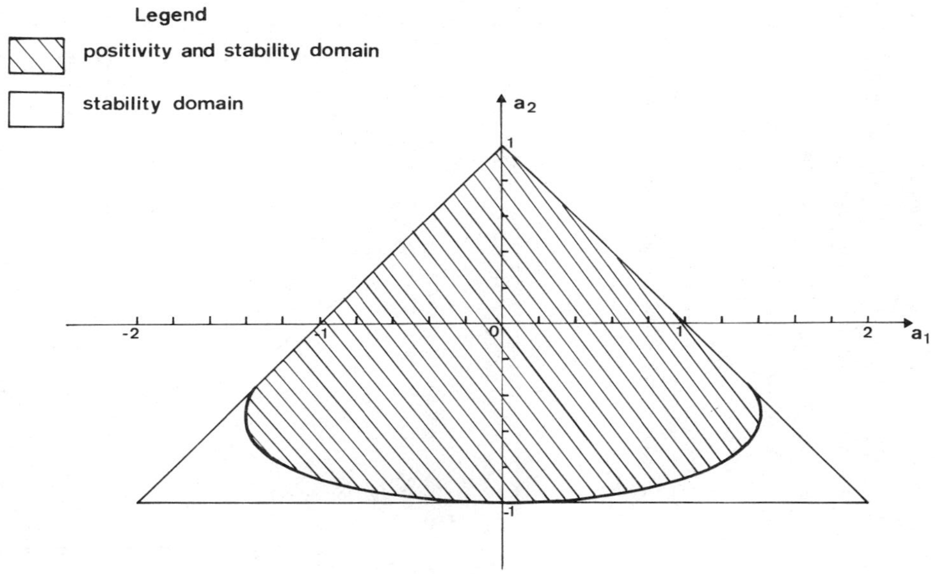

Fig. 5.2-1 Stability and positivity domains in the parameter plane for the discrete transfer function $(1 - a_1 z^{-1} - a_2 z^{-2})^{-1}$.

$$\begin{aligned} 1 + a_1 - a_2 &> 0 \\ 1 - a_1 - a_2 &> 0 \\ 1 + a_2 &> 0 \end{aligned} \tag{5.2-22}$$

To explicitly compute the coefficients d_i one can either apply the transformation $z = (1 + s)/(1 - s)$ and convert the problem into one in the continuous-time domain, or one can apply Lemma B.4-2 (Appendix B) after obtaining a state-space realization for $h(z)$. For this simple example we shall use the first approach.

Applying first the transformation $z = (1 + s)/(1 - s)$ to Eq. (5.2-21) with $r = 0$ (i.e., $d_i \equiv 0$), one obtains

$$h'(s) = \frac{1 + 2s + s^2}{(1 - a_1 - a_2) + s(2 + 2a_2) + s^2(1 + a_1 - a_2)} \tag{5.2-23}$$

and the real part of $h'(s)$ for $s = j\omega$ is given by

$$\text{Re } h'(j\omega) = \frac{(1 - a_1 - a_2) + 2(1 + 3a_2)\omega^2 + (1 + a_1 - a_2)\omega^4}{[1 - a_1 - a_2 - \omega^2(1 + a_1 - a_2)]^2 + 4\omega^2(1 + a_2)} \tag{5.2-24}$$

The real part of $h'(j\omega)$ will be strictly positive real for any real ω, and therefore $h'(s)$ will be strictly positive real if, in addition to the condition of Eq. (5.2-22), a_1 and a_2 satisfy one of the following two conditions:

$$\begin{aligned} &1 + 3a_2 \geq 0 \\ &(1 + 3a_2)^2 - (1 + a_1 - a_2)(1 - a_1 - a_2) < 0 \end{aligned} \tag{5.2-25}$$

The corresponding domain in the a_1-a_2 plane for which strictly real positivity is assured with $r = 0$ is shown in Fig. 5.2-1 by the cross-hatched area [4]. We note that this domain is smaller than the stability domain.

To obtain a strictly positive real transfer function when a_1 and a_2 are outside of this domain but still in the stability domain, one must consider either $r = 1$ or $r = 2$.

For $r = 1$, applying the same procedure as above and taking into account that $-1 < d_1 < 1$ (the numerator should be asymptotically stable), one finds that d_1, a_1, and a_2 must satisfy either the condition

$$1 - d_1a_1 + 3a_2 \geq 0 \tag{5.2-26}$$

or the condition

$$(1 - d_1a_1 + 3a_2)^2 - (1 - d_1^2)(1 + a_1 - a_2)(1 - a_1 - a_2) < 0 \tag{5.2-27}$$

The condition of Eq. (5.2-27) can always be satisfied by a convenient choice of d_1, and in particular it is satisfied for any values of a_1 and a_2 within the stability domain if one chooses

$$d_1 = -0.5a_1 \tag{5.2-28}$$

For $r = 2$, applying the transformation $z = (1 + s)/(1 - s)$, one obtains

$$h'(s) = \frac{1 + d_1 + d_2}{1 - a_1 - a_2} + \frac{(1 + s)\left[\dfrac{2(1 - d_2)}{1 + d_1d_2}\right] + s^2\left[\dfrac{1 + d_2 - d_1}{1 + d_1 + d_2}\right]}{(1 + s)\left[\dfrac{2(1 + a_2)}{1 - a_1 - a_2}\right] + s^2\left[\dfrac{1 + a_1 - a_2}{1 - a_1 - a_2}\right]} \tag{5.2-29}$$

If we choose

$$d_2 = d_1 - 1 \tag{5.2-30}$$

the coefficient of s^2 in the numerator of $h'(s)$ becomes null, and $h'(s)$ has the form of a transfer function already discussed in Sec. 3.3. Therefore so that $h'(s)$ will be strictly positive real one should satisfy the inequality

$$\frac{2(1 - d_2)}{1 + d_1 + d_2} = \frac{2 - d_1}{d_1} \geq \frac{1 + a_1 - a_2}{2(1 + a_2)} \tag{5.2-31}$$

which becomes

$$d_1 \leq \frac{4(1 + a_2)}{3 + a_1 + a_2} \tag{5.2-32}$$

Fourth Step (Return to the Original Structure)

In this last step, it only remains to determine the adaptation law of the form given by Eqs. (5.2-8) and (5.2-9) knowing the adaptation laws given by Eqs. (5.2-19) and (5.2-20) which use ν_k instead of ν_k^0. To do this, we shall try first to find the relation between ν_k^0 and ν_k, and then we shall express the adaptation laws of Eqs. (5.2-19) and (5.2-20) in terms of ν_k^0 instead of ν_k.

The explicit expression of ν_k is obtained from Eq. (5.2-14) using Eq. (5.2-12). One obtains

$$\begin{aligned}\nu_k &= \varepsilon_k + d_1\varepsilon_{k-1} + d_2\varepsilon_{k-2} \\ &= a_1\varepsilon_{k-1} + a_2\varepsilon_{k-2} + [a_1 - \hat{a}_1(k)]\theta_S(k-1) + [a_2 - a_2(k)]\theta_S(k-2) \\ &\quad + [b_1 - \hat{b}_1(k)]\rho(k-1) + d_1\varepsilon_{k-1} + d_2\varepsilon_{k-2}\end{aligned} \tag{5.2-33}$$

Replacing $\hat{a}_1(k)$, $\hat{a}_2(k)$, $\hat{b}_1(k)$ in Eq. (5.2-33) by their expressions given by Eqs. (5.2-10), (5.2-11), (5.2-19), and (5.2-20), one obtains

$$\begin{aligned}\nu_k &= a_1\varepsilon_{k-1} + a_2\varepsilon_{k-2} + [a_1 - \hat{a}_1(k-1) - \alpha_1\nu_k\theta_S(k-1)]\theta_S(k-1) \\ &\quad + [a_2 - a_2(k-1) - \alpha_2\nu_k\theta_S(k-2)]\theta_S(k-2) \\ &\quad + [b_1 - \hat{b}_1(k-1) - \beta_1\nu_k\rho(k-1)]\rho(k-1) + d_1\varepsilon_{k-1} + d_2\varepsilon_{k-2}\end{aligned} \tag{5.2-34}$$

The expression for ν_k^0 is obtained by using Eqs. (5.2-6), (5.2-1), (5.2-2), and (5.2-4):

$$\begin{aligned}\nu_k^0 &= a_1\varepsilon_{k-1} + a_2\varepsilon_{k-2} + [a_1 - \hat{a}_1(k-1)]\theta_S(k-1) \\ &\quad + [a_2 - \hat{a}_2(k-1)]\theta_S(k-2) + [b_1 - \hat{b}_1(k-1)]\rho(k-1) \\ &\quad + d_1\varepsilon_{k-1} + d_2\varepsilon_{k-2}\end{aligned} \tag{5.2-35}$$

Subtracting Eq. (5.2-34) from Eq. (5.2-35), one obtains

$$\nu_k^0 - \nu_k = \sum_{i=1}^{2} \alpha_i\nu_k\theta_S^2(k-i) + \beta_1\nu_k\rho^2(k-1) \tag{5.2-36}$$

which yields

$$\nu_k = \frac{\nu_k^0}{1 + \Sigma_{i=1}^2 \alpha_i \theta_S^2(k - i) + \beta_1 \rho^2(k - 1)} \tag{5.2-37}$$

Using Eq. (5.2-37), the adaptation laws of Eqs. (5.2-10), (5.2-11), (5.2-19), and (5.2-20) become

$$\hat{a}_i(k) = \hat{a}_i(k - 1) + \frac{\alpha_i \theta_S(k - i)}{1 + \Sigma_{i=1}^2 \alpha_i \theta_S^2(k - i) + \beta_1 \rho^2(k - 1)} \nu_k^0, \quad i = 1, 2 \tag{5.2-38}$$

$$\hat{b}_1(k) = \hat{b}_1(k - 1) + \frac{\beta_1 \rho(k - 1)}{1 + \Sigma_{i=1}^2 \alpha_i \theta_S^2(k - i) + \beta_1 \rho^2(k - 1)} \nu_k^0 \tag{5.2-39}$$

Note that ν_k^0 can be computed using Eqs. (5.2-1), (5.2-2), and (5.2-6). One obtains

$$\nu_k^0 = \theta_M(k) - \sum_{i=1}^{2} \hat{a}_i(k - 1)\theta_S(k - i) - \hat{b}_1(k - 1)\rho(k - 1) + \sum_{i=1}^{2} d_i \varepsilon_{k-i} \tag{5.2-40}$$

The adaptation delay at the instant k [if one directly uses the measurement of $\theta_M(k)$] is equal to the time necessary for computing ν_k^0 and for multiplication of ν_k^0 with the weighting factors appearing in Eqs. (5.2-38) and (5.2-39), which are already computed at k - 1.

5.3 DISCRETE MODEL REFERENCE ADAPTIVE SYSTEMS DESCRIBED BY DIFFERENCE EQUATIONS

Even though the design of such discrete MRAS's are direct generalizations of the example presented in the previous section, we shall emphasize here the various types of algorithms which can be obtained for problems of classes A and B.

5.3-1 Adaptation Algorithms for Parallel Model Reference Adaptive Systems

Consider the parallel MRAS described in Sec. 2.1.2 and restated here for the discrete-time case:

The *reference model*:

$$\theta_M(k) = \sum_{i=1}^{n} a_i\theta_M(k - i) + \sum_{i=0}^{m} b_i\rho(k - i)$$

$$= \underline{p}^T\underline{x}_{k-1} \tag{5.3-1}$$

where

$$\underline{p}^T = [a_1, \ldots, a_n, b_0, \ldots, b_m] \tag{5.3-2}$$

$$\underline{x}_{k-1}{}^T = [\theta_M(k - 1), \ldots, \theta_M(k - n), \rho(k), \ldots, \rho(k - m)] \tag{5.3-3}$$

$\underline{p}$ is the parameter vector, $\theta_M(k)$ is the model output at instant k, and $\rho(k)$ is the model input at instant k. [The reference model is characterized by the discrete transfer function $h_M(z) = (\Sigma_{i=0}^{m} b_i z^{-i})/(1 - \Sigma_{i=1}^{n} a_i z^{-i})$].

The *parallel adjustable system*:

$$\theta_S(k) = \sum_{i=1}^{n} \hat{a}_i(k)\theta_S(k - i) + \sum_{i=0}^{m} \hat{b}_i(k)\rho(k - i)$$

$$= \hat{\underline{p}}^T(k)\underline{y}_{k-1} = [\hat{\underline{p}}^I(k) + \hat{\underline{p}}^P(k)]^T\underline{y}_{k-1} \tag{5.3-4}$$

$$\theta_S{}^0(k) = [\hat{\underline{p}}^I(k - 1)]^T\underline{y}_{k-1} \tag{5.3-5}$$

where

$$\hat{\underline{p}}^T(k) = [\hat{a}_1(k), \ldots, \hat{a}_n(k), \hat{b}_0(k), \ldots, \hat{b}_m(k)] \tag{5.3-6}$$

$$\underline{y}_{k-1}{}^T = [\theta_S(k - 1), \ldots, \theta_S(k - n), \rho(k), \ldots, \rho(k - m)] \tag{5.3-7}$$

and $\theta_S{}^0(k)$ and $\theta_S(k)$ are the a priori and a posteriori outputs, respectively, of the adjustable system at instant k.

The *generalized error*:

$$\text{a priori:} \quad \varepsilon_k{}^0 = \theta_M(k) - \theta_S{}^0(k) \tag{5.3-8}$$

$$\text{a posteriori:} \quad \varepsilon_k = \theta_M(k) - \theta_S(k) \tag{5.3-9}$$

The *adaptation algorithm*:

$$\nu_k^0 = \varepsilon_k^0 + \sum_{i=1}^{n} d_i \varepsilon_{k-i} \tag{5.3-10}$$

$$\nu_k = \varepsilon_k + \sum_{i=1}^{n} d_i \varepsilon_{k-i} \tag{5.3-11}$$

$$\underline{\hat{p}}(k) = \underline{\hat{p}}^I(k) + \underline{\hat{p}}^P(k) \tag{5.3-12}$$

$$\underline{\hat{p}}^I(k) = \underline{\hat{p}}^I(k-1) + \phi_1(\nu_k^0) = \sum_{\ell=0}^{k} \phi_1(\nu_\ell^0) + \underline{\hat{p}}^I(-1) \tag{5.3-13}$$

$$\underline{\hat{p}}^P(k) = \phi_2(\nu_k^0) \tag{5.3-14}$$

In Eqs. (5.3-12) and (5.3-13), $\underline{\hat{p}}^I(k)$ corresponds to the part of the adaptation algorithm which provides the memory of the adaptation mechanism and $\underline{\hat{p}}^P(k)$ is a transient term which vanishes when $\nu_k^0 = 0$ [$\lim_{k\to\infty} \nu_k^0 = 0$ implies that $\lim_{k\to\infty} \underline{\hat{p}}(k) = \lim_{k\to\infty} \underline{\hat{p}}^I(k)$]. In other words, $\underline{\hat{p}}^P(k)$ represents the memoryless part of the adaptation algorithm.

As in the example considered in Sec. 5.2 when deriving explicit forms for the adaptation algorithms of Eqs. (5.3-13) and (5.3-14), in the intermediate steps, we shall use a modified adaptation algorithm of the form

$$\underline{\hat{p}}^I(k) = \underline{\hat{p}}^I(k-1) + \underline{\phi}_1'(\nu_k) \tag{5.3-15}$$

$$\underline{\hat{p}}^P(k) = \underline{\phi}_2'(\nu_k) \tag{5.3-16}$$

and the relations between ν_k^0 and ν_k will be established, allowing us to convert the algorithms of the form given by Eqs. (5.3-15) and (5.3-16) to the form given by Eqs. (5.3-13) and (5.3-14). The structures of $\underline{\phi}_1'(\nu_k)$ and $\phi_2'(\nu_k)$, which will be used in the various adaptation algorithms to be presented next, have been suggested by the solutions for the inequality considered in Appendix D, Section D2. For adaptation problems belonging to class A, one has the following algorithm:

THEOREM 5.3-1 (class A algorithm). The parallel MRAS described by Eqs. (5.3-1) to (5.3-14) is globally asymptotically stable if the following adaptation algorithm is used:

1. $$\hat{\underline{p}}(k) = \hat{\underline{p}}^I(k) + \hat{\underline{p}}^P(k) \qquad (5.3\text{-}17)$$

2. $$\hat{\underline{p}}^I(k) = \hat{\underline{p}}^I(k-1) + \frac{G\underline{y}_{k-1}}{1 + \underline{y}_{k-1}^T(G + G'_{k-1})\underline{y}_{k-1}}\nu_k^0 \qquad (5.3\text{-}18)$$

3. $$\hat{\underline{p}}^P(k) = \frac{G'_{k-1}\underline{y}_{k-1}}{1 + \underline{y}_{k-1}^T(G + G'_{k-1})\underline{y}_{k-1}}\nu_k^0 \qquad (5.3\text{-}19)$$

4. $$G'_k + \frac{1}{2}G \geq 0 \quad \text{for all } k \qquad (5.3\text{-}20)$$

5. $$\nu_k^0 = \theta_M(k) - [\hat{\underline{p}}^I(k-1)]^T\underline{y}_{k-1} + \sum_{i=1}^{n} d_i\varepsilon_{k-1} \qquad (5.3\text{-}21)$$

where G is an arbitrary positive definite matrix, G'_k is a constant or time-varying matrix satisfying condition 4 and $d_1, \ldots, d_n$ are selected such that

6. $$h(z) = \frac{1 + \Sigma_{i=1}^n d_i z^{-i}}{1 - \Sigma_{i=1}^n a_i z^{-i}} \qquad (5.3\text{-}22)$$

is a strictly positive real, discrete transfer function.

Concerning the algorithm given by Theorem 5.3-1, one can make the following comments:

1. If
$$G'_k = G' \neq 0 \qquad (5.3\text{-}23)$$
(where G' is a constant matrix) the algorithm is of the proportional + integral type with constant adaptation gains. For
$$G'_k \equiv 0 \qquad (5.3\text{-}24)$$
the algorithm is of the integral type.
2. As shown experimentally in Refs. 5 and 6, choosing G'_k as a negative definite matrix while verifying condition 4 could

improve the convergence of the parameters. On the other hand, if G'_k is a positive definite matrix, this will improve the convergence of the generalized error, but high values of the elements of G'_k can slow down the convergence of the parameters [5].

For adaptation problems belonging to class B, one has two algorithms, which will be presented next; the first is of the integral adaptation type, and the second is of the integral + proportional adaptation type [6].

THEOREM 5.3-2 (class B algorithm--integral adaptation). The parallel MRAS described by Eqs. (5.3-1) to (5.3-14) is globally asymptotically stable if the following algorithm is used:

1. $$\hat{\underline{p}}(k) = \hat{\underline{p}}^I(k) \tag{5.3-25}$$

2. $$\hat{\underline{p}}^I(k) = \hat{\underline{p}}^I(k-1) + \frac{G_{k-1}\underline{y}_{k-1}}{1 + \underline{y}_{k-1}^T G_{k-1}\underline{y}_{k-1}} \nu_k^0 \tag{5.3-26}$$

3. $$G_k = G_{k-1} - \frac{1}{\lambda}\frac{G_{k-1}\underline{y}_{k-1}\underline{y}_{k-1}^T G_{k-1}}{1 + (1/\lambda)\underline{y}_{k-1}^T G_{k-1}\underline{y}_{k-1}}, \quad G_0 > 0,\ \lambda > 0.5 \tag{5.3-27}$$

4. $$\nu_k^0 = \theta_M(k) - [\hat{\underline{p}}^I(k-1)]^T \underline{y}_{k-1} + \sum_{i=1}^{n} d_i \varepsilon_{k-i} \tag{5.3-28}$$

where G_0 is an arbitrary positive definite matrix and $d_1, \ldots, d_n$ are selected such that the discrete transfer function

5. $$h'(z) = \frac{1 + \Sigma_{i=1}^n d_i z^{-i}}{1 - \Sigma_{i=1}^n a_i z^{-1}} - \frac{1}{2\lambda} \tag{5.3-29}$$

is strictly positive real.

Notes:

1. From Eq. (5.3-27) and the condition that G_0 is a positive definite matrix, it can be seen that

$$G_k \leq G_{k-1}$$

(i.e., given a vector $\underline{f}$, one has $\underline{f}^T G_k \underline{f} \leq \underline{f}^T G_{k-1} \underline{f}$). Therefore G_k has the significance of a decreasing integral adaptation gain, and the coefficient λ allows us to weight the rate of the decrease.

2. The value of λ most used in practice is $\lambda = 1$, for which Eq. (5.3-27) becomes

$$G_k = G_{k-1} - \frac{G_{k-1}\underline{y}_{k-1}\underline{y}_{k-1}^T G_{k-1}}{1 + \underline{y}_{k-1}^T G_{k-1}\underline{y}_{k-1}}$$

and Eq. (5.3-29) becomes

$$h'(z) = \frac{1 + \Sigma_{i=1}^{n} d_i z^{-i}}{1 - \Sigma_{i=1}^{n} a_i z^{-i}} - \frac{1}{2}$$

3. When λ tends toward infinity ($\lambda \to +\infty$) the decreasing integral adaptation gain G_k tends toward a constant gain, and one obtains Theorem 5.3-1 particularized for the case of integral adaptation.
4. The positivity condition required by Theorem 5.3-2 is stronger than the condition required by Theorem 5.3-1 since the term $-(1/2\lambda)$ has been added to the transfer function considered in Theorem 5.3-1. This is directly related to the fact that one uses a decreasing integral adaptation gain instead of a constant adaptation gain [when $\lambda \to +\infty$, Eq. (5.3-29) tends toward Eq. (5.3-22)].

It is also possible to give a proportional + integral type of adaptation algorithm for class B problems similar to that given in Theorem 5.3-1 for class A problems, as will be shown next. However, for this type of algorithm there does not exist a rigorous proof for the global asymptotic stability of the resulting MRAS.*

Proportional + Integral Adaptation
Algorithm (Class B)

For the parallel MRAS described by Eqs. (5.3-1) to (5.3-14) the following algorithm can be used:

*For a modified algorithm allowing a rigorous proof see Ref. 7.

1. $$\hat{\underline{p}}(k) = \hat{\underline{p}}^I(k) + \hat{\underline{p}}^P(k) \tag{5.3-30}$$

2. $$\hat{\underline{p}}^I(k) = \hat{\underline{p}}^I(k-1) + \frac{G_{k-1}\underline{y}_{k-1}}{1 + \underline{y}_{k-1}^T(G_{k-1} + G'_{k-1})\underline{y}_{k-1}} \nu_k^0 \tag{5.3-31}$$

3. $$\hat{\underline{p}}^P(k) = \frac{G'_{k-1}\underline{y}_{k-1}}{1 + \underline{y}_{k-1}^T(G_{k-1} + G'_{k-1})\underline{y}_{k-1}} \nu_k^0 \tag{5.3-32}$$

4. $$G_k = G_{k-1} - \frac{1}{\lambda} \frac{G_{k-1}\underline{y}_{k-1}\underline{y}_{k-1}^T G_{k-1}}{1 + (1/\lambda)\underline{y}_{k-1}^T G_{k-1}\underline{y}_{k-1}}, \quad G_0 > 0,\ \lambda > 0.5 \tag{5.3-33}$$

5. $$G'_k = \alpha G_k, \quad \alpha \geq -0.5 \tag{5.3-34}$$

6. $$\nu_k^0 = \theta_M(k) - [\hat{\underline{p}}^I(k-1)]^T \underline{y}_{k-1} + \sum_{i=1}^{n} d_i \varepsilon_{k-i} \tag{5.3-35}$$

where G_0 is an arbitrary positive definite matrix and $d_1, \ldots, d_n$ are selected such that the discrete transfer function

7. $$h'(z) = \frac{1 + \Sigma_{i=1}^n d_i z^{-i}}{1 - \Sigma_{i=1}^n a_i z^{-i}} - \frac{1}{2\lambda} \tag{5.3-36}$$

is strictly positive real.

Notes:

1. When λ tends toward infinity ($\lambda \to +\infty$), one obtains Theorem 5.3-1 for the case $G'_k = \alpha G$ (proportional + integral adaptation).
2. A proof for the asymptotic stability of the MRAS using the above algorithm can be given for $|\alpha| \ll 1$. However, extensive experimentation, reported in Refs. 6 and 8 for $\lambda = 1$, has shown that in the various examples considered the adaptive system converges for any value of $\alpha > -0.5$. See Ref. 7 for a reformulation of the algorithm allowing a global asymptotic stability proof for various α.
3. It was observed experimentally that, as in the case of the proportional + integral adaptation with constant gains, a negative value for α ($-0.5 < \alpha < 0$) will improve the convergence of the parameters, while a positive value of α

will improve the convergence of the generalized error. However, a high positive value of α has as a countereffect of slowing down the convergence of the parameters [6, 9].

The adaptation algorithms for class B problems (integral and integral + proportional) will subsequently be used in Secs. 7.7 and 8.5, where experimental results will validate their successful application.

5.3-2 Convergence Proofs

We shall proceed now to the proofs of Theorems 5.3-1 and 5.3-2 as well as to the proof of the asymptotic stability for the parallel MRAS which uses the proportional + integral adaptation algorithm of Eqs. (5.3-30) to (5.3-36) in the case $|\alpha| << 1$.

Proof of Theorem 5.3-1

First Step. The objective of this first step is to obtain an equivalent feedback representation of the parallel MRAS having a linear feedforward block characterized by the transfer function of Eq. (5.3-22).

Following along the same lines as in the example considered in Sec. 5.2 and going from k to k + 1, from Eqs. (5.3-1), (5.3-4), (5.3-9), and (5.3-11) one obtains

$$\varepsilon_{k+1} = \underline{a}^T \underline{e}_k + [\underline{p} - \hat{\underline{p}}(k+1)]^T \underline{y}_k \tag{5.3-37}$$

and

$$\nu_{k+1} = \varepsilon_{k+1} + \underline{d}^T \underline{e}_k \tag{5.3-38}$$

where

$$\underline{a}^T = [a_1, \ldots, a_n] \tag{5.3-39}$$

$$\underline{e}_k^T = [\varepsilon_k, \ldots, \varepsilon_{k-n+1}] \tag{5.3-40}$$

$$\underline{d}^T = [d_1, \ldots, d_n] \tag{5.3-41}$$

Also, using Eqs. (5.3-5), (5.3-8), and (5.3-10), one obtains

$$\varepsilon_{k+1}^{0} = \underline{a}^T \underline{e}_k + [\underline{p} - \hat{\underline{p}}^I(k)]^T \underline{y}_k \tag{5.3-42}$$

$$\nu_{k+1}^{0} = \varepsilon_{k+1}^{0} + \underline{d}^T \underline{e}_k \tag{5.3-43}$$

From Eqs. (5.3-37), (5.3-38), (5.3-42), and (5.3-43) and also using the expressions for $\underline{p}^I(k + 1)$ and $\underline{p}^P(k + 1)$ derived from Eqs. (5.3-18) and (5.3-19) of Theorem 5.3-1, one can express ν_{k+1} in terms of ν_{k+1}^{0}. Subtracting Eq. (5.3-43) from Eq. (5.3-38), one has

$$\nu_{k+1} - \nu_{k+1}^{0} = \varepsilon_{k+1} - \varepsilon_{k+1}^{0} \tag{5.3-44}$$

By substituting for ε_{k+1} and ε_{k+1}^{0} their expressions given by Eqs. (5.3-37) and (5.3-42), respectively, Eq. (5.3-44) becomes

$$\nu_{k+1} - \nu_{k+1}^{0} = [\hat{\underline{p}}^I(k) - \hat{\underline{p}}(k + 1)]^T \underline{y}_k \tag{5.3-45}$$

By now using Eqs. (5.3-18) and (5.3-19), Eq. (5.3-45) becomes

$$\nu_{k+1} - \nu_{k+1}^{0} = -\frac{\underline{y}_k^T (G + G_k')\underline{y}_k}{1 + \underline{y}_k^T (G + G_k')\underline{y}_k} \nu_{k+1}^{0} \tag{5.3-46}$$

from which

$$\nu_{k+1} = \frac{\nu_{k+1}^{0}}{1 + \underline{y}_k^T (G + G_k')\underline{y}_k} \tag{5.3-47}$$

By using ν_{k+1} instead of ν_{k+1}^{0}, at instant k + 1 Eqs. (5.3-18) and (5.3-19) become

$$\hat{\underline{p}}^I(k + 1) = \underline{p}^I(k) + G\underline{y}_k \nu_{k+1} \tag{5.3-48}$$

$$\hat{\underline{p}}^P(k + 1) = G_k' \underline{y}_k \nu_{k+1} \tag{5.3-49}$$

Now defining the auxiliary variable

$$\underline{r}_k \triangleq [\hat{\underline{p}}^I(k) - \underline{p}] \tag{5.3-50}$$

and using the notation

$$\omega_{k+1} \triangleq -\omega_{k+1}^{1} \triangleq \underline{y}_k^T [\hat{\underline{p}}(k + 1) - \underline{p}] = [\hat{\underline{p}}(k + 1) - \underline{p}]^T \underline{y}_k \tag{5.3-51}$$

one obtains, from Eqs. (5.3-37), (5.3-38), (5.3-48), and (5.3-49), the following equivalent feedback system:

$$\varepsilon_{k+1} = \underline{a}^T \underline{e}_k + \omega_{k+1}^1 \tag{5.3-52}$$

$$\nu_{k+1} = \varepsilon_{k+1} + \underline{d}^T \underline{e}_k \tag{5.3-53}$$

$$\underline{r}_{k+1} = \underline{r}_k + G\underline{y}_k \nu_{k+1} \tag{5.3-54}$$

$$\omega_{k+1} = -\omega_{k+1}^1 = \underline{y}_k^T \underline{r}_k + \underline{y}_k^T [G + G_k'] \underline{y}_k \nu_{k+1} \tag{5.3-55}$$

where Eqs. (5.3-52) and (5.3-53) define a linear, time-invariant feedforward block characterized by the transfer function of Eq. (5.3-22) and Eqs. (5.3-54) and (5.3-55) define a nonlinear, time-varying feedback block.

Second Step. Since the adaptation algorithm has memory due to its structure [see Eq. (5.3-18)], one can directly study the hyperstability of the resulting equivalent feedback system. To apply the hyperstability Theorem C-4 (Appendix C) to the equivalent feedback system of Eqs. (5.3-52) to (5.3-55), we should show that under the conditions of Theorem 5.3-1 the equivalent feedback block given by Eqs. (5.3-54) and (5.3-55) satisfies the inequality

$$\sum_{k=0}^{k_1} \omega_{k+1} \nu_{k+1} \geq -\gamma_0^2 \quad \text{for all } k_1 \geq 0 \tag{5.3-56}$$

and any input/output sequence.

To check the inequality of Eq. (5.3-56) there are two possibilities: Use a generalization of the relation given in Eq. (5.2-18) [3], or use the property of *positive* systems to satisfy this type of inequality [see Appendix B, Sec. B.5 and in particular Eq. (B.5-19), and Appendix D, Sec. D.2]. One observes that Eqs. (5.3-54) and (5.3-55) provide a state-space realization of the relation between ν_{k+1} and ω_{k+1}. Therefore the block defined by Eqs. (5.3-54) and (5.3-55) is positive if it verifies one of the positivity lemmas for discrete, time-varying systems given in Appendix B, Sec. B.5.

The positivity properties of a system being independent of the input/output sequence, one may define a new input/output sequence:

$$u_k \triangleq \nu_{k+1}, \quad v_k \triangleq \omega_{k+1} \tag{5.3-57}$$

Then Eqs. (5.3-54) and (5.3-55) become

$$\underline{r}_{k+1} = \underline{r}_k + G\underline{y}_k u_k \tag{5.3-58}$$

$$v_k = \underline{y}_k^T \underline{r}_k + \frac{1}{2} \underline{y}_k^T G \underline{y}_k u_k + \underline{y}_k^T (\frac{1}{2}G + G_k') \underline{y}_k u_k \tag{5.3-59}$$

This block can be decomposed into two parallel blocks receiving the same input, one which has the output

$$v_k^1 = \underline{y}_k^T \underline{r}_k + \frac{1}{2} \underline{y}_k^T G \underline{y}_k u_k \tag{5.3-60}$$

and the second with the output

$$v_k^2 = \underline{y}_k^T (\frac{1}{2}G + G_k') \underline{y}_k u_k \tag{5.3-61}$$

If each of these two blocks satisfies an input/output inequality of the form

$$\sum_{k=0}^{k_1} v_k^i u_k \geq -\gamma_i^2, \quad i = 1, 2, \; k_1 \geq 0 \tag{5.3-62}$$

then the inequality of Eq. (5.3-56) will also be satisfied since it is equal to the sum of the two inequalities given in Eq. (5.3-62) [note that the use of Lemma C-8 (Appendix C), which is applicable to the parallel combinations of two blocks each satisfying the inequality of Eq. (5.3-62), leads to the same conclusion].

Therefore, to satisfy the inequality of Eq. (5.3-62) each of these two blocks should be a *positive* block. Since the block with input u_k and output v_k^2 does not involve any dynamics, it will be positive if the equivalent gain $\underline{y}_k^T (\frac{1}{2}G + G_k') \underline{y}_k$ is positive or null for any k. This holds if condition 4 of the Theorem 5.3-1 is satisfied (i.e., $\frac{1}{2}G + G_k'$ should be a positive definite or positive semi-definite matrix for any k).

It now remains to show that the block given by Eqs. (5.3-58) and (5.3-60) is a positive system verifying, for example, Lemma B.5-1 or B.5-2. Making the substitutions in Eqs. (B.5-1) and B.5-2), one finds

$$\begin{aligned} &\underline{x}_k = \underline{r}_k, \quad A_k = I, \quad B_k = G\underline{y}_k \\ &C_k = \underline{y}_k^T, \quad J_k = \frac{1}{2} \underline{y}_k^T G \underline{y}_k \end{aligned} \tag{5.3-63}$$

Since G is a positive definite matrix, one can choose $P = G^{-1}$, and Eqs. (B.5-5), (B.5-6), (B.5-7), and (B.5-8) will be satisfied for $Q_k = 0$, $S_k = 0$, and $R_k = 0$, and therefore, since Lemma B.5-1 implies Lemma B.5-4 and Eq. (B.5-19)(see Appendix B, Sec. B.5), one has

$$\sum_{k=0}^{k_1} v_k{}^1 u_k \geq -\frac{1}{2}\underline{r}_0{}^T G^{-1} \underline{r}_0 = -\gamma_1{}^2 \quad \text{for all } k_1 \geq 0 \qquad (5.3\text{-}64)$$

One concludes that the equivalent feedback block given by Eqs. (5.3-54) and (5.3-55) under the conditions of Theorem 5.3-1 satisfies the inequality of Eq. (5.3-56).

Third Step. Since the feedback block of the equivalent system satisfies the conditions for the applicability of the hyperstability Theorem C-4, one concludes that the equivalent feedback system of Eqs. (5.3-52) to (5.3-56) will be asymptotically hyperstable if the transfer function of Eq. (5.3-22) is *strictly positive real*. But this is exactly condition 6 of the Theorem 5.3-1. Since the equivalent feedback system is asymptotically hyperstable, the original parallel MRAS using the adaptation algorithm of Theorem 5.3-1 will be *globally asymptotically stable*.

Proof of Theorem 5.3-2

First Step. The objective of the first step is to obtain an equivalent feedback representation of the parallel MRAS considered such that the linear feedforward block of the equivalent feedback system can be characterized by the transfer function of Eq. (5.3-29).

Following a development similar to that used for the proof of Theorem 5.3-1 [Eqs. (5.3-37) to (5.3-55)], one finds the following equivalent feedback system:

$$\varepsilon_{k+1} = \underline{a}^T \underline{e}_k + \omega_{k+1}{}^1 \qquad (5.3\text{-}65)$$

$$\nu_{k+1} = \varepsilon_{k+1} + \underline{d}^T \underline{e}_k \qquad (5.3\text{-}66)$$

$$\underline{r}_{k+1} = \underline{r}_k + G_k \underline{y}_k \nu_{k+1} \qquad (5.3\text{-}67)$$

$$\omega_{k+1} = -\omega_{k+1}{}^1 = \underline{y}_k{}^T \underline{r}_k + \underline{y}_k{}^T G_k \underline{y}_k \nu_{k+1} \qquad (5.3\text{-}68)$$

where $\underline{r}_k$ and ω_{k+1} are defined by Eqs. (5.3-50) and (5.3-51), respectively. However, the linear feedforward block given by Eqs. (5.3-65) and (5.3-66) is not characterized by the discrete transfer function of Eq. (5.3-29) but by that of Eq. (5.3-22). To obtain a linear feedforward block which is characterized by the discrete transfer function of Eq. (5.3-29), an additional transformation of the system of Eqs. (5.3-65) to (5.3-68) must be considered.

By defining the new variable

$$\nu^*_{k+1} \triangleq \nu_{k+1} - \frac{1}{2\lambda}\,\omega^1_{k+1} = \nu_{k+1} + \frac{1}{2\lambda}\,\omega_{k+1} \tag{5.3-69}$$

Eqs. (5.3-65) to (5.3-68) become

$$\varepsilon_{k+1} = \underline{a}^T\underline{e}_k + \omega^1_{k+1} \tag{5.3-70}$$

$$\nu^*_{k+1} = \varepsilon_{k+1} + \underline{d}^T\underline{e}_k - \frac{1}{2\lambda}\,\omega^1_{k+1} \tag{5.3-71}$$

$$\underline{r}_{k+1} = \underline{r}_k + G_k\underline{y}_k\left(\nu^*_{k+1} - \frac{1}{2\lambda}\,\omega_{k+1}\right) \tag{5.3-72}$$

$$\omega_{k+1} = -\omega^1_{k+1} = \underline{y}_k^T\underline{r}_k + \underline{y}_k^T G_k\underline{y}_k\left(\nu^*_{k+1} - \frac{1}{2\lambda}\,\omega_{k+1}\right) \tag{5.3-73}$$

In this form, the equivalent feedback system representation of the parallel MRAS considered has a linear feedforward block defined by Eqs. (5.3-70) and (5.3-71) which is characterized by the discrete transfer function of Eq. (5.3-29) and a nonlinear time-varying feedback block defined by Eqs. (5.3-72) and (5.3-73).

Note that in comparison to the structure of the equivalent feedback block of Eqs. (5.3-67) and (5.3-68) the feedback block defined by Eqs. (5.3-72) and (5.3-73) contains in addition a negative local feedback through a gain equal to $1/2\lambda$, which is essential to the next step of the proof.

Second Step. As for the proof of Theorem 5.3-1, in order to apply the hyperstability Theorem C-4 (Appendix C) to the equivalent feedback system of Eqs. (5.3-70) to (5.3-73) we should show that under the conditions of Theorem 5.3-2 the equivalent feedback block defined by Eqs. (5.3-72) and (5.3-73) with input ν^*_{k+1} and output ω_{k+1} satisfies the inequality

$$\sum_{k=0}^{k_1} \omega_{k+1}\nu^*_{k+1} \geq -\gamma_0^2 \quad \text{for all } k_1 \geq 0 \tag{5.3-74}$$

and any input/output sequence.

To show this, we shall first decompose the feedback block of Eqs. (5.3-72) and (5.3-73) into two blocks connected by feedback. We shall then evaluate an inequality of the form of Eq. (5.3-74) for each of these two blocks, and then using the relations between the inputs and outputs of these blocks and ν^*_{k+1} and ω_{k+1}, respectively, we shall evaluate the left-hand term of the inequality of Eq. (5.3-74).

Since the input/output property characterized by Eq. (5.3-74) must be independent of the input/output sequence, one may define a new input/output sequence as

$$u_k^* \triangleq \nu^*_{k+1}, \quad v_k \triangleq \omega_{k+1} \tag{5.3-75}$$

Then Eqs. (5.3-72) and (5.3-73) become

$$\underline{r}_{k+1} = \underline{r}_k + G_k\underline{y}_k(u_k^* - \frac{1}{2\lambda} v_k) \tag{5.3-76}$$

$$v_k = \underline{y}_k^T \underline{r}_k + \underline{y}_k^T G_k \underline{y}_k (u_k^* - \frac{1}{2\lambda} v_k) \tag{5.3-77}$$

The system defined by Eqs. (5.3-76) and (5.3-77) can furthermore be decomposed into two blocks connected by negative feedback, as shown in Fig. 5.3-1. The first block (feedforward) is defined by

$$\underline{r}_{k+1} = \underline{r}_k + G_k\underline{y}_k u_k^1 \tag{5.3-78}$$

$$v_k^1 = v_k = \underline{y}_k^T \underline{r}_k + \underline{y}_k^T G_k \underline{y}_k u_k^1 \tag{5.3-79}$$

and the second block (feedback) is defined by

$$v_k^2 = \frac{1}{2\lambda} v_k = \frac{1}{2\lambda} u_k^2 \tag{5.3-80}$$

where

$$u_k^1 = u_k^* - \frac{1}{2\lambda} v_k = u_k^* - v_k^2 \tag{5.3-81}$$

By using the relations given by Eqs. (5.3-79) to (5.3-81) and the notations of Eq. (5.3-75), the inequality of Eq. (5.3-74) can be expressed as

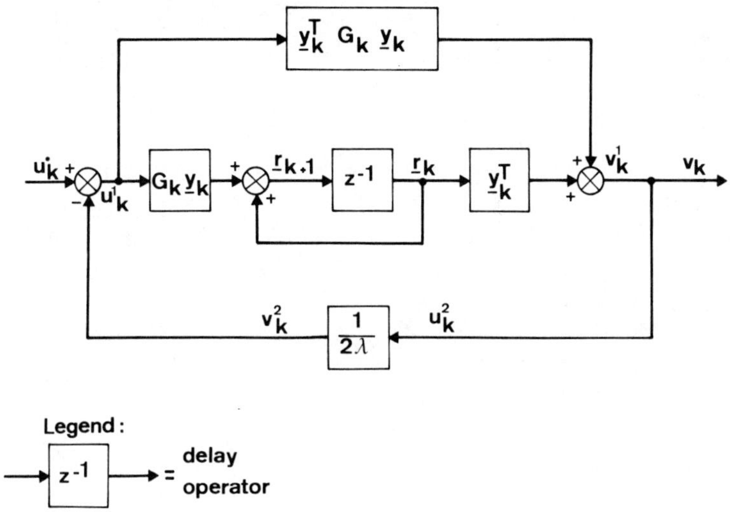

Fig. 5.3-1 Decomposition of the equivalent feedback block of Eqs. (5.3-76) and (5.3-77) into two blocks connected by negative feedback.

$$\sum_{k=0}^{k_1} v_k u_k^* = \sum_{k=0}^{k_1} v_k (u_k^1 + v_k^2)$$
$$= \sum_{k=0}^{k_1} v_k^1 u_k^1 + \sum_{k=0}^{k_1} v_k^2 u_k^2 \geq -\gamma_0^2 \quad \text{for all } k_1 \geq 0 \tag{5.3-82}$$

In fact, the left-hand side of the inequality of Eq. (5.3-82) has been decomposed into two sums of input/output products, each corresponding to one of the blocks in which the system of Eqs. (5.3-76) and (5.3-77) has been decomposed. We shall separately evaluate each of these sums, and then we shall show that the inequality of Eq. (5.3-82) is satisfied.

Using Eqs. (5.3-79) and (5.3-80), one has

$$\sum_{k=0}^{k_1} v_k^2 u_k^2 = \frac{1}{2\lambda} \sum_{k=0}^{k_1} (v_k)^2 = \frac{1}{2\lambda} \sum_{k=0}^{k_1} [\underline{r}_k^T \underline{y}_k \underline{y}_k^T \underline{r}_k$$
$$+ 2u_k^1 (\underline{y}_k^T G_k \underline{y}_k) \underline{y}_k^T \underline{r}_k + u_k^1 (\underline{y}_k^T G_k \underline{y}_k)^2 u_k^1] \tag{5.3-83}$$

We shall now proceed to the evaluation of the sum of the input/output products for the block with input u_k^1 and output v_k^1 which is defined by Eqs. (5.3-78) and (5.3-79). To do this we shall use two lemmas, discussed below.

Consider a discrete, linear, time-varying system given by

$$\underline{x}_{k+1} = A_k\underline{x}_k + B_k\underline{u}_k \tag{5.3-84}$$

$$\underline{v}_k = C_k\underline{x}_k + J_k\underline{u}_k \tag{5.3-85}$$

where $\underline{x}_k$ is the n-dimensional state vector and $\underline{u}_k$ and $\underline{v}_k$ are m-dimensional vectors representing the input and the output, respectively.

LEMMA 5.3-1 [6]. If the time-varying symmetric matrices P_k, Q_k, and R_k and a time-varying matrix S_k exist such that

$$A_k^T P_{k+1} A_k - P_k = -Q_k \tag{5.3-86}$$

$$C_k - B_k^T P_{k+1} A_k = S_k^T \tag{5.3-87}$$

$$J_k + J_k^T - B_k^T P_{k+1} B_k = R_k \tag{5.3-88}$$

then the sum of the input/output inner products for the system of Eqs. (5.3-84) and (5.3-85) can be expressed by

$$\sum_{k=k_0}^{k_1} \underline{v}_k^T \underline{u}_k = \frac{1}{2}\underline{x}_{k_1+1}^T P_{k_1+1}\underline{x}_{k_1+1} - \frac{1}{2}\underline{x}_{k_0}^T P_{k_0}\underline{x}_{k_0} + \frac{1}{2}\sum_{k=k_0}^{k_1} (\underline{x}_k^T Q_k \underline{x}_k + 2\underline{u}_k^T S_k^T \underline{x}_k + \underline{u}_k^T R_k \underline{u}_k) \tag{5.3-89}$$

The proof of Lemma 5.3-1 closely follows the derivation of Lemma B.5-4 from Lemma B.5-1 [see Appendix B, Eqs. (B.5-16) to (B.5-18)] and it is omitted. Note that if, in addition, in Lemma 5.3-1 P_k is positive definite, Q_k is positive semidefinite, and Q_k, S_k, and R_k also satisfy Eq. (B.5-8), the system of Eqs. (5.3-84) and (5.3-85) is *positive* (see Lemma B.5-1).

In addition to Lemma 5.3-1 we shall also use a relation between G_k and G_{k+1} derived from Eq. (5.3-27) by applying the *matrix inversion lemma*, given next.

LEMMA 5.3-2 (matrix inversion lemma) [1]. By considering a nonsingular (n x n)-dimensional matrix P, a nonsingular (m x m)-dimensional matrix R, and an (n x m)-dimensional matrix H (of maximum rank), the following identity holds:

$$(P^{-1} + H^T R^{-1} H)^{-1} = P - PH^T(R + HPH^T)^{-1} HP \qquad (5.3\text{-}90)$$

Using the matrix inversion lemma, from Eq. (5.3-27) one obtains

$$G_{k+1}^{-1} = G_k^{-1} + \frac{1}{\lambda} \underline{y}_k \underline{y}_k^T \qquad (5.3\text{-}91)$$

Now identifying the terms of Eqs. (5.3-84) and (5.3-85) in the system of Eqs. (5.3-78) and (5.3-79), one finds

$$\begin{aligned} &\underline{x}_k = \underline{r}_k, \quad A_k = I, \quad B_k = G_k \underline{y}_k \\ &C_k = y_k^T, \quad J_k = \underline{y}_k^T G_k \underline{y}_k \end{aligned} \qquad (5.3\text{-}92)$$

To express the sum of the input/output products for the system of Eqs. (5.3-78) and (5.3-79) in the form of Eq. (5.3-89), one should find the matrices P_k, Q_k, S_k, and R_k leading to the verification of Eqs. (5.3-86) to (5.3-88), which in this case become

$$P_{k+1} - P_k = -Q_k \qquad (5.3\text{-}93)$$

$$\underline{y}_k^T - \underline{y}_k^T G_k P_{k+1} = S_k^T \qquad (5.3\text{-}94)$$

$$2\underline{y}_k^T G_k \underline{y}_k - \underline{y}_k^T G_k P_{k+1} G_k \underline{y}_k = R_k \qquad (5.3\text{-}95)$$

Choosing

$$\begin{aligned} &P_k = G_k^{-1} \\ &P_{k+1} = G_{k+1}^{-1} = G_k^{-1} + \frac{1}{\lambda} \underline{y}_k y_k^T \end{aligned} \qquad (5.3\text{-}96)$$

and using the relation [derived from Eq. (5.3-91)]

$$G_k G_{k+1}^{-1} = I + \frac{1}{\lambda} G_k \underline{y}_k \underline{y}_k^T \qquad (5.3\text{-}97)$$

from Eqs. (5.3-93), (5.3-94), and (5.3-95) one obtains

$$Q_k = -\frac{1}{\lambda}\underline{y}_k\underline{y}_k^T, \quad S_k^T = -\frac{1}{\lambda}(\underline{y}_k^T G_k \underline{y}_k)\underline{y}_k^T$$
$$R_k = \underline{y}_k^T G_k \underline{y}_k - \frac{1}{\lambda}(\underline{y}_k^T G_k \underline{y}_k)^2 \tag{5.3-98}$$

Therefore, one can write

$$\sum_{k=0}^{k_1} v_k^1 u_k^1 = \frac{1}{2}\underline{r}_{k_1+1}^T G_{k_1+1}^{-1}\underline{r}_{k_1+1} - \frac{1}{2}\underline{r}_0^T G_0^{-1}\underline{r}_0$$
$$- \frac{1}{2\lambda}\sum_{k=0}^{k_1}[\underline{r}_k^T\underline{y}_k\underline{y}_k^T\underline{r}_k + 2u_k^1(\underline{y}_k^T G_k \underline{y}_k)\underline{y}_k^T\underline{r}_k$$
$$+ u_k^1(\underline{y}_k^T G_k \underline{y}_k)^2 u_k^1] + \frac{1}{2}\sum_{k=0}^{k_1} u_k^1(\underline{y}_k^T G_k \underline{y}_k)u_k^1 \tag{5.3-99}$$

Equation (5.3-99) by itself does not satisfy an inequality of the form of Eq. (5.3-74). However, summing up Eqs. (5.3-83) and (5.3-99), one obtains

$$\sum_{k=0}^{k_1} v_k u_k = \sum_{k=0}^{k_1} v_k^1 u_k^1 + \sum_{k=0}^{k_1} v_k^2 u_k^2 = \frac{1}{2}\underline{r}_{k_1+1}^T G_{k_1+1}^{-1}\underline{r}_{k_1+1}$$
$$- \frac{1}{2}\underline{r}_0^T G_0^{-1}\underline{r}_0 + \frac{1}{2}\sum_{k=0}^{k_1} u_k^1(\underline{y}_k^T G_k \underline{y}_k)u_k^1$$
$$\geq -\frac{1}{2}\underline{r}_0^T G_0^{-1}\underline{r}_0 = -\gamma_0^2 \tag{5.3-100}$$

since $G_{k_1+1}^{-1}$, G_0^{-1}, and G_k are positive definite matrices. One thus concludes that the equivalent feedback block of Eqs. (5.3-72) and (5.3-73) under the conditions of Theorem 5.3-2 satisfies the inequality of Eq. (5.3-74).

Third Step. Now applying the hyperstability Theorem C-4 (Appendix C), one concludes that the equivalent feedback system of Eqs. (5.3-70) to (5.3-73) will be asymptotically hyperstable if the discrete transfer function of Eq. (5.3-29) is *strictly positive real*. But this is exactly condition 5 of the Theorem 5.3-2. Therefore the parallel MRAS using the adaptation algorithm given in the Theorem 5.3-2 will be *globally asymptotically stable*.

We shall present now a proof for the asymptotic stability of parallel MRAS's when using the proportional + integral adaptation algorithm given in Eqs. (5.3-30) to (5.3-36) for the case $|\alpha| \ll 1$. The proof closely follows that of the Theorem 5.3-2, and for this reason only the main points will be mentioned.

From Eqs. (5.3-1) to (5.3-14) and Eqs. (5.3-30) to (5.3-36) one obtains the equivalent feedback system described by

$$\varepsilon_{k+1} = \underline{a}^T\underline{e}_k + \omega_{k+1}^1 \quad (5.3\text{-}101)$$

$$\nu_{k+1}^* = \varepsilon_{k+1} + \underline{d}^T\underline{e}_k - \frac{1}{2\lambda}\omega_{k+1}^1 \quad (5.3\text{-}102)$$

$$\underline{r}_{k+1} = \underline{r}_k + G_k\underline{y}_k(\nu_{k+1}^* - \frac{1}{2\lambda}\omega_{k+1}) \quad (5.3\text{-}103)$$

$$\omega_{k+1} = -\omega_{k+1}^1 = \underline{y}_k^T\underline{r}_k + \underline{y}_k^T(G_k + G_k')\underline{y}_k(\nu_{k+1}^* - \frac{1}{2\lambda}\omega_{k+1}) \quad (5.3\text{-}104)$$

This system differs from that of Eqs. (5.3-70) to (5.3-73) only in the last equation [compare Eqs. (5.3-73) and (5.3-104)]. The feedback system of Eqs. (5.3-101) to (5.3-104) will be asymptotically hyperstable if the discrete transfer function of Eq. (5.3-36) is strictly positive real, as required by the algorithm, but this supposes that the feedback block defined by Eqs. (5.3-103) and (5.3-104) satisfies the inequality of Eq. (5.3-74). We shall show this next for the case when $|\alpha| \ll 1$.

Following a procedure similar to that in the proof of Theorem 5.3-2, the feedback block given by Eqs. (5.3-103) and (5.3-104) can be further decomposed into a feedforward block defined by

$$\underline{r}_{k+1} = \underline{r}_k + G_k\underline{y}_k u_k^1 \quad (5.3\text{-}105)$$

$$v_k^1 = v_k = \underline{y}_k^T\underline{r}_k + \underline{y}_k^T(G_k + G_k')\underline{y}_k u_k^1 \quad (5.3\text{-}106)$$

and a feedback block defined by Eq. (5.3-80), the relations between u_k^1, u_k^*, and v_k^2 being defined by Eq. (5.3-81). With this decomposition, the inequality of Eq. (5.3-74) takes on the form given by Eq. (5.3-82). In this case the second term on the left-hand side of the inequality of Eq. (5.3-83) becomes [also using Eq. (5.3-34)]

$$\sum_{k=0}^{k_2} v_k^2 u_k^2 = \frac{1}{2\lambda} \sum_{k=0}^{k_1} (v_k)^2$$

$$= \frac{1}{2\lambda} \sum_{k=0}^{k_1} [\underline{r}_k^T \underline{y}_k \underline{y}_k^T \underline{r}_k + 2(1+\alpha) u_k^1 (\underline{y}_k^T G_k \underline{y}_k) \underline{y}_k^T \underline{r}_k$$

$$+ (1+\alpha)^2 u_k^1 (\underline{y}_k^T G_k \underline{y}_k)^2 u_k^1] \qquad (5.3\text{-}107)$$

If one assumes that $|\alpha| \ll 1$, then Eq. (5.3-107) can be approximated by Eq. (5.3-83).

On the other hand, using a development similar to that in the proof of Theorem 5.3-2, for the feedforward block of Eqs. (5.3-105) and (5.3-106) one obtains the following expression:

$$\sum_{k=0}^{k_1} v_k^1 u_k^1 = \frac{1}{2} \underline{r}_{k_1+1}^T G_{k+1}^{-1} \underline{r}_{k_1+1} - \frac{1}{2} \underline{r}_0^T G_0^{-1} \underline{r}_0$$

$$- \frac{1}{2\lambda} \sum_{k=0}^{k_1} [\underline{r}_k^T \underline{y}_k \underline{y}_k^T \underline{r}_k + 2u_k^1 (\underline{y}_k^T G_k \underline{y}_k) \underline{y}_k^T \underline{r}_k$$

$$+ u_k^1 (\underline{y}_k^T G_k \underline{y}_k)^2 u_k^1] + (\frac{1}{2} + \alpha) \sum_{k=0}^{k_1} u_k^1 (\underline{y}_k^T G_k \underline{y}_k) u_k^1 \qquad (5.3\text{-}108)$$

Summing up Eqs. (5.3-108) and (5.3-83) [the latter being an approximation of Eq. (5.3-107) for $|\alpha| \ll 1$], one obtains

$$\sum_{k=0}^{k_1} v_k u_k = \sum_{k=0}^{k_1} v_k^1 u_k^1 + \sum_{k=0}^{k_1} v_k^2 u_k^2$$

$$\approx \frac{1}{2} \underline{r}_{k_1+1}^T G_{k_1+1}^{-1} \underline{r}_{k_1+1} - \frac{1}{2} \underline{r}_0^T G_0^{-1} \underline{r}_0 + (\frac{1}{2} + \alpha) \sum_{k=0}^{k_1} u_k^1 (\underline{y}_k^T G_k \underline{y}_k) u_k^1$$

$$\geq -\frac{1}{2} \underline{r}_0^T G_0^{-1} \underline{r}_0 = -\gamma_0^2 \qquad (5.3\text{-}109)$$

if

$$\alpha + \frac{1}{2} \geq 0 \qquad (5.3\text{-}110)$$

which is condition 5 of the algorithm, already satisfied under the hypothesis $|\alpha| \ll 1$.

5.3-3 *Computation of the Coefficients* d_i

Theorem 5.3-1 requires the selection of a set of coefficients d_i, $i = 1, 2, \ldots, n$, such that the discrete transfer function

$$h(z) = \frac{1 + \Sigma_{i=1}^{n} d_i z^{-i}}{1 - \Sigma_{i=1}^{n} a_i z^{-i}} \tag{5.3-111}$$

is strictly positive real. Theorem 5.3-2, as well as the proportional + integral adaptation algorithm for class B problems, requires that the discrete transfer function

$$h'(z) = \frac{1 + \Sigma_{i=1}^{n} d_i z^{-i}}{1 - \Sigma_{i=1}^{n} a_i z^{-i}} - \frac{1}{2\lambda}, \quad \lambda > 0.5 \tag{5.3-112}$$

be strictly positive real. In both Eqs. (5.3-111) and (5.3-112), the a_i, $i = 1, 2, \ldots, n$, are the coefficients of the denominator of the discrete transfer function characterizing the reference model.

For particular situations, depending on the values of a_i, the discrete transfer functions of Eqs. (5.3-111) and (5.3-112) can be strictly positive real for $d_i = 0$, $i = 1, 2, \ldots, n$, as shown in the example discussed in Sec. 5.2 or in Ref. 4. We shall next discuss the computation of the coefficients d_i in the general case.

To explicitly compute the coefficients d_i one can give state realizations for $h(z)$ and $h'(z)$ and then apply Lemma B.4-2, Appendix B (see also Theorem B.4-1). Such a state realization for $h(z)$ of Eq. (5.3-111) has the form

$$\underline{x}_{k+1} = A\underline{x}_k + \underline{b}u_k \tag{5.3-113}$$

$$v_k = (\underline{d} + \underline{a})^T \underline{x}_k + u_k \tag{5.3-114}$$

where

$$A = \begin{bmatrix} 0 & 1 & 0 & \cdots & 0 \\ \vdots & & \ddots & \ddots & \\ \vdots & & & \ddots & 0 \\ 0 & & & & 1 \\ a_n & a_{n-1} & & \cdots & a_1 \end{bmatrix}, \quad \underline{b} = \begin{bmatrix} 0 \\ \vdots \\ \vdots \\ 0 \\ 1 \end{bmatrix}$$

$$\underline{d}^T = [d_n \cdots d_1], \quad \underline{a}^T = [a_n \cdots a_1] \tag{5.3-115}$$

and for h'(z) of Eq. (5.3-112) has the form

$$\underline{x}_{k+1} = A\underline{x}_k + \underline{b}u_k \tag{5.3-116}$$

$$v_k = (\underline{d} + \underline{a})^T \underline{x}_k + (1 - \frac{1}{2\lambda})u_k \tag{5.3-117}$$

where A, $\underline{b}$, $\underline{d}$, and $\underline{a}$ are also given by Eq. (5.3-115).

To obtain a strictly positive real transfer function one uses Lemma B.4-2 (Appendix B). A simplification in the computation of the coefficients d_i occurs if we first select Q' in Eq. (B.4-16) such that $P = [P_{ij}]$, which is the solution of the Lyapunov equation

$$A^T PA - P = -Q', \quad Q' > 0 \tag{5.3-118}$$

has a value for P_{nn} of either

$$P_{nn} \le 2 \tag{5.3-119}$$

for the case of the state-space realization of Eqs. (5.3-113) and (5.3-114) or of

$$P_{nn} \le 2 - \frac{1}{\lambda}, \quad \lambda > 0.5 \tag{5.3-120}$$

for the case of the state realization of Eqs. (5.3-116) and (5.3-117).

If $P_{nn} \le 2$, the system of Eqs. (5.3-113) and (5.3-114) can be decomposed into two parallel systems. One is

$$\underline{x}_{k+1} = A\underline{x}_k + \underline{b}u_k \tag{5.3-121}$$

$$v_k^1 = (\underline{d} + \underline{a})^T x_k + \frac{1}{2}P_{nn}u_k \tag{5.3-122}$$

The other system is characterized by a constant gain,

$$v_k^2 = (1 - \frac{1}{2}P_{nn})u_k \tag{5.3-123}$$

and it will be strictly positive real as long as the inequality of Eq. (5.3-119) is satisfied; its output will be null when $P_{nn} = 2$.

If $P_{nn} \le 2 - (1/\lambda)$, the system of Eqs. (5.3-116) and (5.3-117) can also be decomposed into two parallel systems, one described by Eqs. (5.3-121) and (5.3-122) and the other one characterized by a constant gain and given by

$$v_k^2 = (1 - \frac{1}{2\lambda} - \frac{1}{2}P_{nn})u_k \tag{5.3-124}$$

This last system will be strictly positive real as long as the inequality of Eq. (5.3-120) is satisfied; its output will be null when $P_{nn} = 2 - (1/\lambda)$.

If the system described by Eqs. (5.3-121) and (5.3-122) is characterized by a strictly positive real transfer function, then the parallel combination of the two systems, with one system being characterized by a positive (or null) constant gain [Eq. (5.3-123) for h(z) and Eq. (5.3-124) for h'(z)] and the other by a strictly positive real transfer function, will also be characterized by a strictly positive real transfer function.

The system described by Eqs. (5.3-121) and (5.3-122) will be characterized by a strictly positive real transfer matrix if the following two equations are satisfied:

$$A^T PA - P = -Q, \quad Q > 0 \tag{5.3-125}$$

$$P_{nn}\underline{a}^T + [0, P_{n1}, \ldots, P_{n,n-1}] = \underline{a}^T + \underline{d}^T \tag{5.3-126}$$

These two equations are obtained from Eqs. (B.4-16) and (B.4-17) using Eqs. (5.3-115), (5.3-121), and (5.3-122), and Eq. (B.4-18) is identically null for the system of Eqs. (5.3-121) and (5.3-122) [K in Eqs. (B.4-17) and (B.4-18) is null]. Therefore, from Eqs. (5.3-125) and (5.3-126), one obtains

$$\begin{aligned} d_i &= (P_{nn} - 1)a_i + P_{n,n-i}, \quad i = 1, \ldots, n - 1 \\ d_n &= (P_{nn} - 1)a_n \end{aligned} \tag{5.3-127}$$

If, furthermore, $P_{nn} = 1$, then

$$\begin{aligned} d_i &= P_{n,n-i}, \quad i = i, \ldots, n - 1 \\ d_n &= 0 \end{aligned} \tag{5.3-128}$$

One sees also that if $P_{nn} = 1$ and in addition $P_{n,n-i} = 0$ for i = 1, 2, ..., n - 1, then $d_i = 0$ for i = 1, 2, ..., n in Eq. (5.3-127) or (5.3-128), and therefore the transfer functions h(z) and h'(z) (for $\lambda \geq 1$) will be strictly positive real with $d_i = 0$ for i = 1, 2, ..., n.

In practice one can also use the following approximation: One can decompose the system of Eqs. (5.3-113) and (5.3-114) or of Eqs. (5.3-116) and (5.3-117) into two parallel blocks, one described by

$$\underline{x}_{k+1} = A\underline{x}_k + \underline{b}u_k \tag{5.3-129}$$

$$v_k^{\,1} = (\underline{d} + \underline{a})^T \underline{x}_k \tag{5.3-130}$$

which is characterized by the transfer function

$$h_1(z) = \frac{\Sigma_{i=1}^{n} (a_i + d_i) z^{-1}}{1 - \Sigma_{i=1}^{n} a_i z^{-1}} \tag{5.3-131}$$

and a second which has either a unit transfer function for the case of Eqs. (5.3-113) and (5.3-114) or a gain equal to $1 - (1/2\lambda)$ for the case of Eqs. (5.3-116) and (5.3-117). The sum of both will be strictly positive real if for any z, on $|z| = 1$, $|h_1(z)| < 1$ for the first case and $|h_1(z)| < 1 - (1/2\lambda)$ for the second case. These last conditions can be fulfilled if

$$d_i = -a_i + \varepsilon_i \tag{5.3-132}$$

where $0 < \varepsilon_i << 1$.

One can also inquire whether the coefficients d_i may be computed recursively without selecting them a priori in order to satisfy condition 6 of Theorem 5.3-1 or condition 5 of Theorem 5.3-2. This approach gains importance in the case where the parameters a_i, $i = 1, \ldots, n$, are unknown or only partially known (such a situation occurs in identification problems when the reference model represents the process to be identified). The answer to the above question is yes, and we shall next show a possibility for recursive computation of the coefficients d_i.

One considers again Eqs. (5.3-1) to (5.3-14) describing the parallel MRAS in which Eqs. (5.3-10) and (5.3-11) are replaced by

$$\nu_k^{\,0} = \varepsilon_k^{\,0} + \sum_{i=1}^{n} \hat{d}_i^{\,I}(k-1)\varepsilon_{k-1} = \varepsilon_k^{\,0} + [\hat{\underline{d}}^I(k-1)]^T \underline{e}_{k-1} \tag{5.3-133}$$

and

$$\nu_k = \varepsilon_k + \hat{\underline{d}}^T(k)\underline{e}_{k-1} \tag{5.3-134}$$

respectively, where

$$\underline{\hat{d}}^T(k) = [\hat{d}_1(k) \cdots \hat{d}_n(k)] \tag{5.3-135}$$

$$\underline{e}_k^T = [\varepsilon_{k-1} \cdots \varepsilon_{k-n}] \tag{5.3-136}$$

We now define an extended adjustable parameter vector

$$\begin{aligned}[\underline{\hat{p}}_e(k)]^T &= [\underline{\hat{p}}(k)^T, -\underline{\hat{d}}(k)^T] \\ &= [\hat{a}_1(k) \cdots \hat{a}_n(k), \hat{b}_0(k) \cdots \hat{b}_m(k), \\ &\quad -\hat{d}_1(k) \cdots -\hat{d}_n(k)]\end{aligned} \tag{5.3-137}$$

and the extended observation vector

$$\begin{aligned}\underline{\tilde{y}}_{k-1}^T &= [\underline{y}_{k-1}^T, \underline{e}_{k-1}^T] \\ &= [\theta_S(k-1) \cdots \theta_S(k-n), \rho(k) \cdots \rho(k-m), \\ &\quad \varepsilon_{k-1} \cdots \varepsilon_{k-n}]\end{aligned} \tag{5.3-138}$$

For this type of MRAS one has the following counterparts of Theorems 5.3-1 and 5.3-2.

THEOREM 5.3-3 (class A algorithm). The parallel MRAS described by Eqs. (5.3-1) to (5.3-9), Eqs. (5.3-133) and (5.3-134), and Eqs. (5.3-12) to (5.3-14) is globally asymptotically stable if the following algorithm is used.:

1. $$\underline{\hat{p}}_e(k) = \underline{\hat{p}}_e^I(k) + \underline{\hat{p}}_e^P(k) \tag{5.3-139}$$

2. $$\underline{\hat{p}}_e^I(k) = \underline{\hat{p}}_e^I(k-1) + \frac{\tilde{G}\underline{\tilde{y}}_{k-1}}{1 + \underline{\tilde{y}}_{k-1}^T(\tilde{G} + \tilde{G}'_{k-1})\underline{\tilde{y}}_{k-1}} \nu_k^0, \quad G > 0 \tag{5.3-140}$$

3. $$\underline{\hat{p}}_e^P(k) = \frac{\tilde{G}'_{k-1}\underline{\tilde{y}}_{k-1}}{1 + \underline{\tilde{y}}_{k-1}^T(\tilde{G} + \tilde{G}'_{k-1})\underline{\tilde{y}}_{k-1}} \nu_k^0 \tag{5.3-141}$$

4. $$G'_k + \frac{1}{2}G \geq 0 \quad \text{for all } k \tag{5.3-142}$$

5. $$\begin{aligned}\nu_k^0 &= \theta_M(k) - [\underline{\hat{p}}^I(k-1)]^T\underline{y}_{k-1} + [\underline{\hat{d}}^I(k-1)]^T\underline{e}_{k-1} \\ &= \theta_M(k) - [\underline{\hat{p}}_e^I(k-1)]^T\underline{\tilde{y}}_{k-1}\end{aligned} \tag{5.3-143}$$

where G is an arbitrary positive definite matrix.

THEOREM 5.3-4 (class B algorithm) [10]. The parallel MRAS described by Eqs. (5.3-1) to (5.3-9), Eqs. (5.3-133) and (5.3-134), and Eqs. (5.3-12) to (5.3-14) is globally asymptotically stable if the following algorithm is used:

1. $\hat{\underline{p}}_e(k) = \hat{\underline{p}}_e^I(k)$ (5.3-144)

2. $\hat{\underline{p}}_e^I(k) = \hat{\underline{p}}_e^I(k-1) + \dfrac{\tilde{G}_{k-1}\tilde{\underline{y}}_{k-1}}{1 + \tilde{\underline{y}}_{k-1}^T \tilde{G}_{k-1}\tilde{\underline{y}}_{k-1}} \nu_k^0$ (5.3-145)

3. $\tilde{G}_k = \tilde{G}_{k-1} - \dfrac{1}{\lambda} \dfrac{\tilde{G}_{k-1}\tilde{\underline{y}}_{k-1}\tilde{\underline{y}}_{k-1}^T \tilde{G}_{k-1}}{1 + (1/\lambda)\tilde{\underline{y}}_{k-1}^T \tilde{G}_{k-1}\tilde{\underline{y}}_{k-1}}, \quad G_0 > 0, \ \lambda > 0.5$ (5.3-146)

4. $\nu_k^0 = \theta_M(k) - [\hat{\underline{p}}_e^I(k-1)]^T \tilde{\underline{y}}_{k-1}$ (5.3-147)

where G_0 is an arbitrary positive definite matrix.

Note that the above two adaptation algorithms *do not require the a priori selection of the coefficients* d_i as in the algorithms of Theorems 5.3-1 and 5.3-2. In exchange, the dimension of the adjustable parameter vector is augmented (2n + m + 1 instead of n + m + 1), and, of course, the number of basic operations to be done at each step is also augmented (the vector $\tilde{\underline{y}}_{k-1}$ has a greater dimension than $\underline{y}_{k-1}$). Theorem 5.3-4 can be extended to obtain a proportional + integral type of adaptation algorithm for class B problems in a way similar to Theorem 5.3-2.

The proofs of Theorems 5.3-3 and 5.3-4 closely follow those of Theorems 5.3-1 and 5.3-2. The fact that the condition of strictly real positivity of a transfer function has disappeared comes from the fact that using these algorithms, the linear feedforward block of the equivalent feedback system will not have any dynamics and will always be characterized by a positive gain.

For the case of Theorem 5.3-3, from Eqs. (5.3-1), (5.3-4) and (5.3-9) one obtains

$$\varepsilon_{k+1} = \underline{a}^T \underline{e}_k + [\underline{p} - \hat{\underline{p}}(k+1)]^T \underline{y}_k \qquad (5.3\text{-}148)$$

where

$$\underline{a}^T = [a_1 \cdots a_n] \qquad (5.3\text{-}149)$$

Substituting Eq. (5.3-148) into Eq. (5.3-134), one obtains

$$\nu_{k+1} = [\underline{a} + \hat{\underline{a}}(k + 1)]^T\underline{e}_k + [\underline{p} - \hat{\underline{p}}(k + 1)]^T\underline{y}_k \quad (5.3\text{-}150)$$

Now defining the extended parameter vector

$$\underline{p}_e^T = [\underline{p}^T, \underline{a}^T] = [a_1 \cdots a_n, b_0 \cdots b_m, a_1 \cdots a_n] \quad (5.3\text{-}151)$$

Eq. (5.3-150) can be written, also using Eqs. (5.3-137) and (5.3-138), as

$$\nu_{k+1} = [\underline{p}_e - \hat{\underline{p}}_e(k + 1)]^T\tilde{\underline{y}}_k = \omega_{k+1}^1 \quad (5.3\text{-}152)$$

Then, similar to the proof of Theorem 5.3-1, one obtains

$$\tilde{\underline{r}}_{k+1} = \tilde{\underline{r}}_k + \tilde{G}\tilde{\underline{y}}_k\nu_{k+1} \quad (5.3\text{-}153)$$

$$\omega_{k+1} = -\omega_{k+1}^1 = \tilde{\underline{y}}_k^T\tilde{\underline{r}}_k + \tilde{\underline{y}}_k^T(\tilde{G} + \tilde{G}_k')\tilde{\underline{y}}\nu_{k+1} \quad (5.3\text{-}154)$$

where

$$\tilde{\underline{r}}_k = [\hat{\underline{p}}_e^I(k) - \underline{p}_e] \quad (5.3\text{-}155)$$

Equations (5.3-152) to (5.3-154) define an equivalent feedback system. The linear feedforward block defined by Eq. (5.3-152) corresponds to a block with a positive unit gain (which, of course, is strictly positive real), and the feedback block has a structure similar to that of Eqs. (5.3-54) and (5.3-55) except that it is of higher dimension. Therefore the feedback block of Eqs. (5.3-153) and (5.3-154) will also satisfy the Popov inequality, and since the equivalent feedforward block is characterized by a positive gain, the equivalent feedback system will be asymptotically hyperstable.

In a similar way one can prove Theorem 5.3-4. In this case the linear feedforward block of the equivalent feedback system will be described by

$$\nu_{k+1}^* = \left(1 - \frac{1}{2\lambda}\right)[\underline{p}_e - \hat{\underline{p}}_e(k + 1)]^T\tilde{\underline{y}}_k = \left(1 - \frac{1}{2\lambda}\right)\omega_{k+1}^1 \quad (5.3\text{-}156)$$

and the feedback block will have a structure similar to that of Eqs. (5.3-72) and (5.3-73) in which $\underline{r}_k$, G_k, $\underline{y}_k$ have been replaced by $\tilde{\underline{r}}_k$, $\tilde{\underline{G}}_k$, $\tilde{\underline{y}}_k$. The linear feedforward block of Eq. (5.3-156) is always characterized in this case by a positive gain since $\lambda > 0.5$. For the details of the proof and experimental results see [10].

Note: Equation (5.3-150) suggests that an approximation of the algorithms given in Theorems 5.3-3 and 5.3-4 can be considered, and this will lead to a reduction of the number of basic operations at each step. This approximation consists of choosing in Eq. (5.3-133)

$$\underline{\hat{d}}^I(k) = -\underline{\hat{a}}^I(k) \tag{5.3-157}$$

and of adapting $\underline{\hat{p}}(k)^T = [\hat{a}_1(k) \cdots \hat{a}_n(k), \hat{b}_0(k) \cdots \hat{b}_m(k)]$ using the algorithms of Theorem 5.3-1 or 5.3-2. This approximation is based on the fact that if $\underline{\hat{p}}(k)$ tends toward $\underline{\hat{p}}$, $\underline{\hat{a}}_k(k)$ will tend toward $\underline{\hat{a}}$ and ν_k will tend toward zero.

5.3-4 *Adaptation Algorithms for Series-Parallel Model Reference Adaptive Systems*

The equations of the adaptive system in this case are similar to those of the parallel case except that the adjustable system is described by

$$\theta_S(k) = \sum_{i=1}^{n} \hat{a}_i(k)\theta_M(k - i) + \sum_{i=0}^{m} \hat{b}_i(k)\rho(k - i)$$

$$= [\underline{\hat{p}}(k)]^T \underline{x}_{k-1} \tag{5.3-158}$$

and

$$\theta_S^{\,0}(k) = [\underline{\hat{p}}^I(k - 1)]^T \underline{x}_{k-1} \tag{5.3-159}$$

For series-parallel MRAS's one has the following adaptation algorithms, which are counterparts of the algorithms given for the parallel case.

THEOREM 5.3-5 (class A algorithm). The series-parallel MRAS described by Eqs. (5.3-1), (5.3-158), and (5.3-159) and Eqs. (5.3-8) to (5.3-14) is globally asymptotically stable if the following adaptation algorithm is used

1. $$\underline{\hat{p}}(k) = \underline{\hat{p}}^I(k) + \underline{\hat{p}}^P(k) \tag{5.3-160}$$

2. $$\underline{\hat{p}}^I(k) = \underline{\hat{p}}^I(k - 1) + \frac{G\underline{x}_{k-1}}{1 + \underline{x}_{k-1}^T(G + G'_{k-1})\underline{x}_{k-1}} \varepsilon_k^{\,0} \tag{5.3-161}$$

3. $$\underline{p}^P(k) = \frac{G'_{k-1}\underline{x}_{k-1}}{1 + \underline{x}_{k-1}^T(G + G'_{k-1})\underline{x}_{k-1}} \varepsilon_k^0 \qquad (5.3\text{-}162)$$

4. $$G'_k + \frac{1}{2}G \geq 0 \quad \text{for all } k \qquad (5.3\text{-}163)$$

5. $$\varepsilon_k^0 = \theta_M(k) - [\hat{\underline{p}}^I(k-1)]^T\underline{x}_{k-1} \qquad (5.3\text{-}164)$$

where G is an arbitrary positive definite matrix and G'_k is a constant or a time-varying matrix satisfying condition 4.

THEOREM 5.3-6 (class B algorithm--integral adaptation). The series-parallel MRAS described by Eqs. (5.3-1), (5.3-158), and (5.3-159) and Eqs. (5.3-8) to (5.3-14) is globally asymptotically stable if the following adaptation algorithms is used:

1. $$\hat{\underline{p}}(k) = \hat{\underline{p}}^I(k) \qquad (5.3\text{-}165)$$

2. $$\hat{\underline{p}}^I(k) = \hat{\underline{p}}^I(k-1) + \frac{G_{k-1}\underline{x}_{k-1}}{1 + \underline{x}_{k-1}^T G_{k-1}\underline{x}_{k-1}} \varepsilon_k^0 \qquad (5.3\text{-}166)$$

3. $$G_k = G_{k-1} - \frac{1}{\lambda}\frac{G_{k-1}\underline{x}_{k-1}\underline{x}_{k-1}^T G_{k-1}}{1 + (1/\lambda)\underline{x}_{k-1}^T G_{k-1}\underline{x}_{k-1}}, \quad G_0 > 0,\ \lambda > 0.5 \qquad (5.3\text{-}167)$$

4. $$\varepsilon_k^0 = \theta_M(k) - [\hat{\underline{p}}^I(k-1)]^T\underline{x}_{k-1} \qquad (5.3\text{-}168)$$

where G_0 is an arbitrary positive definite matrix.

A proportional + integral adaptation algorithm for series-parallel discrete MRAS's can also be derived. Such an algorithm is presented in Table 7.3-1.

Comparing the adaptation algorithms for the series-parallel case with those for the parallel case, one can see that they have a similar structure, but in the series-parallel case

1. $\underline{y}_{k-1}$, which contains the measurements of the output of the adjustable systems at instants k - 1, ..., k - n, is replaced by $\underline{x}_{k-1}$, which contains the measurements of the model outputs at instants k - 1, ..., k - n.

2. ν_k^0 is replaced by ε_k^0, which is the a priori output generalized error. This means that in Eq. (5.3-10) [or equivalently in Eqs. (5.3-21) and (5.3-28)] all the coefficients d_i, i = 1, 2, ..., n, have been set to zero.
3. No condition upon the positive real character of a discrete transfer function is required.

The reason for these changes clearly emerges when writing the equivalent feedback system for the series-parallel MRAS and then comparing it with the equivalent feedback system for the parallel MRAS.

5.4 DISCRETE MODEL REFERENCE ADAPTIVE SYSTEMS DESCRIBED BY STATE-SPACE EQUATIONS

5.4-1 *Algorithms for Parallel Model Reference Adaptive Systems*

We shall next discuss the algorithms belonging to class A.

Consider the parallel MRAS described by

The *reference model*:

$$\underline{x}_{k+1} = A_M\underline{x}_k + B_M\underline{u}_k \tag{5.4-1}$$

The *parallel adjustable system*:

$$\underline{y}_{k+1}^0 = A_S(k)\underline{y}_k + B_S(k)\underline{u}_k \tag{5.4-2}$$

$$\underline{y}_{k+1} = A_S(k + 1)\underline{y}_k + B_S(k + 1)\underline{u}_k \tag{5.4-3}$$

The *generalized state error*:

$$\underline{e}_k^0 = \underline{x}_k - \underline{y}_k^0 \tag{5.4-4}$$

$$\underline{e}_k = \underline{x}_k - \underline{y}_k \tag{5.4-5}$$

The *adaptation algorithm*:

$$\underline{v}_k^0 = D\underline{e}_k^0 \tag{5.4-6}$$

$$\underline{v}_k = D\underline{e}_k \tag{5.4-7}$$

$$A_S(k + 1) = \sum_{\ell=0}^{k} \Phi_1(\underline{v}, k, \ell) + \Phi_2(\underline{v}, k) + A_S(0) \tag{5.4-8}$$

$$B_S(k + 1) = \sum_{\ell=0}^{k} \Psi_1(\underline{v}, k, \ell) + \Psi_2(\underline{v}, k) + B_S(0) \qquad (5.4\text{-}9)$$

where $\underline{x}$ and $\underline{y}$ are state vectors (n-dimensional), $\underline{u}$ is the input vector (m-dimensional), A_M and B_M are constant matrices of appropriate dimension and $A_S(k + 1)$ and $B_S(k + 1)$ are time-varying matrices of appropriate dimension, $\Phi_1(\underline{v}, k, \ell)$ and $\Psi_1(\underline{v}, k, \ell)$ are discrete matrix functionals of $\underline{v}$, $\Phi_2(\underline{v}, k)$ and $\Psi_2(\underline{v}, k)$ are matrix functions of $\underline{v}$, $\underline{y}_{k+1}^0$ is the *a priori* state vector of the adjustable system computed with the values of the adjustable parameters at instant k, and $\underline{y}_{k+1}$ is the *a posteriori* state vector of the adjustable system at instant k + 1 (after the adaptation has acted).

The design follows the basic steps encountered in the continuous-time case, including the point emphasized in the example presented in Sec. 5.2 concerning the use of $\underline{v}_{k+1}^0$ for computing $A_S(k + 1)$ and $B_S(k + 1)$.

One has the following basic result [11, 12].

THEOREM 5.4-1. The discrete-time parallel MRAS described by Eqs. (5.4-1) to (5.4-9) is globally asymptotically stable (in $\underline{e}$-space) if

1. $\Phi_1(\underline{v}, k, \ell)$, $\Phi_2(\underline{v}, k)$, $\Psi_1(\underline{v}, k, \ell)$, and $\Psi_2(\underline{v}, k)$ are given by

$$\Phi_1(\underline{v}, k, \ell) = F_A(k - \ell)\underline{v}_{\ell+1}[G_A\underline{y}_\ell]^T, \quad \ell \le k \qquad (5.4\text{-}10)$$

$$\Phi_2(\underline{v}, k) = F_A'(k)\underline{v}_{k+1}[G_A'(k)\underline{y}_k]^T \qquad (5.4\text{-}11)$$

$$\Psi_1(\underline{v}, k, \ell) = F_B(k - \ell)\underline{v}_{\ell+1}[G_B\underline{u}_\ell]^T, \quad \ell \le k \qquad (5.4\text{-}12)$$

$$\Psi_2(\underline{v}, k) = F_B'(k)\underline{v}_{k+1}[G_B'(k)\underline{u}_k]^T \qquad (5.4\text{-}13)$$

where $F_A(k - \ell)$ and $F_B(k - \ell)$ are *positive definite discrete kernels* whose z-transforms are *positive real transfer matrices with a pole at* $z = 1$, G_A and G_B are *constant positive definite matrices*, and $F_A'(k)$, $F_B'(k)$, $G_A'(k)$, and $G_B'(k)$ are *time-varying positive definite (or semidefinite) matrices for all* $k \ge 0$.

2. The transfer matrix

$$H(z) = Dz(zI - A_M)^{-1} = D + DA_M(zI - A_M)^{-1} \quad (5.4\text{-}14)$$

is strictly positive real.

Noted below are several particular cases of Theorem 5.4-1 that are often used in its application.

Integral Adaptation.

$$\Phi_1(\underline{v}, k, \ell) = F_A\underline{v}_{\ell+1}[G_A\underline{y}_\ell]^T \quad (5.4\text{-}15)$$

$$\Psi_1(\underline{v}, k, \ell) = F_B\underline{v}_{\ell+1}[G_B\underline{u}_\ell]^T \quad (5.4\text{-}16)$$

where F_A, F_B, G_A, and G_B are positive definite constant matrices.

In addition to the integral adaptation, the following adaptation algorithm can also be considered.

Proportional Adaptation.

$$\Phi_2(\underline{v}, k) = F'_A\underline{v}_{k+1}[G'_A\underline{y}_k]^T \quad (5.4\text{-}17)$$

$$\Psi_2(\underline{v}, k) = F'_B\underline{v}_{k+1}[G'_B\underline{u}_k]^T \quad (5.4\text{-}18)$$

where F'_A, F'_B, G'_A, and G'_B are positive definite (or semidefinite) constant matrices. (This leads to a proportional + integral adaptation algorithm.)

The explicit computation of the matrix D in Eq. (5.4-14) assuring the satisfaction of condition 2 of Theorem 5.4-1 is discussed in Sec. 5.4-2. However, to effectively apply the results of Theorem 5.4-1, we should express $\underline{v}_{\ell+1}$ and $\underline{v}_{k+1}$ in Eqs. (5.4-10) to (5.4-13) or in Eqs. (5.4-15) to (5.4-18) using the measurements available prior to the adaptation; they are $\underline{v}_{\ell+1}{}^0$ and $\underline{v}_{k+1}{}^0$.

For the case of proportional + integral adaptation using a procedure similar to that in Sec. 5.2, one obtains (for $G'_A = G_A$ and $G'_B = G_B$)

$$\underline{v}_{k+1} = [I + DN(k)]^{-1}\underline{v}_{k+1}{}^0 \quad (5.4\text{-}19)$$

where

$$N(k) = (F_A + F'_A)\underline{y}_k{}^T G_A\underline{y}_k + (F_B + F'_B)\underline{u}_k{}^T G_B\underline{u}_k \quad (5.4\text{-}20)$$

The proof of Theorem 5.4-1 follows the procedures shown in Secs. 4.1 and 5.2, and it is omitted [11, 12].

Class B algorithms for parallel MRAS's described by state-space equations can be obtained by directly extending the results for the single-input/single-output case by the method indicated in Ref. 6.

5.4-2 *Computation of the Matrix D*

To explicitly compute the matrix D which will assure that

$$H(z) = D + DA_M(zI - A_M)^{-1} \tag{5.4-21}$$

is strictly positive real as required by Theorem 5.4-1, one can use Lemma B.4-2 after defining a state-space realization for H(z).

The problem of computing D can be drastically simplified if one restricts D to be a positive definite matrix. If this is done, the transfer matrix H(z) given by Eq. (5.4-21) will be strictly positive real if the associated transfer matrix

$$H_1(z) = \frac{1}{2}D + DA_M(zI - A_M)^{-1} \tag{5.4-22}$$

is strictly positive real because $H(z) = \frac{1}{2}D + H_1(z)$ and $\frac{1}{2}D$ is also a strictly positive real transfer matrix if D is positive definite. So that $H_1(z)$ given by Eq. (5.4-22) will be strictly positive real, the system

$$A_M{}^T PA_M - P = -LL^T - Q = -Q' \tag{5.4-23}$$

$$PA_M = DA_M - K^T L^T \tag{5.4-24}$$

$$K^T K = D - P \tag{5.4-25}$$

should admit a solution for P which is positive definite for an arbitrary positive definite matrix Q' (Lemma B.4-2). By choosing D = P, the system of Eqs. (5.4-23), (5.4-24), and (5.4-25) will be satisfied (with K = 0) if, of course, the eigenvalues of A_M are in $|z| < 1$.

Note: As in the SISO case, the need for an a priori selection of the matrix D can be removed by extending the results of Theorem 5.3-3. In this case, Eqs. (5.4-6) and (5.4-7) should be replaced by

$$\underline{v}_k^0 = \underline{e}_k^0 + \hat{D}^I(k - 1)\underline{e}_{k-1} \qquad (5.4\text{-}26)$$

$$\underline{v}_k = \underline{e}_k + \hat{D}(k)\underline{e}_{k-1} \qquad (5.4\text{-}27)$$

5.5 DESIGN OF DISCRETE-TIME MODEL REFERENCE ADAPTIVE SYSTEMS USING ONLY INPUT AND OUTPUT MEASUREMENTS

All the methods presented in Sec. 4.2 concerning the design of continuous-time MRAS's have their direct counterpart in the case of discrete-time MRAS's. On the other hand, the designs presented in Secs. 5.2 and 5.3 for single-input/single-output discrete-time MRAS's described by difference equations require only input and output measurements at instant k plus the previous measurements up to k - n [one needs ε_k, ε_{k-1}, ..., ε_{k-n}, $\theta_S(k - 1)$, ..., $\theta_S(k - n)$, $\rho(k - 1)$, ..., $\rho(k - m)$, $m \le n$]. These previous measurements are obtained by merely passing the input and outputs through discrete-time, state-variable filters formed by a cascade of n delay operators [for $\theta_M(k)$ and $\theta_S(k)$] or of m delay operators [for $\rho(k)$]. In fact, the designs presented in Secs. 5.2 and 5.3 for constant adaptation gains are a direct counterpart of the design presented in Sec. 4.2-2 if one considers that $\varepsilon_{k-(n-1)}$ (when $d_n = 0$) in Eq. (5.3-11) is the equivalent of the filtered generalized output error ε_f and $\varepsilon_{k-(n-2)}$, ..., ε_{k-1}, ε_k are the equivalent of $\dot{\varepsilon}_f$, ..., $\varepsilon_f^{(n-1)}$. The design presented in Secs. 5.2 and 5.3 which uses state-variable filters formed by a cascade of delay operators can be straightforwardly extended to the case of state-variable filters characterized by other transfer functions. (See Problem 5-3.)

It therefore remains to examine the case when the *adjustable system is in state-variable form and one has access only to the inputs and outputs of the reference model and of the adjustable system.* (Such a situation occurs, for example, when building adaptive observers; see Chap. 8.) Furthermore, one assumes that a state-variable filter cannot be tolerated on the input to the adjustable system. In the continuous-time case, the designs concerning this type of problem have been presented in Secs. 4.2-3,

4.2-4, and 4.2-5. The principles of these designs are directly applicable to the discrete case, and we shall emphasize only the specific particularities of the discrete-time case.

First, the design presented in Sec. 4.2-5 based on the use of a nonminimal state realization of the adjustable system can be directly applied in the discrete case in connection with the designs given in Sec. 5.4 (see Problem 5-8). We shall next present the counterpart of the designs given in Secs. 4.2-3 and 4.2-4.

Case 1 (the counterpart of the design given in Sec. 4.2-3). The basic idea of this design is to make the output of the state-space realization of the adjustable system zero-state equivalent to that obtained when an input/output description by a difference equation is used. To achieve this, as in the continuous case, supplementary adaptation signals will be applied to the state-space realization of the adjustable system. The transient adaptation signals in this case have very simple expressions.

The particularity of the design in the discrete case is that the state-variable filters are formed by a cascade of delay operators if the adjustable system in the input/output form is described by a difference equation. The transient adaptation signals in this case have very simple expressions.

Case 2 (the counterpart of the design given in Sec. 4.2-4). The objective of this design is to directly use the generalized output error ε_k for the implementation of the adaptation algorithm instead of $\nu_k = \varepsilon_k + \Sigma_{i=1}^{n} d_i\varepsilon_{k-i}$. The derivation of the adaptation algorithm proceeds along the same lines as in the continuous time case, and the methodology will be illustrated in Chap. 8.

5.6 CONCLUDING COMMENTS

We have seen in this chapter how the general methodology based on the use of hyperstability and positivity concepts is applied to the design of discrete-time MRAS's. We wish to emphasize the following aspects:

1. The basic steps of the methodology developed for continuous-time MRAS's can also be recognized in the discrete case.
2. The last step of the design, which is the return from the asymptotically stable equivalent feedback system to the original system, is slightly more complex than in the continuous case. We must, in addition, circumvent the presence of the inherent delay of one sample caused by the discrete nature of the adaptation process.
3. The problem of circumventing the one sample adaptation delay is solved by introducing the a priori variables. These are variables depending on the adjustable parameters, and they are calculated (or measured) at instant k, but they depend on the values of the adjustable parameters at instant k - 1. The adaptation algorithms which will give the values of the parameters at instant k will depend on these a priori variables.
4. Adaptation algorithms using time-varying adaptation gains have been derived. They can easily be implemented on digital computers since the gains at each step are calculated recursively. These types of algorithms provide in many situations better performances than those using constant adaptation gains. Several types of such algorithms exist [7].
5. The design developed for continuous-time MRAS's when only input and output measurements are available can be extended to the discrete case (see Sec. 5.5).

PROBLEMS

5.1 Check the positive real (or strictly positive real) character (if any) of the following discrete transfer functions:

$$\frac{1}{1 - az^{-1}}, \quad \frac{z^{-1}}{1 - az^{-1}}, \quad \frac{1 + dz^{-1}}{1 - az^{-1}} - \frac{1}{2}$$

($|a| < 1$ in all three cases.)

5.2 Find the positivity domains in the a_1-a_2 plane for the discrete transfer function

$$h(z) = \frac{1 + d_1 z^{-1} + d_2 z^{-2}}{1 - a_1 z^{-1} - a_2 a^{-2}}$$

For each of the two cases (a) $d_1 = 0.5$, $d_2 = 0$ and (b) $d_1 = 0.5$, $d_2 = d_1 - 1$. Compare the positivity domains obtained in the two cases. Repeat the same problem for $d_1 = -0.5$, 0.8, -0.8, 0.2, and -0.2.

5.3 Find adaptation algorithms for a parallel MRAS described by

The reference model:

$$\theta_M(k) = \sum_{i=1}^{n} a_i \theta_M(k - i) + \sum_{i=0}^{m} b_i \rho(k - i)$$

The filtered output of the reference model:

$$\theta_{Mf}(k) = \sum_{i=1}^{n-1} c_i \theta_{Mf}(k - i) + \theta_M(k - r)$$

The filtered input:

$$\rho_f(k) = \sum_{i=1}^{n-1} c_i \rho_f(k - i) + \rho(k - r)$$

The parallel adjustable model:

$$\theta_S'(k) = \sum_{i=1}^{n} \hat{a}_i(k) \theta_S'(k - i) + \sum_{i=0}^{m} \hat{b}_i(k) \rho_f(k - i)$$

The filtered output generalized error:

$$\varepsilon_k{}^f = \theta_{Mf}(k) - \theta_S'(k)$$

5.4 What is the formulation of Theorems 5.3-1 and 5.3-2 for the case when the parallel adjustable system described by Eq. (5.3-4) is replaced by

$$\theta_S(k) = \hat{p}(k)^T \underline{y}_{k-1} + \underline{\ell}^T \underline{e}_{k-1}$$

where $\underline{\ell}$ is a constant vector $\underline{\ell}^T = [\ell_1, \ldots, \ell_n]$ and $\underline{e}_{k-1}{}^T = [\varepsilon_{k-1}m \ldots, \varepsilon_{k-n}]$?

5.5 Show that if in Theorem 5.3-1 one replaces the constant integral adaptation gain G by a time-varying adaptation gain G_k given by

$$G_{k+1} = G_k + \underline{y}_k \underline{y}_k^T \quad (G_0 > 0)$$

one still obtains a globally asymptotically stable parallel MRAS. (Hint: Follow the proof of Theorem 5.3-1 and make use of Lemma B.5-2 or B.5-3 instead of Lemma B.5-1.)

5.6 Consider the second-order reference model

$$\theta_M(k) = a_1\theta_M(k - 1) + a_2\theta_M(k - 2) + b_1\rho(k - 1) + b_2\rho(k - 2)$$

and the parallel adjustable system

$$\underline{y}_{k+1} = \begin{bmatrix} \hat{a}_1(k + 1) & 1 \\ \hat{a}_2(k + 1) & 0 \end{bmatrix} \underline{y}_k + \begin{bmatrix} \hat{b}_1(k + 1) \\ \hat{b}_2(k + 1) \end{bmatrix} (k)$$

$$+ \begin{bmatrix} u_a^1(k + 1) \\ 0 \end{bmatrix} + \begin{bmatrix} u_b^1(k + 1) \\ 0 \end{bmatrix}$$

$$\theta_S(k) = \underline{c}^T \underline{y}_k [1 \;\; 0] \begin{bmatrix} y_{1,k} \\ y_{2,k} \end{bmatrix} = y_{1,k}$$

Develop the complete adaptation algorithms (integral and integral + proportional) using only input and output measurements, as discussed in Sec. 5.5 (cases 1 and 2).

5.7 Extend the designs developed in Problem 5.6 for the case of an MRAS with a series-parallel adjustable model described by

$$\underline{y}_{k+1} = \begin{bmatrix} \hat{a}_1(k+1) & 1 \\ \hat{a}_2(k+1) & 0 \end{bmatrix} \begin{bmatrix} \theta_M(k) \\ y_{2,k} \end{bmatrix} + \begin{bmatrix} \hat{b}_1(k+1) \\ \hat{b}_2(k+1) \end{bmatrix} \rho_k$$

$$+ \begin{bmatrix} u_a{}^1(k+1) \\ 0 \end{bmatrix} + \begin{bmatrix} u_b{}^1(k+1) \\ 0 \end{bmatrix}$$

$$\theta_S(k) = [1, 0]\underline{y}_k$$

where $\underline{y}_k{}^T = [y_{1,k}, y_{2,k}]$ and $\theta_M(k)$ and $\theta_S(k)$ are the outputs of the reference model and of the adjustable system, respectively.

5.8 Extend in the discrete case the design discussed in Sec. 4.2-5 which uses a nonminimal state realization of the adjustable system. Consider parallel and series-parallel adjustable models.

REFERENCES

1. J. M. Mendel, *Discrete Techniques of Parameter Estimation. The Equation Error Formulation,* Marcel Dekker, New York, 1973.
2. P. Kudva and K. S. Narendra, An identification procedure for discrete multivariable systems, IEEE Trans. Autom. Control, *AC-19,* no. 5, 546-549 (1974).
3. I. D. Landau, Synthesis of hyperstable discrete model reference *ence on Circuits and Systems,* Nov. 1971. Univ. of Santa Clara, 1972.
4. L. Ljung, On positive real transfer functions and the converg- ence of some recursive schemes, IEEE Trans. Autom. Control, *AC-22,* no. 4, 539-551 (1977).
5. I. D. Landau. Sur une méthode de synthèse des systèmes adaptatifs avec modele utilisés pour la commande et l'identification d'une classe de procédés physiques. Centre National de Ra Recherche Scientifique: A.O. 8495 Thèse d'Etat ès Sciences, Université de Grenoble, June 1973.

6. I. D. Landau, Unbiased recursive identification using model reference adaptive techniques, IEEE Trans. Autom. Control, *AC-21*, no. 2, 194-202 (1976), and *AC-23*, no. 1, 97-99 (1978).
7. I. D. Landau and H. M. Silveira. A stability theorem with applications to adaptive control in Proceedings of 17th IEEE Control and Decision Conf., San Diego, January 1979. IEEE, N.Y., 1979.
8. G. Bethoux, Approche unitaire des méthodes d'idendification et de commande adaptative des procédeés dynamiques, Thèse 3me Cycle, Institut National Polytechnique de Grenoble, Grenoble, July 1976.
9. I. D. Landau and G. Bethoux, Algorithms for discrete time model reference adaptive systems, in *Proceedings of the Sixth IFAC Congress,* Part ID, 58.4.1-58.4.11, Boston, Aug. 1975. International Federation of Automatic Control, distributed by Instruments Society of America, Pittsburgh.
10. I. D. Landau, Elimination of the real positivity condition in the design of parallel MRAS, IEEE Trans. on Autom. Control, *AC-23,* no. 6, 1978.
11. I. D. Landau, Algorithmes d' adaptation hyperstables pour une classe de systèmes adaptatifs echantillonnés, C. R. Acad. Sci. Paris, Series A, *275*, 1391-1394 (1972).
12. I. D. Landau, Design of discrete model reference adaptive systems using the positivity concept, *Proceedings of the Third IFAC Symposium on Sensitivity Adaptivity and Optimality,* Ischia, 1973, pp. 307-314. Instruments Society of America, Pittsburgh, 1973.

Chapter Six

ADAPTIVE MODEL-FOLLOWING CONTROL SYSTEMS

6.1 INTRODUCTION

In Chap. 1 (Sec. 1.1), it was emphasized that the direct application of linear optimal control theory often has a limited use because of the difficulty of expressing real design objectives in terms of the *quadratic index of performance*. As already pointed out, one of the most efficient ways to avoid this difficulty is to use so-called *linear model-following control (LMFC) systems*. This type of control system uses a model which specifies the design objectives either only for the computation of the control law (implicit model) or as a real part of the system (explicit model) for generating part of the control law as shown in Fig. 6.1-1 (where K_U is the feedforward gain, K_P is the plant feedback gain, and K_M is the model feedback gain). In both cases, the objective is to minimize the error between the states (or the outputs) of the model and those of the controlled plant. However, these control schemes do not overcome the difficulties related to the uncertainty concerning the values of the plant parameters and to the variation of these parameters during operation.

Several studies concerning the sensitivity of the LMFC system's performance in the presence of plant parameter variations are available [1-3]. The analysis of the performance of various LMFC system designs leads to the conclusion that *in order to realize model-following control systems which assure the desired performance in the presence of*

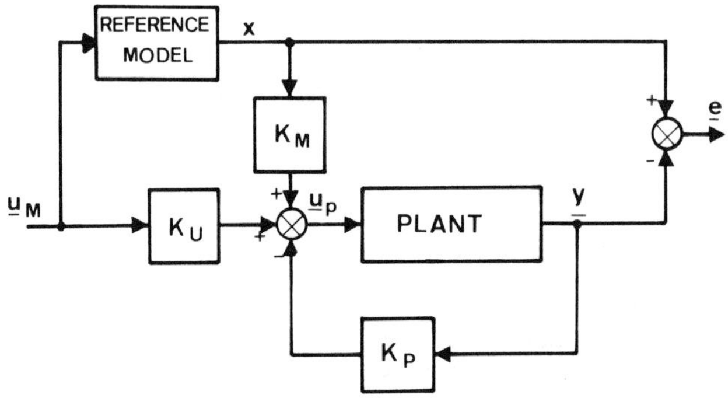

Fig. 6.1-1 Linear model-following control system (with explicit model).

parameter variations and (or) poor knowledge about the parameter values, an adaptive design must be employed.

For the case of LMFC systems which use an explicit model, as shown in Sec. 1.3-3, adaptive model-following control (AMFC) systems appear as a natural extension when an adaptive design must be considered. Adaptive model-following control systems belong to the class of model reference adaptive systems, and under some conditions, the design methods presented in Chap. 4 are applicable. However, to design AMFC systems using model reference adaptive system (MRAS) techniques, there are two requirements: *One must know what kind of linear model-following control system design can effectively lead to an adaptive model-following control system,* and, second, *one must know if for the specific application considered a solution for the linear model-following control design exists.*

As mentioned above, only LMFC systems with an explicit model (represented schematically in Fig. 6.1-1) have as a counterpart AMFC systems with an MRAS-type structure.

Two design methods are currently used for this type of LMFC system. The first method is directed toward the minimization of a quadratic index of performance in terms of both the state error between the plant and the model and of the plant input. [See Sec. 1.1, Eq. (1.1-3).] Classical linear optimal control theory can be applied, and under some conditions an optimal control law can be found. For details, see Refs. 4 to 6.

However, the design problem for an LMFC system can also be formulated in another way: Given a model and a plant, do there exist three matrices K_M, K_P, and K_U (see Fig. 6.1-1) such that for null initial conditions *perfect model following* (or *perfect model matching*) occurs (for any input $\underline{u}_M$ belonging to the class of piece-wise continuous vector functions). This formulation is the essence of the second approach. Perfect model following can be expressed, for example, by the condition that the transfer matrices of the model and of the controlled plant be identical or by the condition that the generalized state error vector $\underline{e} = \underline{x} - \underline{y}$ (or the generalized output error vector) and its derivatives be null for any input $\underline{u}_M$. Another way to express perfect model following is to define a new system with an augmented state vector $\hat{\underline{x}}^T = [\underline{x}^T, \underline{y}^T]$, input $\underline{u}_M$, and output $\underline{e}$ and then show that input $\underline{u}_M$ has no effect upon output $\underline{e}$.

It turns out that in the case where solutions exist for K_M, K_P, K_U leading to perfect model following, the design is easier to compute and leads to smooth control. Experimental results reported in Ref. 7 show that even when such a solution does not exist, better results can be obtained if one modifies the model such that perfect model following can be achieved, instead of trying to minimize the quadratic index of performance in terms of generalized state error $\underline{e} = \underline{x} - \underline{y}$ and plant input $\underline{u}_P$ using the initial model.

Since in the case of perfect model following we shall have $\underline{e} = \underline{0}$ and since when an asymptotically stable MRAS is used one has $\lim_{t\to\infty} \underline{e} = \underline{0}$, one concludes that perfect model-following design can be extended to AMFC systems. *The objective in the latter case is to asymptotically tend toward perfect model following.* But *this will be possible to achieve only if such a solution exists, the role of the adaptation mechanism being only to find the solution* (more exactly, to converge to the solution).

In designing AMFC systems, we shall therefore assume that a solution for perfect model following exists for any values of the plant parameters in a given range of possible variations.

Since the perfect model-following assumption is a basic one, even though this book is concerned with adaptive control design and not with linear control design, we shall give a brief review of various methods for checking the possible existence of perfect model following and some indications how to modify a model in order to satisfy the conditions for perfect model following. This will be the topic of Sec. 6.2 as well as of Sec. 6.4-1.

In Sec. 6.3, we shall discuss the design of AMFC using state measurements. This will be done for two types of applications:

1. Adaptive model following in the presence of an external input (tracking)
2. Adaptive state variable controller (elimination of state disturbances)

Then in Sec. 6.4 we shall discuss the case of AMFC systems using only input and output measurements.

In Sec. 6.5, some practical aspects of the implementation of AMFC systems will be discussed. We shall conclude this chapter with the presentation of two case studies: the adaptive control of a dc electrical drive and an aircraft longitudinal control problem.

For other applications of AMFC systems, we urge the reader to consider the references mentioned in Sec. 1.3-5.

6.2 LINEAR MODEL-FOLLOWING CONTROL SYSTEMS (THE PERFECT MODEL-FOLLOWING CONTROL PROBLEM)

Consider the LMFC system represented in Fig. 6.1-1, which can be described by the following equations:

The reference model:

$$\dot{\underline{x}} = A_M\underline{x} + B_M\underline{u}_M \tag{6.2-1}$$

The plant to be controlled:

$$\dot{\underline{y}} = A_P\underline{y} + B_P\underline{u}_P \tag{6.2-2}$$

The plant control input:

$$\underline{u}_P = -K_P\underline{y} + K_M\underline{x} + K_U\underline{u}_M \tag{6.2-3}$$

In these equations, $\underline{x}$ is the model state vector (n-dimensional), $\underline{y}$ is the plant state vector (n-dimensional), $\underline{u}_M$ is the model vector input (m-dimensional), $\underline{u}_P$ is the plant vector input (m_1-dimensional), and A_M, B_M, A_P, B_P, K_M, K_P, and K_U are constant matrices of appropriate dimension. The pairs (A_P, B_P), (A_M, B_M) are stabilizable, and, furthermore, A_M is a Hurwitz matrix (i.e., the reference model is asymptotically stable).

One defines the error between the states of the model and those of the plant (the generalized state error vector) as

$$\underline{e} = \underline{x} - \underline{y} \tag{6.2-4}$$

As pointed out in Sec. 6.1, three basic approaches for perfect model following have been considered in the literature:

1. Algebraic conditions on the error equation [8, 9]
2. Algebraic conditions on the augmented system with the state vector $\hat{\underline{x}}^T = [\underline{x}^T, \underline{y}^T]$, input $\underline{u}_M$, and output $\underline{e}$ [10, 11]
3. Transfer matrix matching [12, 13]

We shall next briefly consider the first approach. (The third approach will be considered in Sec. 6.4-1.)

Algebraic Conditions on the Error Equation [8, 9, 14]

Subtracting Eq. (6.2-2) from Eq. (6.2-1) and also using Eq. (6.2-3), one obtains the following vector differential equation:

$$\dot{\underline{e}} = (A_M - B_P K_M)\underline{e} + [A_M - A_P + B_P(K_P - K_M)]\underline{y} + (B_M - B_P K_U)\underline{u}_M \tag{6.2-5}$$

To have perfect model following, one must assure that for any $\underline{u}_M$, piecewise continuous, and $\underline{e}(0) = \underline{0}$ we shall have $\underline{e}(t) = \underline{x} - \underline{y} = 0$, $\dot{\underline{e}}(t) = \dot{\underline{x}} - \dot{\underline{y}} = \underline{0}$. One concludes immediately that this can be achieved if

$$[A_M - A_P + B_P(K_P - K_M)]\underline{y} + (B_M - B_P K_U)\underline{u}_M = \underline{0} \quad \text{for all}$$
$$\underline{y} \in R^n, \ \underline{u}_M \in C^m \tag{6.2-6}$$

and in addition to this, the remaining unforced system

$$\dot{\underline{e}} = (A_M - B_P K_M)\underline{e} \tag{6.2-7}$$

must be asymptotically stable, which implies that the matrix $(A_M - B_P K_M)$ must be a Hurwitz matrix.

So that Eq. (6.2-6) holds for any $\underline{y}$ and $\underline{u}_M$, one must have

$$A_M - A_P + B_P(K_P - K_M) = 0 \tag{6.2-8}$$

$$B_M - B_P K_U = 0 \tag{6.2-9}$$

or

$$B_P(K_M - K_P) = A_M - A_P \tag{6.2-10}$$

$$B_P K_U = B_M \tag{6.2-11}$$

Before discussing the solutions for K_M, K_P, and K_U which lead to the verification of Eqs. (6.2-10) and (6.2-11), let us make the observation that in Eq. (6.2-10) K_M and K_P have an equal and opposite influence. This signifies that in order to obtain perfect model following what is important is not the values of K_M and K_P individually but their difference. This has three consequences:

1. The feedback gain matrix K_P can be computed without relying upon the perfect model-following condition of Eq. (6.2-10). For example, it can be computed using linear optimal control in order to minimize the effect of state disturbances acting on the plant. However, K_P must be such that the resulting matrix K_M which leads to the satisfaction of Eq. (6.2-10) assures also that the matrix $(A_M - B_P K_M)$ is a Hurwitz matrix.
2. If some of the states of the plant are not available, perfect model following (if a solution exists) can still be achieved by a convenient choice of K_M [1] (one replaces the missing plant states in the control law by the model states).
3. If all the states of the plant are available and K_P has not been fixed in advance, K_M is not necessary.

After these preliminary observations, let us look to the conditions assuring the existence of solutions for K_M, K_P, and K_U leading to verification of Eqs. (6.2-10) and (6.2-11). The problem is not an easy one, because in most cases B_P is not a nonsingular square matrix. Using the result concerning the existence of a solution for the matrix equation

$$M \cdot X = N \tag{6.2-12}$$

which states that a solution X for Eq. (6.2-12) exists if and only if [15]

$$\text{rank } M = \text{rank}[M, N] \tag{6.2-13}$$

one concludes from Eqs. (6.2-10) and (6.2-11) that perfect model following can be achieved if [9]

$$\text{rank } B_P = \text{rank}[B_P, (A_M - A_P)] = \text{rank}[B_P, B_M] \tag{6.2-14}$$

This means in fact that a solution exists if the column vectors of the difference matrix $(A_M - A_P)$ and of the matrix B_M are linearly dependent on the column vectors of the matrix B_P.

A class of solutions can be obtained by the use of the Penrose pseudo-inverse of B_P, denoted by B_P^+ (we assume that $m_1 < n$ and that the left pseudo-inverse exists; for $m > n$, one must use the right pseudo-inverse).[1] If B_P^+ exists, by left-multiplying Eqs. (6.2-10) and (6.2-11) one obtains

$$K_M - K_P = B_P^+(A_M - A_P) \tag{6.2-15}$$

$$K_U = B_P^+ B_M \tag{6.2-16}$$

Introducing the expressions $K_M - K_P$ and K_U given by Eqs. (6.2-15) and (6.2-16) into Eqs. (6.2-10) and (6.2-11), one obtains sufficient conditions for the existence of solutions for perfect model-following control:

$$(I - B_P B_P^+)(A_M - A_P) = 0 \tag{6.2-17}$$

$$(I - B_P B_P^+)B_M = 0 \tag{6.2-18}$$

---

[1] $B_P^+ = (B_P^T B_P)^{-1} B_P^T$ is called the left Penrose pseudo-inverse of B_P, which exists if $B_P^T B_P$ is a nonsingular matrix. B_P^+ has the property that $B_P^+ B_P = I$.

The conditions of Eqs. (6.2-17) and (6.2-18) are known as Erzberger's conditions for perfect model following [8, 14]. If the conditions of Eqs. (6.2-17) and (6.2-18) are satisfied, then $K_M - K_p$ and K_U can be computed directly using Eqs. (6.2-15) and (6.2-16).

If B_p is a nonsingular square matrix, then $B_p^+ = B_p^{-1}$, and Eqs. (6.2-17) and (6.2-18) are always verified, but this is very seldom the case. When B_p is a singular or a rectangular matrix $B_pB_p^+ \neq I$ and so that Eqs. (6.2-17) and (6.2-18) will be verified, the matrix $I - B_pB_p^+$ must be orthogonal to B_M and to $A_M - A_p$.

The perfect model-following conditions of Eqs. (6.2-17) and (6.2-18) are essentially related to the structure of A_M, A_p, B_M, and B_p and not to the values of their elements [7, 8].

Note that Erzberger's conditions of Eqs. (6.2-17) and (6.2-18) and the rank condition of Eq. (6.2-14) are always satisfied if the reference model and the plant have a similar Luenberger-type controllable canonical structure [16].

We now turn to the question, How can we modify a given model in order to satisfy the perfect model-following conditions? At this time, no general procedure to do this is available. The basic idea to be considered is to define a model having a structure for which perfect model following can be achieved and which gives a good approximation of the desired model.

6.3 ADAPTIVE MODEL-FOLLOWING CONTROL SYSTEMS DESCRIBED BY STATE-SPACE EQUATIONS

6.3-1 *Introduction*

As pointed out in Sec. 6.1, in spite of their advantages, LMFC systems do not overcome the problem of uncertainty in or variation of plant parameters. For these situations, AMFC systems should be considered. If such a solution can be used (when perfect model following can be achieved), it has the advantage of not requiring an explicit identification of the plant parameters, and, as will be shown next, the adaptation laws have an explicit form, which of course does not require the real-time solution of a set of linear or nonlinear equations.

Two basic implementations of AMFC systems can be considered: (1) with parameter adaptation [Fig. 6.3-1(a)] and (2) with signal-synthesis adaptation [Fig. 6.3-1(b)]. Mainly, the objective of these AMFC systems is to assure that the generalized state error vector tends toward zero when the plant parameters differ from their nominal values for any piecewise continuous input vector function $\underline{u}_M$. From similar considerations discussed in Sec. 2.1, one can show that the two configurations shown in Figs. 6.3-1(a) and (b) are equivalent. For the AMFC system with parameter adaptation, the plant input can be expressed by [see Fig. 6.3-1(a)]

$$\underline{u}_P = -K_P(\underline{e}, t)\underline{y} + K_U(\underline{e}, t)\underline{u}_M + K_M\underline{x} \tag{6.3-1}$$

where $K_P(\underline{e}, t)$ and $K_U(\underline{e}, t)$ are time-variable matrices depending on the generalized error vector $\underline{e} = \underline{x} - \underline{y}$ and K_M is a constant matrix. But $K_P(\underline{e}, t)$ and $K_U(\underline{e}, t)$ can also be expressed as

$$K_P(\underline{e}, t) = K_P - \Delta K_P(\underline{e}, t) \tag{6.3-2}$$

$$K_U(\underline{e}, t) = K_U + \Delta K_U(\underline{e}, t) \tag{6.3-3}$$

where K_U and K_P are constant matrices designed for some specific plant parameter values. With this decomposition, one can write

$$\underline{u}_P = \underline{u}_{P1} + \underline{u}_{P2} \tag{6.3-4}$$

where

$$\underline{u}_{P1} = -K_P\underline{y} + K_M\underline{x}_M + K_U\underline{u}_M \tag{6.3-5}$$

$$\underline{u}_{P2} = \Delta K_P(\underline{e}, t)\underline{y} + \Delta K_U(\underline{e}, t)\underline{u}_M \tag{6.3-6}$$

The plant input $\underline{u}_{P1}$ represents the linear control [see Sec. 6.2, Eq. (6.2-3)] and the plant input $\underline{u}_{P2}$ is the contribution of the adaptive loop. The corresponding block diagram is represented in Fig. 6.3-1(b). *In this form, the adaptation mechanism appears in the form of a supplementary feedback loop which will improve the performance of the LMFC systems.* Since the two configurations are equivalent, for the remainder of this section we shall consider only AMFC systems with signal-synthesis adaptation.

From the structural point of view, one can also consider two types of AMFC systems: (1) with a parallel reference model and

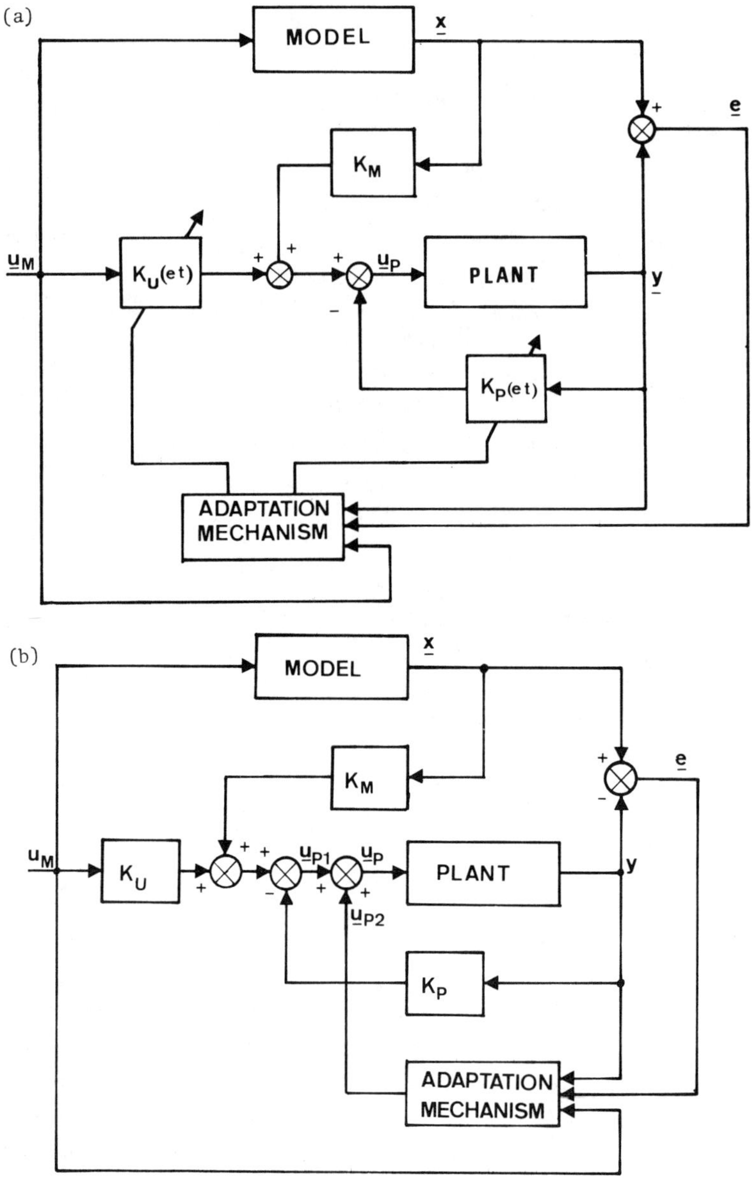

Fig. 6.3-1 Parallel AMFC systems. (a) With parameter adaptation. (b) With signal-synthesis adaptation.

(2) with a series-parallel reference model. The first structure is used for tracking where the desired trajectory is defined by the model and the input acting on it. The second structure used for elimination of the plant state disturbances is an *adaptive state-variable controller*. A combination of both structures is possible.

6.3-2 *Parallel Adaptive Model-Following Control Systems*

A parallel AMFC system with signal-synthesis adaptation [Fig. 6.3-1(b)] is described by the following equations:

The reference model:

$$\underline{\dot{x}} = A_M\underline{x} + B_M\underline{u}_M \tag{6.3-7}$$

The plant to be controlled:

$$\underline{\dot{y}} = A_P\underline{y} + B_P\underline{u}_{P1} + B_P\underline{u}_{P2} \tag{6.3-8}$$

The generalized state error:

$$\underline{e} = \underline{x} - \underline{y} \tag{6.3-9}$$

The linear control signal:

$$\underline{u}_{P1} = -K_P\underline{y} + K_M\underline{x} + K_U\underline{u}_M \tag{6.3-10}$$

The adaptation signal:

$$\underline{u}_{P2} = \Delta K_P(\underline{e}, t)\underline{y} + \Delta K_U(\underline{e}, t)\underline{u}_M \tag{6.3-11}$$

In these equations $\underline{x}$ and $\underline{y}$ are n-dimensional vectors, $\underline{u}_M$ is an m-dimensional vector function, $\underline{u}_{P1}$ and $\underline{u}_{P2}$ are m_1-dimensional vectors, and A_M, B_M, A_P, B_P, K_M, K_P, K_U, ΔK_P, and ΔK_U are matrices of appropriate dimension.

The objective of the adaptation mechanism which will generate the two time-varying matrices $\Delta K_P(\underline{e}, t)$, $\Delta K_U(\underline{e}, t)$ is to assure that the generalized state error $\underline{e}$ goes to zero under certain conditions.

The complete structure of the parallel AMFC system assuring an asymptotic perfect model following is given in Table 6.3-1 later in this section. The derivation of this structure is shown next.

The AMFC system of the form described by Eqs. (6.3-7) to (6.3-11) is, in fact, a special form of the basic parallel MRAS configuration described by state-space equations [see Sec. 4.1, Eqs. (4.1-1) to (4.1-7)] if in Eq. (6.3-7) we replace $\underline{u}_M$ by $\underline{u}$ and we rewrite Eq. (6.3-8) using Eqs. (6.3-10) and (6.3-11). One then obtains

$$\dot{\underline{y}} = [A_p - B_pK_p + B_pK_M + B_p\,\Delta K_p(e, t)]\underline{y} + B_p[K_U + \Delta K_U(e, t)]\underline{u} + B_pK_M\underline{e} \tag{6.3-12}$$

Comparing Eq. (6.3-12) with Eq. (4.1-2), one identifies

$$A_p - B_pK_p + B_pK_M + B_p\,\Delta K_p(\underline{e}, t) = A_S(\underline{v}, t) \tag{6.3-13}$$

$$B_p[K_U + \Delta K_U(\underline{e}, t)] = B_S(\underline{v}, t) \tag{6.3-14}$$

and the term $B_pK_M\underline{e}$ is similar to those introduced in Eq. (4.1-39) (when A_M has a pole at s = 0). This similarity suggests that we also use a similar adaptation mechanism with

$$\underline{v} = D\underline{e} \tag{6.3-15}$$

and

$$\begin{aligned} A_S(\underline{v}, t) &= B_p[\textstyle\int_0^t \Phi_1(\underline{v}, t, \tau)\,d\tau + \Phi_2(\underline{v}, t)] + A_S(0) \\ &= \textstyle\int_0^t \Phi_1'(\underline{v}, t, \tau)\,d\tau + \Phi_2'(\underline{v}, t) + A_S(0) \end{aligned} \tag{6.3-16}$$

$$\begin{aligned} B_S(\underline{v}, t) &= B_p[\textstyle\int_0^t \Psi_1(\underline{v}, t, \tau)\,d\tau + \Psi_2(\underline{v}, t)] + B_S(0) \\ &= \textstyle\int_0^t \Psi_1'(\underline{v}, t, \tau)\,d\tau + \Psi_2'(\underline{v}, t) + B_S(0) \end{aligned} \tag{6.3-17}$$

where

$$A_S(0) = A_p + B_p[K_M - K_p + \Delta K_p(0)] \tag{6.3-18}$$

and

$$B_S(0) = B_p[K_U + \Delta K_U(0)] \tag{6.3-19}$$

Introducing Eqs. (6.3-16) and (6.3-17) into Eqs. (6.3-13) and (6.3-14), respectively, one obtains

$$\Delta K_P(\underline{e}, t) = \Delta K_P(\underline{v}, t) = \int_0^t \Phi_1(\underline{v}, t, \tau)\, d\tau + \Phi_2(\underline{v}, t) + \Delta K_P(0) \tag{6.3-20}$$

and

$$\Delta K_U(\underline{e}, t) = \Delta K_U(\underline{v}, t) = \int_0^t \Psi_1(\underline{v}, t, \tau)\, d\tau + \Psi_2(\underline{v}, t) + \Delta K_U(0) \tag{6.3-21}$$

where Φ_1 and Ψ_1 are m_1 x n and m_1 x n matrices, respectively, which denote a nonlinear time-varying relation between $\Delta K_P(\underline{v}, t)$ and $\Delta K_U(\underline{v}, t)$ and the values of $\underline{v}$ for $0 \le \tau \le t$, and Φ_2 and Ψ_2 are matrices of the same dimensions which denote a nonlinear time-varying relation between $\Delta K_P(\underline{v}, t)$ and $\Delta K_U(\underline{v}, t)$ and the values of $\underline{v}(t)$ having the property that $\Phi_2(0, t) = 0$ and $\Psi_2(0, t) = 0$ for all t. The integral terms in Eqs. (6.3-20) and (6.3-21) will assure the memory of the adaptation mechanism, while Φ_2 and Ψ_2 are transient terms vanishing at the end of the adaptation process when $\underline{v} = 0$ (or $\underline{e} = 0$).

The hypotheses under which the design of the AMFC system is accomplished are as follows:

1. A_M, B_M, A_P, and B_P belong to the class of matrices which satisfy the perfect model-following conditions (i.e., there exists a solution for K_M, K_P, and K_U which leads to perfect model following).
2. The hypotheses of the *basic* case (see Sec. 2.2) for the design of MRAS's are fulfilled, which also means that A_P and B_P are assumed to be constant during the adaptation process.

These assumptions are necessary for an exact analytic development of the design procedure. As several experimental results have shown (see Sec. 6.7), the design which will be presented can be also used when A_P is a nonstationary matrix. The design to be developed below allows one to determine D, Φ_1, Ψ_1, Φ_2, and Ψ_2 so that the AMFC system will be globally asymptotically stable, and therefore $\lim_{t\to\infty} \underline{e}(t) = \underline{0}$ for all initial conditions $\underline{x}(0)$, $\underline{y}(0)$, $K_U + \Delta K_U(0)$, and $K_P - \Delta K_P(0)$ and all piecewise continuous input vector functions $\underline{u}_M$.

The design method proceeds in four steps as for the parallel MRAS's considered in Sec. 4.1, and we shall mention next only the first step (equivalent representation as a feedback system).

Combining Eqs. (6.3-7) to (6.3-11), one can write

$$\dot{\underline{e}} = (A_M - B_P K_M)\underline{e} + [A_M - A_P + B_P(K_P - K_M) - B_P\ \Delta K_P(\underline{v}, t)]\underline{y}$$
$$+ [B_M - B_P K_U - B_P\ \Delta K_U(\underline{v}, t)]\underline{u}_M \qquad (6.3\text{-}22)$$

Since one assumes that a solution for the perfect model-following problem exists, according to Eqs. (6.2-10) and (6.2-11) one can write

$$A_M - A_P = B_P(K_M - K_P^{\,0}) \qquad (6.3\text{-}23)$$
$$B_M = B_P K_U^{\,0}$$

where $K_P^{\,0}$ and $K_U^{\,0}$ are the unknown values of K_P and K_U assuring the perfect model following, and Eq. (6.3-22) becomes

$$\dot{\underline{e}} = (A_M - B_P K_M)\underline{e} + B_P[K_P - K_P^{\,0} - \Delta K_P(\underline{v}, t)]\underline{y}$$
$$+ B_P[K_U^{\,0} - K_U - \Delta K_U(\underline{v}, t)]\underline{u}_M \qquad (6.3\text{-}24)$$

Equation (6.3-24) together with Eqs. (6.3-15), (6.3-20), and (6.3-21) define an equivalent feedback system described by the following equations:

$$\dot{\underline{e}} = (A_M - B_P K_M)\underline{e} + B_P \underline{w}_1 \qquad (6.3\text{-}25)$$

$$\underline{v} = D\underline{e} \qquad (6.3\text{-}26)$$

$$\underline{w} = -\underline{w}_1 = [\Delta K_P(\underline{v}, t) + K_P^{\,0} - K_P]\underline{y} + [\Delta K_U(\underline{v}, t) - K_U^{\,0} + K_U]\underline{u}_M$$
$$= [\textstyle\int_0^t \Phi_1(\underline{v}, t, \tau)\, d\tau + \Phi_2(\underline{v}, t) + \Delta K_P^{\,0}]\underline{y}$$
$$+ [\textstyle\int_0^t \Psi_1(\underline{v}, t, \tau)\, d\tau + \Psi_2(\underline{v}, t) + \Delta K_U^{\,0}]\underline{u}_M \qquad (6.3\text{-}27)$$

where

$$\Delta K_P^{\,0} = \Delta K_P(0) - K_P + K_P^{\,0} \qquad (6.3\text{-}28)$$
$$\Delta K_U^{\,0} = \Delta K_U(0) + K_U - K_U^{\,0}$$

The equivalent feedback system of Eqs. (6.3-25) to (6.3-27) can be partitioned into a linear time-invariant feedforward block described by Eqs. (6.3-25) and (6.3-26) and a nonlinear time-varying feedback block described by Eq. (6.3-27). This equivalent feedback system will be asymptotically hyperstable if it satisfies the conditions of Theorem 4.1-1.

The results of the design for the case of integral + proportional adaptation are summarized in Table 6.3-1.

Discussion of the Results

1. In comparison with other types of adaptive control systems, the implementation of these adaptation laws does not require the real-time solution of a set of linear or nonlinear equations. The aid of a computer is required in the design stage for the computation of the parameters of the adaptation mechanism.
2. It is important to note that in the adaptation law (a) the integral terms which contain $\underline{v}$ provide the memory of the adaptation mechanism, and (b) the proportional terms on $\underline{v}$ are introduced in order to accelerate the reduction of the generalized error $\underline{e}$ at the beginning of the adaptation process. They are also beneficial in the presence of parameter variations with zero steady-state value and in the presence of state disturbances (see Secs. 6.6 and 6.7 and Refs. 14 and 17).
3. As one can see, the implementation of the adaptive laws requires only summators, multipliers, and integrators. These elements are associated in order to realize canonical structures of the following form: multiplier → PI (proportional + integral) amplifier → multiplier. (See also Fig. 4.1-2.) These basic components of the nonlinear part of the adaptation mechanism generate the adaptation signal $\underline{u}_{p2}$ from the signal $\underline{v}$. The signal $\underline{v}$ is obtained from the generalized error $\underline{e}$ by the gain matrix D [14].
4. The ratio between the values of the proportional gain and the values of the integral gain in the PI amplifiers has an important influence on the speed of reduction of the

Table 6.3-1 Continuous-Time Adaptive Model-Following Control Systems

	Parallel AMFC (tracking)	Series-parallel AMFC (state regulation)	Observations
Reference model	$\dot{\underline{x}} = A_M\underline{x} + B_M\underline{u}$	$\dot{\underline{x}} = A_M\underline{y}$	dim $\underline{x}$ = dim $\underline{y}$ A_M = Hurwitz matrix (A_P, B_P) = stabilizable (A_M, B_P) = stabilizable
Plant to be controlled	$\dot{\underline{y}} = A_P\underline{y} + B_P\underline{u}_{P1} + B_P\underline{u}_{P2}$	$\dot{\underline{y}} = A_P\underline{y} + B_P\underline{u}_{P1} + B_P\underline{u}_{P2}$	
Generalized state error vector	$\underline{e} = \underline{x} - \underline{y}$	$\underline{e} = \underline{x} - \underline{y}$	
Linear control signal	$\underline{u}_{P1} = -K_P\underline{y} + K_M\underline{x} + K_U\underline{u}_M$	$\underline{u}_{P1} = -K_P\underline{y} - K\underline{e}$	
Adaptation signal	$\underline{u}_{P2} = \Delta K_P(\underline{e}, t)\underline{y} + \Delta K_U(\underline{e}, t)\underline{u}_M$	$\underline{u}_{P2} = \Delta K_P(\underline{e}, t)\underline{y}$	
Perfect model-following conditions	$(I - B_PB_P^+)(A_M - A_P) = 0$ $(I - B_PB_P^+)B_M = 0$ or $\text{rank } B_P = \text{rank}[B_P, A_M - A_P]$ $= \text{rank}[B_P, B_M]$	$(I - B_PB_P^+)(A_M - A_P) = 0$ or $\text{rank } B_P = \text{rank}[B_P, A_M - A_P]$	

Adaptation law (integral + proportional)	$\underline{v} = D\underline{e}$ $\Delta K_P(\underline{e}, t) = \int_0^t F\underline{v}(G\underline{y})^T \, d\tau + F'\underline{v}(G\underline{y})^T + \Delta K_P(0)$ $\Delta K_U(\underline{e}, t) = \int M\underline{v}(N\underline{u}_M)^T \, d\tau + M'\underline{v}(N\underline{u}_M)^T + \Delta K_U(0)$ $F, M, G, N > 0; \; F', M' \geq 0$	$\underline{v} = D\underline{e}$ $\Delta K_P(e, t) = \int_0^t F\underline{v}(G\underline{y})^T \, d\tau + F'\underline{v}(G\underline{y})^T + \Delta K_P(0)$ $F, G > 0; \; F' \geq 0$	
Stability condition	$D(sI - A_M + B_P K_M)^{-1} B_P$ = strictly positive real transfer matrix, or $D = B_P^T P, \quad P > 0$ where P satisfies $(A_M - B_P K_M)^T P + P(A_M - B_P K_M) = -Q, \quad Q > 0$	$D(sI - B_P K)^{-1} B_P$ = strictly positive real transfer matrix, or $D = B_P^T P, \quad P > 0$ where P satisfies $(B_P K)^T P + P B_P K = -Q, \quad Q > 0$	If $B_P = B_P^0 R$, where $R > 0$, B_P can be replaced by B_P^0. Q is an arbitrary positive definite matrix.

state distance ($||e||$). Experimental results have shown that a high ratio leads to a high speed reduction of the model plant state error, but in counterpart, the control energy is higher (also the parameter adaptation speed is slower) [14, 18]. The simultaneous augmentation of proportional and integral gains improves the speed of adaptation. The gains are limited by the saturations existing in the adaptation loop and by the imperfect characterization of the plant by Eq. (6.3-8) if a reduced order state equation is used [14].

5. For the computation of the linear compensator D, one must solve the Lyapunov equation (see Table 6.3-1, last row) after the choice of a definite positive symmetric matrix Q. In the case of multivariable adaptive control system design the matrix $A_M - B_P K_M$ may have a very large dimension, and experience shows that it is often "ill-conditioned."[2] (This situation occurs, for example, in some aircraft control problems when the phugoid mode and the short period dynamics are considered together. See Sec. 6-7.) Consequently, the direct method for solving the Lyapunov equation often breaks down. On the other hand, the methods based on the knowledge of the eigenvalues and the eigenvectors of matrix $A_M - B_P K_M$ are not very attractive because of computational difficulties. Therefore, it seems that series methods (where the matrix P is obtained as the sum of a series) are the most suitable for almost all of the practical cases. The Jameson method [19] was successfully used to solve the Lyapunov equation [14].
6. The choice of matrix Q (see Table 6.3-1) has a great influence on the relative performance of the various components of the plant model state error. [In fact, the matrix Q characterizes the derivative of the Lyapunov function for the adaptive

[2]The largest and the smallest elements of the matrix differ by several orders of magnitude.

system, which is $\dot{V} = -\underline{e}^T Q \underline{e}$. See Eqs. (6.3-41) and (6.3-43) for $z \equiv 0$]. Experimental results have shown that by choosing Q as a diagonal matrix of a given trace the performance of the jth component of the generalized state error $\underline{e}$ is improved in comparison to the other components by augmenting the term q_{jj} and decreasing the others. (Q works as a weighting matrix.) This is illustrated by the example presented in Sec. 6.7. The multiplication of all the terms of Q by a constant is equivalent to the multiplication of the gains of the PI amplifiers in the nonlinear time-varying part of the adaptation mechanism [14].

7. From the stability conditions it appears that B_p must be known. When B_p is not known but can be estimated within the approximation of an unknown but positive definite matrix, i.e.,

$$B_p = B_p^0 R \tag{6.3-29}$$

where B_p^0 is a known matrix and R is any positive definite matrix, the transfer matrix of the equivalent linear feedforward block becomes (if one also chooses $K_M = 0$)

$$H(s) = D(sI - A_M)^{-1} B_p^0 R \tag{6.3-30}$$

This transfer matrix will be strictly positive real for any positive definite matrix R by choosing

$$D = [B_p^0]^T P \tag{6.3-31}$$

where P is the positive definite matrix solution of the Lyapunov equation:

$$PA_M + A_M^T P = -Q, \quad Q > 0 \tag{6.3-32}$$

Note that if B_p and B_M have the same structure, one can sometimes choose $B_p^0 = B_M$.

8. As pointed out in Sec. 6.2, the matrices K_M and K_p have the same effect for obtaining perfect model following. If K_p is not restricted by other considerations, K_M can be eliminated.

The Case of a Time-Varying Plant

In the analytical design presented previously, it was assumed that A_p and B_p are unknown (with some restrictions on B_p) but constant during the adaptation process. However, it will next be shown that under a few supplementary assumptions the design presented guarantees that the generalized state error vector remains bounded when A_p is time-varying. (This is related to the property of hyperstable feedback systems which have a bounded output for a bounded input [20]). In addition, an estimation of this bound of the generalized state error vector will be obtained.

Consider the plant equation (6.3-8) in which A_p is replaced by

$$A_p(t) = A_p(0) + \Delta A_p(t) \tag{6.3-33}$$

Then Eq. (6.3-24) becomes

$$\begin{aligned}\dot{\underline{e}} &= (A_M - B_pK_M)\underline{e} + B_p[K_p - K_p{}^0 - \Delta K_p(\underline{v}, t)]\underline{y} \\ &\quad + B_p[K_U{}^0 - K_U - \Delta K_U(\underline{v}, t)]\underline{u}_M - B_p\, \Delta K_p{}^0(t)\underline{y}\end{aligned} \tag{6.3-34}$$

where

$$\begin{aligned}A_M - A_p(t) &= A_M - A_p(0) - \Delta A_p(t) \\ &= B_p[K_M - K_p{}^0 - \Delta K_p{}^0(t)]\end{aligned} \tag{6.3-35}$$

and, respectively,

$$\Delta A_p(t) = B_p\, \Delta K_p{}^0(t) \tag{6.3-36}$$

The equivalent feedback system will be described in this case by

$$\dot{\underline{e}} = (A_M - B_pK_M)\underline{e} + B_p\underline{w}_1 + B_p\underline{z}(t) \tag{6.3-37}$$

and Eqs. (6.3-26) and (6.3-27). In Eq. (6.3-37), $\underline{z}(t)$ is given by

$$\underline{z}(t) = -\Delta K_p{}^0(t)\underline{y} \tag{6.3-38}$$

and represents an external input applied to the free feedback system of Eqs. (6.3-25) to (6.3-27). The free feedback system being asymptotically hyperstable, if $\lim_{t\to\infty} \underline{z}(t) = 0$ [which implies that $\lim_{t\to\infty} \Delta A_p(t) = 0$] if $\underline{y} \not\equiv 0$ or $\underline{z}(t) \in L_2$,[3] the AMFC system will

[3] That is, $\int_0^\infty \underline{z}^T(t)\underline{z}(t)\, dt < \infty$.

still assure that $\lim_{t\to\infty} \underline{e} = 0$. If $||\underline{z}(t)|| < a$, one can assure only that $||\underline{v}(t)|| < b$ and, consequently, $||\underline{e}(t)|| < c$, where a, b, and c are positive, finite constants. But $||\underline{z}(t)|| < a$ if the variations of the terms of the matrix A_p are bounded, i.e.,

$$||\Delta A_p(t)|| < \alpha \quad \text{for all } t \geq 0 \tag{6.3-39}$$

and the state of the plant is bounded, i.e.,

$$||\underline{y}|| < \delta \quad \text{for all } t \geq 0 \tag{6.3-40}$$

where α and δ are finite positive constants.

An estimation of the bound of $||\underline{e}||$ can be obtained by applying the methodology indicated in Ref. 21. One considers for the system of Eqs. (6.3-25) to (6.3-27) a Lyapunov function of the type used in Theorem 4.4-1, that is,

$$\begin{aligned} V(\underline{e}, t) = {} & \underline{e}^T P \underline{e} + 2 \int_{-\infty}^{t} \underline{v}^T [\int_{-\infty}^{t'} F\underline{v}(G\underline{y})^T \, d\tau]\underline{y} \, dt' \\ & + 2 \int_{-\infty}^{t} \underline{v}^T F' \underline{v}\underline{y}^T G\underline{y} \, dt' \\ & + 2 \int_{-\infty}^{t} \underline{v}^T [\int_{-\infty}^{t'} M\underline{v}(N\underline{u}_M)^T \, d\tau]\underline{u}_M \, dt' \\ & + 2 \int_{-\infty}^{t} \underline{v}^T M' \underline{v}\underline{u}_M^T N\underline{u}_M \, dt' \end{aligned} \tag{6.3-41}$$

where

$$\underline{v} = B^T P\underline{e} \tag{6.3-42}$$

and P is the positive definite matrix solution of the Lyapunov equation (see Table 6.3-1). Computing next the derivatives of $V(\underline{e}, t)$ along the trajectories of the forced feedback system defined by Eqs. (6.3-25), (6.3-26), and (6.3-37) using the adaptation laws given in Table 6.3-1, one finally obtains

$$\dot{V}(\underline{e}, t) = -\underline{e}^T Q\underline{e} + 2\underline{e}^T PB_p\underline{z} \tag{6.3-43}$$

Now, if $\dot{V}(\underline{e}, t)$ is negative outside a closed region including the origin of $\underline{e}$-space, then all the solutions of Eq. (6.3-37) will enter in this region [21]. From Eq. (6.3-43) one obtains

$$\dot{V}(\underline{e}, t) \le -\lambda_{min}(Q)||\underline{e}||^2 + 2||\underline{e}||\cdot||P||\cdot||B_p\underline{z}||$$
$$= ||\underline{e}||[-\lambda_{min}(Q)||\underline{e}|| + 2||P||\cdot||B_p\underline{z}||] \tag{6.3-44}$$

where $\lambda_{min}(Q)$ is the minimum eigenvalue of the matrix Q and $||\underline{e}||$ the Euclidean norm ($||\underline{e}||^2 = \underline{e}^T\underline{e}$). But from Eqs. (6.3-36) and (6.3-38), one has

$$||B_p\underline{z}|| \le ||\Delta A_p(t)||\cdot||\underline{y}|| \tag{6.3-45}$$

and one concludes from Eq. (6.3-44) that $\dot{V}(\underline{e}, t)$ will be negative for all $||\underline{e}||$ satisfying the inequality

$$||\underline{e}|| > 2||P||\cdot||\Delta A_p(t)||\cdot||\underline{y}||[\lambda_{min}(Q)]^{-1} \tag{6.3-46}$$

Therefore, the upper bound for $||\underline{e}||$ will be

$$||\underline{e}||_{max} < 2||P||\cdot||\Delta A_p(t)||_{max}||\underline{y}||_{max}[\lambda_{min}(Q)]^{-1} \tag{6.3-47}$$

The Effect of a Plant State Disturbance

For this aspect, see Ref. 18 for a theoretical analysis and Ref. 14 for simulation results.

6.3-3 Series-Parallel Adaptive Model-Following Control Systems Used as State-Variable Controllers

One can ask if it is possible to use AMFC system configurations when we are interested in realizing adaptive state variable controllers which act only in the presence of state disturbances. If the objective is to assure certain dynamics for the elimination of the plant state disturbances independently of the values of the plant parameters, the answer is yes. Series-parallel AMFC system configurations could be used successfully for solving this problem. The reference model in this case is defined by

$$\dot{\underline{x}} = A_M\underline{y} \tag{6.3-48}$$

where $\underline{x}$ is the state of the model (i.e., the desired state) and $\underline{y}$ is the state of the plant. The significance of Eq. (6.3-48) is that at each moment the desired slope of variation in $\underline{y}$ (denoted by $\dot{\underline{x}}$) is defined by A_M and by the measured value of the plant state vector $\underline{y}$.

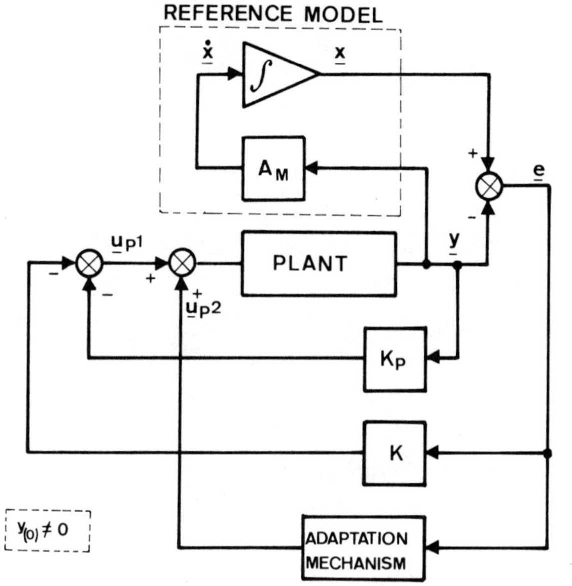

Fig. 6.3-2 Series-parallel AMFC system (adaptive state variable controller).

(We want to force the dynamics of the plant characterized by the matrix A_P to be equal to that of the model.)

The plant is described by

$$\dot{\underline{y}} = A_P\underline{y} + B_P\underline{u}_{P1} + B_P\underline{u}_{P2} \tag{6.3-49}$$

where $\underline{u}_{P1}$ is the linear control signal and $\underline{u}_{P2}$ the adaptation signal. We define

$$\underline{u}_{P1} = -K_P\underline{y} - K\underline{e} \tag{6.3-50}$$

where

$$\underline{e} = \underline{x} - \underline{y} \tag{6.3-51}$$

and K_P and K are gain matrices (the control depends on the plant state and on the difference between the desired state and the real one).

The adaptation signal is given by

$$\underline{u}_{P2} = \Delta K_P(\underline{e}, t)\underline{y} \tag{6.3-52}$$

Table 6.3-2 Discrete-Time Adaptive Model-Following Control Systems[a]

	Parallel AMFC (tracking)	Series-parallel AMFC (state regulation)
Reference model	$\underline{\tilde{x}}_{k+1}^{0} = A_M\underline{\tilde{x}}_k + B_M\underline{u}_M(k)$ $\underline{\tilde{x}}_{k+1} = A_M\underline{\tilde{x}}_k + B_M\underline{u}_M(k) - B_P[\Delta K_P(k+1)$ $- \Delta K_P^{I}(k)]\underline{y}_k - B_P[\Delta K_U(k+1) -$ $- \Delta K_U^{I}(k)]\underline{u}_M(k)$	$\underline{\tilde{x}}_{k+1}^{0} = A_M\underline{y}_k$ $\underline{\tilde{x}}_{k+1} = A_M\underline{y}_k - B_P[\Delta K_P(k+1) - \Delta K_P^{I}(k)]\underline{y}_k$
Plant to be controlled	$\underline{y}_{k+1} = A_P\underline{y}_k + B_P\underline{u}_{P1}(k) + B_P\underline{u}_{P2}(k)$	$\underline{y}_{k+1} = A_P\underline{y}_k + B_P\underline{u}_{P1}(k) + B_P\underline{u}_{P2}(k)$
Generalized state error	$\underline{e}_k^{0} = \underline{\tilde{x}}_k^{0} - \underline{y}_k$ $\underline{e}_k = \underline{\tilde{x}}_k - \underline{y}_k$	$\underline{e}_k^{0} = \underline{\tilde{x}}_k^{0} - \underline{y}_k$ $\underline{e}_k = \underline{\tilde{x}}_k - \underline{y}_k$
Linear control signal	$\underline{u}_{P1}(k) = -K_P\underline{y}_k + K_M\underline{x}_k + K_U\underline{u}_M(k)$	$\underline{u}_{P1}(k) = -K_P\underline{y}_k - K\underline{e}_k$
Adaptation signal	$\underline{u}_{P2}(k) = \Delta K_P^{I}(k)\underline{y}_k + \Delta K_U^{I}(k)\underline{u}_M(k)$	$u_{P2}(k) = \Delta K_P^{I}(k)\underline{y}_k$
Perfect model-following conditions	$(I - B_P B_P^{+})(A_M - A_P) = 0$ $(I - B_P B_P^{+})B_M = 0$ or $\text{rank } B_P = \text{rank } [B_P, A_M - A_P] = \text{rank } [B_P, B_M]$	$(I - B_P B_P^{+})(A_M - A_P) = 0$ or $\text{rank } B_P = \text{rank}[B_P, A_M - A_P]$

Adaptation algorithm (integral + proportional)	$\underline{v}_{k+1}^{0} = D\underline{e}_{k+1}^{0}$ $\underline{v}_{k+1} = \{I + DB_p[(F + F')\underline{y}_k^T G\underline{y}_k + (M + M')\underline{u}_M^T(k)N\underline{u}_M(k)]\}^{-1}\underline{v}_{k+1}^{0}$ $\Delta K_p(k+1) = \Delta K_p^I(k+1) + \Delta K_p^P(k+1)$ $\Delta K_p^I(k+1) = \Delta K_p^I(k) + F\underline{v}_{k+1}[G\underline{y}_k]^T$ $\Delta K_p^P(k+1) = F'\underline{v}_{k+1}[G\underline{y}_k]^T$ $F, G > 0;\ F' \geq 0$ $\Delta K_U(k+1) = \Delta K_U^I(k+1) + \Delta K_U^P(k+1)$ $\Delta K_U^P(k+1) = \Delta K_U^I(k) + M\underline{v}_{k+1}[N\underline{u}_M(k)]^T$ $\Delta K_U^P(k+1) = M'\underline{v}_{k+1}[N\underline{u}_M(k)]^T$ $M, N > 0;\ M' \geq 0$	$\underline{v}_{k+1}^{0} = D\underline{e}_{k+1}^{0}$ $v_{k+1} = [I + DB_p(F + F')\underline{y}_k^T G\underline{y}_k]^{-1}\underline{v}_{k+1}^{0}$ $\Delta K_p(k+1) = \Delta K_p^I(k+1) + K_p^P(k+1)$ $\Delta K_p^I(k+1) = \Delta K_p^I(k) + F\underline{v}_{k+1}[G\underline{y}_k]^T$ $\Delta K_p^P(k+1) = F'\underline{v}_{k+1}[G\underline{y}_k]^T$ $F, G > 0;\ F' \geq 0$
Stability conditions	$Dz(zI - A_M + B_pK_M)^{-1}B_p$ = strictly positive real transfer matrix, or $D = B_p^T P B_p,\quad P > 0$ where P satisfies $(A_M - B_pK_M)^T P(A_M - B_pK_M) - P = -Q,\quad Q > 0$	$Dz(zI - B_pK)^{-1}B_p$ = strictly positive real transfer matrix, or $D = B_p^T P B_p,\quad P > 0$ where P satisfies $(B_pK)^T P B_p K - P = -Q,\quad Q > 0$

[a]Superscript 0 denotes the a priori values.

Because of the use of the plant state vector $\underline{y}$ in the equation of the reference model (see Sec. 2.1), one has a series-parallel configuration.

The complete structure of the series-parallel AMFC system assuring an asymptotic perfect model following is given in Table 6.3-1, and a block diagram is shown in Fig. 6.3-2. This type of structure has been used for voltage and frequency adaptive control in power systems [22].

6.3-4 The Discrete-Time Case

In this section, we shall present the discrete-time formulation of the counterpart of the design presented in Sec. 6-3-2 for parallel and series-parallel AMFC systems. The algorithms for both structures are summarized in Table 6.3-2. The adaptation algorithms of Table 6.3-2 have been successfully used in Ref. 23. The problem formulation is similar to the continuous case. However, in applying the basic design techniques developed in Sec. 5.4 for discrete-time MRAS's described by state equations, we shall be obliged to superimpose auxiliary transient signals on the reference model which do not appear in the continuous case. (These auxiliary transient signals allow us to obtain an equivalent feedback system of similar structure.)

In comparison to the basic model given by Eq. (5.4-1) the equation of the modified reference model contains an additional transient signal (see the equation of $\tilde{\underline{x}}_k$) which vanishes when adaptation is achieved. Since the reference model is asymptotically stable by assumption, $\lim_{k\to\infty} \underline{v}_k = 0$ will also imply that $\lim_{k\to\infty} \tilde{\underline{x}}_k$ (i.e., the state of the modified model will tend toward those of the original model once the adaptation is completed). *However, this approach has an important limitation:* B_p *must be known.* An approximation when B_p is not known is to replace B_p in Table 6.3-2 by a matrix B which falls within the range of variation of B_p.

An extensively tested approximate solution has been proposed by Béthoux and Courtiol [24] when B_p is not known. In this case, the reference model is given by

$$\underline{x}_{k+1} = A_M \underline{x}_k + B_M \underline{u}_M(k) \tag{6.3-53}$$

and the adaptation signal is given by

$$\underline{u}_{P2}(k) = \Delta K_P(k + 1)\underline{y}_k + \Delta K_U(k + 1)\underline{u}_M(k) \tag{6.3-54}$$

which also allows us to apply the design given in Sec. 5.4. However, we must predict $\Delta K_P(k + 1)$ and $\Delta K_U(k + 1)$ at instant k in order to compute $\underline{u}_{P2}(k)$. To obtain this, one replaces $\underline{v}_{k+1}^0$ in Table 6.3-2 by $\underline{v}_k$, and this will be a good prediction of $\underline{v}_{k+1}^0$ if the sampling period is short (one can see, for example, that in the case of an integral adaptation the adjustable parameters in the equations for $\underline{v}_{k+1}^0$ and $\underline{v}_k$ are the same) and B_P is replaced by a matrix B which falls within the range of variation of B_P.

Note that in the design given in Table 6.3-2 an adaptation algorithm with constant adaptation gains has been used. If the plant to be controlled is a linear time-invariant one with unknown parameters, an algorithm with decreasing adaptation gain can be used (see Problem 6.6).

6.4 DESIGN OF ADAPTIVE MODEL-FOLLOWING CONTROL SYSTEMS USING ONLY INPUT AND OUTPUT MEASUREMENTS

6.4-1 Introduction

The problem to be discussed in this section is the design of an AMFC system for a plant which is described by

$$\dot{\underline{y}} = A_P \underline{y} + B_P \underline{u}_P \tag{6.4-1}$$

$$\theta_P = C\underline{x} \tag{6.4-2}$$

where A_P and B_P are constant but unknown matrices and the objective is to assure an asymptotically output perfect model following with respect to a reference model described by

$$\dot{\underline{x}} = A_M \underline{x} + B_M \underline{u}_M \tag{6.4-3}$$

$$\theta_M = C_M \underline{y} \tag{6.4-4}$$

It is assumed that dim $\underline{x}$ = dim $\underline{y}$, that dim θ_M = dim θ_P, and that the use of pure derivatives of θ_P for the implementation of the adaptive laws is not allowed.

Two approaches for solving this problem can be considered. One approach is based on the use of the techniques which are presented in Chap. 8 for building an adaptive observer for the plant given by Eqs. (6.4-1) and (6.4-2). These techniques will allow us to asymptotically observe the states of the plant through an observation model which asymptotically tends toward the following linear observation model:

$$\underline{y}' = A_p'\underline{y}' + B_p'\underline{u}_p \tag{6.4-5}$$

$$\theta_p' = C'\underline{y}' = \theta_p \tag{6.4-6}$$

Then there are two possibilities (either one applies): (1) The design given in Sec. 6.3 where $\underline{y}$ is replaced by the observed state $\underline{y}'$ (and one chooses $C_M = C'$) or, (2) A control law is determined (based on the current parameter estimates provided by the adaptive observer) such that $\theta_{\underline{p}}' = \theta_M$. (The control law plus the observer form an implicit reference model.) For details see Problems 8.3 and 8.4.

The second approach is to consider the plant and the model described by input/output relations in terms of differential operators and to try to apply the methods presented in Sec. 4.2 after the linear design is elucidated. We shall now present the second approach.

6.4-2 *The Linear Model-Following Control Problem*

Single-Input/Single-Output Case

Consider a single-input/single-output linear time-invariant plant described by

$$\hat{N}(p)\theta_p = \hat{M}(p)\mu \tag{6.4-7}$$

where

$$\hat{N}(p) = p^n + \sum_{i=0}^{n-1} \hat{a}_i p^i \tag{6.4-8}$$

is a polynomial of degree n and

$$\hat{M}(p) = \sum_{i=0}^{m} \hat{b}_i p^i \tag{6.4-9}$$

is a Hurwitz polynomial of degree m ($m < n$). θ_p is the plant output, and μ is the plant control input. *The assumption that $\hat{M}(p)$ is a Hurwitz polynomial implies that the zeros of the plant transfer function are in the open left half plane and that they therefore can eventually be cancelled without leading to an unbounded control input.*

Consider also a reference model having the same structure and described by

$$N(p)\theta_M = M(p)\rho \qquad (6.4\text{-}10)$$

where

$$N(p) = p^n + \sum_{i=0}^{n-1} a_i p^i \qquad (6.4\text{-}11)$$

is a Hurwitz polynomial of degree n (the reference model is asymptotically stable) and

$$M(p) = \sum_{i=0}^{m} b_i p^i \qquad (6.4\text{-}12)$$

is a polynomial of degree m [some of the coefficients b_i can be zero, and also the effective degree of M(p) can be lower than m].

The generalized output error is defined as

$$\varepsilon = \theta_M - \theta_P \qquad (6.4\text{-}13)$$

The objective of the design is to determine the plant control input μ such that for any ρ and $\varepsilon(0) = 0$ one has $\varepsilon(t) = 0$ for all t. In this case, *output perfect model following* is achieved. In addition to this objective, the design should recognize the constraint that the derivatives of θ_P and ρ cannot be used but that we can use the filtered derivatives of θ_P and ρ obtained from state-variable filters up to the degree of the state-variable filter. Since we wish to shift n poles, it is necessary to feed back n - 1 derivatives of θ_P (if available). This suggests that we use state-variable filters of degree n - 1 characterized by the transfer function

$$H_F(s) = \frac{1}{C(s)} = \frac{1}{s^{n-1} + \Sigma_{i=0}^{n-2} c_i s^i} \qquad (6.4\text{-}14)$$

where C(s) is a Hurwitz polynomial (i.e., the state-variable filter is asymptotically stable). One can now define the filtered values of θ_P and ρ, denoted by θ_{Pf} and ρ_f, as

$$p^{n-1} + \left(\sum_{i=0}^{n-1} c_i p^i\right)\theta_{Pf} = C(p)\theta_{Pf} = \theta_P \tag{6.4-15}$$

$$C(p)\rho_f = \rho \tag{6.4-16}$$

Since the plant control input is assumed to be a function of θ_P and ρ (and their filtered derivatives), perfect model following will be achieved if the transfer function $L[\theta_p]$ $L[\rho]$ of the plant plus the controller is equal to that of the reference model, given by M(s)/N(s). However, since we shall use state-variable filters of degree n - 1 as part of the controller, n - 1 poles will be added. So in order to end with only n poles, a pole-zero cancellation must be considered. But the most we can hope for is to introduce a physically realizable dynamic compensator at the plant input. With these thoughts in mind, the controller configuration shown in Fig. 6.4-1 can be considered, where

$$B(s) = \sum_{i=0}^{n-1} B_i s^i \tag{6.4-17}$$

$$D(s) = \sum_{i=0}^{n-1} D_i s^i \tag{6.4-18}$$

$$C^*(s) = \sum_{i=0}^{n-1} C_i^* s^i \tag{6.4-19}$$

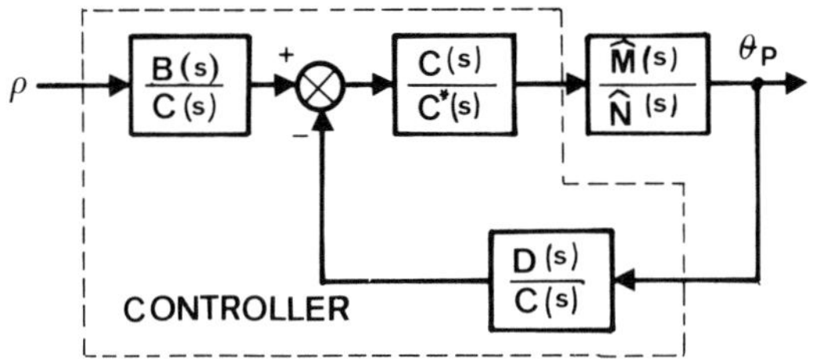

Fig. 6.4-1 Linear Model-Following Control System using only input and output measurements.

and the design problem can be reformulated as follows: Find B(s), D(s), and $C^*(s)$ such that

$$\frac{\mathcal{L}[\theta_p]}{\mathcal{L}[\rho]} = \frac{B(s)\hat{M}(s)}{C^*(s)\hat{N}(s) + D(s)\hat{M}(s)} = \frac{M(s)}{N(s)} \tag{6.4-20}$$

(Note that the coefficients B_i, C_i^*, and D_i are real scalars.)

A solution to this problem can be obtained if one defines

$$B(s) = H_0(s)M(s) \tag{6.4-21}$$

and

$$C^*(s) = H^*(s)\hat{M}(s) \tag{6.4-22}$$

where H(s) and $H^*(s)$ are n - m - 1 Hurwitz polynomials of the form

$$H_0(s) = s^{n-m-1} + \sum_{i=0}^{n-m-2} h_i s^i \tag{6.4-23}$$

$$H^*(s) = s^{n-m-1} + \sum_{i=0}^{n-m-2} h_i^* s^i \tag{6.4-24}$$

Using B(s) and $C^*(s)$ defined by the Eqs. (6.4-21) to (6.4-22), one obtains

$$\frac{\mathcal{L}[\theta_p]}{\mathcal{L}[\rho]} = \frac{H_0(s)M(s)}{H^*(s)\hat{N}(s) + D(s)} \tag{6.4-25}$$

and the condition of Eq. (6.4-20) becomes

$$H^*(s)\hat{N}(s) + D(s) = H_0(s)N(s) \tag{6.4-26}$$

One observes that the choice of the coefficients of $H_0(s)$ is free as long as $H_0(s)$ remains a Hurwitz polynomial (the nonobservable or noncontrollable states will be asymptotically stable) and that for a given $H_0(s)$ the problem always has a solution. However, the dynamic compensator $C(s)/C^*(s)$ in Fig. 6.4-1 located on the input of the plant (inside the feedback loop) can be replaced by an equivalent one, as shown in Fig. 6.4-2. The transfer function of this dynamic compensator is

$$\frac{C(s)}{\hat{b}_m C(s) + \hat{B}(s)} = \frac{C(s)}{C^*(s)} \tag{6.4-27}$$

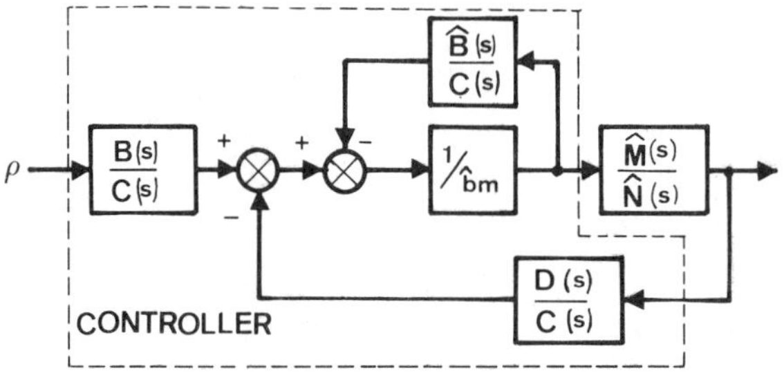

Fig. 6.4-2 Equivalent configuration for the LMFC system shown in Fig. 6.4-1.

and it will be equivalent to that given in Fig. 6.4-1 if

$$\hat{b}_m C(s) + \hat{B}(s) = C^*(s) = H^*(s)\hat{M}(s) \tag{6.4-28}$$

Using the controller structure shown in Fig. 6.4-2 and defining the filtered value of the plant control input μ_f as

$$C(p)\mu_f = \mu \tag{6.4-29}$$

one can express the plant control input as

$$\mu = \left[\frac{1}{\hat{b}_m}\left(-\sum_{i=0}^{n-2}\hat{B}_i p^i\right)\mu_f + \left(\sum_{i=0}^{n-1} B_i p^i\right)\rho_f - \left(\sum_{i=0}^{n-1} D_i p^i\right)\theta_{pf}\right]$$

$$= \left(\sum_{i=0}^{n-2} k_i^{1} p^i\right)\mu_f + \left(\sum_{i=0}^{n-1} k_i^{2} p^i\right)\rho_f + \left(\sum_{i=0}^{n-1} k_i^{3} p^i\right)\theta_{Pf} \tag{6.4-30}$$

Using the expression for B(s) given by Eq. (6.4-21), one can obtain an equivalent scheme for the control system of Fig. 6.4-2, as shown in Fig. 6.4-3, which will be useful later. Using this equivalent scheme, one can write

$$H_0(p)^{-1}\mu = \left(\sum_{i=0}^{n-2} k_i^{1} p^i\right)H_0^{-1}(p)\mu_f + \left(\sum_{i=0}^{n-1} k_i^{2} p^i\right)H_0^{-1}(p)\rho_f$$

$$+ \left(\sum_{i=0}^{n-1} k_i^{3} p^i\right)H_0^{-1}(p)\theta_{Pf} \tag{6.4-31}$$

The expressions for the plant control input μ can be simplified when n and m have particular values. For the case m = 0, $H_0(p)$ can be chosen as

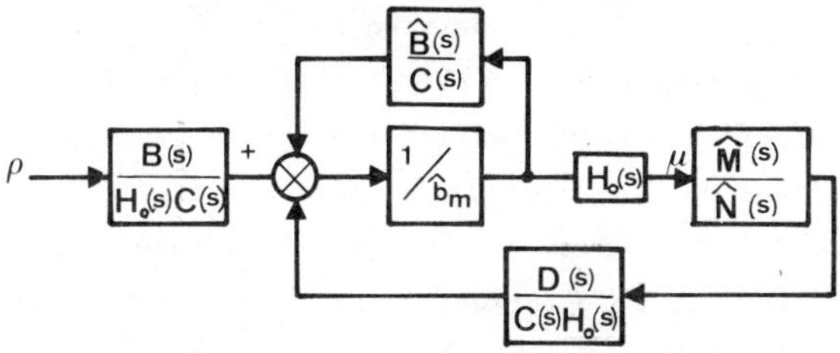

Fig. 6.4-3 Another equivalent configuration for the LMFC system shown in Fig. 6.4-1 or 6.4-2.

$$H_0(p) = C(p) = p^{n-1} + \sum_{i=0}^{n-1} c_i p^i \tag{6.4-32}$$

and B(s) becomes, respectively,

$$B(s) = b_0 C(s) \tag{6.4-33}$$

For the case n - m = 1, one has

$$H_0(s) = 1, \quad B(s) = M(s) \tag{6.4-34}$$

and Eqs. (6.4-30) and (6.4-31) are identical.

6.4-3 *The Adaptive Model-Following Control Problem*

In this section, we shall consider that the AMFC problem is a direct extension of the linear design discussed in the previous section for the case when the parameters $\hat{a}_i$ and $\hat{b}_i$ in Eq. (6.4-7) are unknown and constant during the adaptation process. Since we shall consider the design of an MRAS where only input and output measurements are available for the implementation of the adaptive law, the design methods presented in Sec. 4.2 should be considered. In the present case, the adjustable system is formed by the plant plus the controller defined by Eq. (6.4-31) in which k_i^1, k_i^2, and k_i^3 are now adjustable parameters. The objective of the design is to determine the control μ [and, consequently, the adaptation law for $k_i^1(t)$, $k_i^2(t)$, and $k_i^3(t)$] such that the output generalized error tends toward zero for any input ρ and any values (unknown) of parameters $\hat{b}_i$ and $\hat{a}_i$ satisfying the hypothesis of the linear design. Since the structure of the adjustable plant is defined and state-variable filters should be placed on the output of the plant (i.e., of the adjustable system) in order to obtain the filtered derivatives of θ_p, we shall try to use the design developed either in Sec. 4.2-3 or 4.2-4 (see also Sec. 4.2-6). However, the algorithms developed in Sec. 4.2-3 and 4.2-4 for the computation of the transient adaptation signals cannot be directly used (except in particular cases) since we do not have access inside the adjustable system in the same manner as in the structures considered in these previous sections. Nevertheless, these algorithms give hints for solving the present problem.

In designing AMFC systems using only input and output measurements we distinguish three basic cases depending on the difference between the denominator (n) and the degree of the numerator (m) of the *frozen* plant transfer function:

Case 1: $n - m = 1$. In this case no additional transient adaptation signal is necessary.

Case 2: $n - m = 2$. In this case an additional transient adaptation signal is applied to the input of the controlled plant and the designs given either in Sec. 4.2-3 or 4.2-4 are directly applicable.

Case 3: $n - m \geq 3$. In this case an additional transient adaptation signal is applied to the reference model. However, the design principles emphasized in Sec. 4.2-6 are still applicable.

The results of the design for the various cases are summarized in Table 6.4-1.

Next we shall present the derivation of the adaptation laws for the general case ($n - m \geq 3$).

The General Case ($n - m \geq 3$)

The plant to be controlled is described by

$$\left(p^n + \sum_{i=0}^{n-1} \hat{a}_i p^i\right)\theta_p = \left(\sum_{i=0}^{m} \hat{b}_i p^i\right)\mu, \quad m \geq 1 \tag{6.4-35}$$

where $\hat{a}_i$ and $\hat{b}_i$ are unknown but constant during adaptation and, furthermore, the plant zeros are in the open left half plane. The reference model is described by

$$\left(p^n + \sum_{i=0}^{n-1} a_i p^i\right)\theta_M = \left(\sum_{i=0}^{m} b_i p^i\right)\rho - \left(p^{n-1} + \sum_{i=0}^{n-2} c_i p^i\right)k(t)\mu_0 \tag{6.4-36}$$

where the a_i are such that the reference model is asymptotically stable and some of the coefficients b_i can be null. The coefficients c_i are such that the polynomial

$$C(p) = p^{n-1} + \sum_{i=0}^{n-2} c_i p^i \tag{6.4-37}$$

is a Hurwitz polynomial. The operator C(p) also serves to define a state-variable filter with the transfer function $C(s)^{-1}$. The additional transient adaptation signal μ_0 is a transient signal characterized by

$$\mu_0\Big|_{\varepsilon \equiv 0} = 0 \tag{6.4-38}$$

where ε is the generalized output error defined by

$$\varepsilon = \theta_M - \theta_P \tag{6.4-39}$$

k(t) in Eq. (6.4-36) is an additional adjustable gain. Both μ_0 and k(t) will be determined as part of the design. Since we have access to the internal structure of the reference model, the implementation of the second term on the right-hand side of Eq. (6.4-36) does not require any derivatives of $[k(t)\mu_0]$.

One also introduces a set of filtered values for θ_M, θ_P, θ_S, ε, ρ, and μ defined by

$$C(p)\theta_{Mf} = \theta_M, \quad C(p)\theta_{Pf} = \theta_P, \quad C(p)\varepsilon_f = \varepsilon$$
$$C(p)\mu_f, = \mu, \quad C(p)\rho_f = \rho \tag{6.4-40}$$

and a linear compensator acting on the signal ε_f and defined by

$$\nu = \left(p^{n-1} + \sum_{i=0}^{n-2} d_i p^i\right)\varepsilon_f \tag{6.4-41}$$

We will next derive the equivalent feedback system. Subtracting Eq. (6.4-35) from Eq. (6.4-36) and also using Eq. (6.4-40), one obtains

$$\left(p^n + \sum_{i=0}^{n-1} a_i p^i\right)\varepsilon_f = \left(\sum_{i=0}^{m} b_i p^i\right)\rho_f - k(t)\mu_0 - \left(\sum_{i=0}^{m} \hat{b}_i p^i\right)\mu_f$$
$$- \left(\sum_{i=0}^{n-1} \Delta a_i p^i\right)\theta_{Pf} = w_1 \qquad (6.4\text{-}42)$$

where $\Delta a_i = a_i - \hat{a}_i$. Now introducing the linear operator

$$H_0(p) = p^{n-m-1} + \sum_{i=0}^{n-m-2} h_i p^i \qquad (6.4\text{-}43)$$

and using the relations

$$\left(\sum_{i=0}^{m} b_i p^i\right)\rho_f = \left(\sum_{i=0}^{m} b_i p^i\right)\left(p^{n-m-1} + \sum_{i=0}^{n-m-2} h_i p^i\right)H_0^{-1}(p)\rho_f$$
$$= \left(\sum_{i=0}^{n-1} B_i p^i\right)H_0^{-1}(p)\rho_f \qquad (6.4\text{-}44)$$

$$\left(\sum_{i=0}^{m} \hat{b}_i p^i\right)\mu_f = \hat{b}_m H_0^{-1}(p)\mu + \left(\sum_{i=0}^{n-2} B_i' p^i\right)H_0^{-1}(p)\mu_f \qquad (6.4\text{-}45)$$

and the relation [25]

$$\left(\sum_{i=0}^{n-1} \Delta a_i p^i\right)\theta_{Pf} = \left(\sum_{i=0}^{n-1} D_i p^i\right)H_0(p)^{-1}\theta_{Pf} + \left(\sum_{i=0}^{n-2} G_i p^i\right)H_0(p)^{-1}\mu_f \qquad (6.4\text{-}46)$$

from Eq. (6.4-42) one obtains (after adding and subtracting the term $\hat{b}_m\mu_0$)

$$w = -w_1 = \hat{b}_m[H_0(p)^{-1}\mu + \mu_0] + [k(\nu, t) - \hat{b}_m]\mu_0$$
$$+ \left[\sum_{i=0}^{n-2} (B_i' + G_i)p^i\right]H_0(p)^{-1}\mu_f - \left(\sum_{i=0}^{n-1} B_i p^i\right)H_0(p)^{-1}\rho_f$$
$$+ \left(\sum_{i=0}^{n-1} D_i p^i\right)H_0(p)^{-1}\theta_{Pf} \qquad (6.4\text{-}47)$$

Using for $H_0(p)^{-1}\mu + \mu_0$ and expression similar to that given for $H_0(p)^{-1}\mu$ in Eq. (6.4-31) (which corresponds to the linear case) but with adjustable coefficients k_i^1, k_i^2, and k_i^3, one has

$$H_0(p)^{-1}\mu + \mu_0 = \left[\sum_{i=0}^{n-2} k_i^{\,1}(\nu, t)p^i\right]H_0(p)^{-1}\mu_f + \left[\sum_{i=0}^{n-1} k_i^{\,2}(\nu, t)p^i\right]H_0(p)^{-1}\rho_f + \left[\sum_{i=0}^{n-1} k_i^{\,3}(\nu, t)p^i\right]H_0(p)^{-1}\theta_{Pf} \tag{6.4-48}$$

and introducing Eq. (6.4-48) into Eq. (6.4-47), we obtain an expression of the form

$$w = -w_1 = \sum_{j=1}^{3n} f_j(\nu, t)z_j \tag{6.4-49}$$

From this point, the design methodology emphasized in Sec. 4.2-6 can be applied, and the results are summarized in Table 6.4-1.

Next, to explicitly obtain μ and μ_0, one must multiply both sides of Eq. (6.4-48) by the operator $H_0(p)$ given in Eq. (6.4-43). In so doing, one will have to effectuate products of the form $H_0(p)[k_i(\nu, t)H_0(p)^{-1}q_i(t)]$, and we must express the results using only $k_i(\nu, t)$ and $\dot{k}_i(\nu, t)$, which are directly available. One obtains

$$\mu = \left[\sum_{i=0}^{n-2} k_i^{\,1}(\nu, t)p^i\right]\mu_f + \left[\sum_{i=0}^{n-1} k_i^{\,2}(\nu, t)p^i\right]\rho_f + \left[\sum_{i=0}^{n-1} k_i^{\,3}(\nu, t)p^i\right]\theta_{Pf} \tag{6.4-50}$$

and

$$\mu_0 = H_0^{-1}(p)\left\{\sum_{i=0}^{n-2}\sum_{j=0}^{n-m-2} H_0^{\,j}(p)[\dot{k}_i^{\,1}(\nu, t)p^{j+i}H_0^{-1}(p)\mu_f] + \sum_{i=0}^{n-1}\sum_{j=0}^{n-m-2} H_0^{\,j}(p)[\dot{k}_i^{\,2}(\nu, t)p^{j+i}H_0^{-1}(p)\rho_f] + \sum_{i=0}^{n-1}\sum_{j=0}^{n-m-2} H_0^{\,j}(p)[\dot{k}_i^{\,3}(\nu, t)p^{j+i}H_0^{-1}(p)\theta_{Pf}]\right\} \tag{6.4-51}$$

where

$$H_0^{\,j}(p) = p^{n-m-2-j} + \sum_{i=0}^{n-m-3-j} h_{i+j+1}p^i \tag{6.4-52}$$

In practice μ_0 can be directly obtained by computing the right-hand side of Eq. (6.4-48) from which one subtracts $H^0(p)^{-1}\mu$ obtained from Eq. (6.4-50).

Table 6.4-1 Adaptive Model-Following Control of Single-Input/Single-Output Plants

	$n - m \geq 3$	$n - m \leq 2$	Observations
Reference model	$\left(p^n + \sum_{i=0}^{n-1} a_i p^i\right)\theta_M = \left(\sum_{i=0}^{m} b_i p^i\right)\rho - \left(p^{n-1} + \sum_{i=0}^{n-1} c_i p^i\right)k(t)\mu_0$ (ρ = external input)	$\left(p^n + \sum_{i=0}^{n-1} a_i p^i\right)\theta_P = \left(\sum_{i=0}^{m} b_i p^i\right)\rho$ (ρ = external input)	Poles and zeros of the reference model are in the left half plane; μ_0 = transient adaptation signal.
Plant to be controlled	$\left(p^n + \sum_{i=0}^{n-1} \hat{a}_i p^i\right)\theta_P = \left(\sum_{i=0}^{m} \hat{b}_i p^i\right)\mu$	$\left(p^n + \sum_{i=0}^{n-1} \hat{a}_i p^i\right)\theta_P = \left(\sum_{i=0}^{m} \hat{b}_i p^i\right)(\mu' + \mu'')$ $\mu' + \mu'' = \mu$	$n - m = 1$; $\mu'' = 0$; μ = control signal. Zeros of the plant are in the left half plane.
Generalized output error	$\varepsilon = \theta_M - \theta_P$	$\varepsilon = \theta_M - \theta_P$	
State-variable filter	$C(s)^{-1} = \left(s^{n-1} + \sum_{i=0}^{n-2} c_i s^i\right)^{-1}$	$C(s)^{-1} = \left(s^{n-1} + \sum_{i=0}^{n-2} c_i s^i\right)^{-1}$	$C(s)$ = Hurwitz polynomial.
Filtered variables	$C(p)\theta_{Mf} = \theta_M$, $C(p)\theta_{Pf} = \theta_P$, $C(p)\varepsilon_f = \varepsilon$, $C(p)\mu_f = \mu$, $C(p)\rho_f = \rho$		
Plant control signal (linear + adaptive)	$\mu = \left[\sum_{i=0}^{n-2} k_i^1(\nu, t)p^i\right]\mu_f + \left[\sum_{i=0}^{n-1} k_i^2(\nu, t)p^i\right]\rho_f + \left[\sum_{i=0}^{n-1} k_i^3(\nu, t)p^i\right]\theta_{Pf}$	$\mu' = \left[\sum_{i=0}^{n-2} k_i^1(\nu, t)p^i\right]\mu_f + \left[\sum_{i=0}^{n-1} k_i^2(\nu, t)p^i\right]\rho_f + \left[\sum_{i=0}^{n-1} k_i^3(\nu, t)p^i\right]\theta_{Pf}$	$m = 0$; if $H_0(p) = C(p)$, $[\Sigma_{i=0}^{n-1} k_i^2(\nu, t)p^i]\rho_f$ is replaced by $k_0^2(\nu, t)\rho$.

Transient adaptation signal	μ_0 = Eq. (6.4-51) $H_0(p) = p^{n-m-1} + \sum_{i=0}^{n-m-2} h_i p^i$ $H_0^j(p) = p^{n-m-2-j} + \sum_{i=0}^{n-m-3-j} h_{i+j+1} p^i$	$\mu'' = \left[\sum_{i=0}^{n-2} \dot{k}_i^1(\nu, t)p^i\right] H_0^{-1}(p)\mu_f + \left[\sum_{i=0}^{n-1} \dot{k}_i^2(\nu, t)p^i\right] H_0^{-1}(p)\rho_f + \left[\sum_{i=0}^{n-1} \dot{k}_i^3(\nu, t)p^i\right] H_0^{-1}(p)\theta_{Pf}$ $H_0(p) = p^{n-m-1} + \sum_{i=0}^{n-m-2} h_i p^i$	$H_0(p)$ = Hurwitz polynomial; $n - m = 1$; $\mu'' = 0$. $H_0(p)$ can be equal to $C(p)$. $m = 0$. If $H_0(p) = C(p)$; $[\Sigma_{i=0}^{n-1} k_i(\nu, t)p^i]H_0^{-1}(p)\rho_f$ is replaced by $k_0^2(\nu, t)\rho$ and Eq. (6.4-51) by (6.4-54)
Adaptation law	$\dot{k}(t) = \alpha_{3n}\nu\mu_0$ $\alpha_j > 0;\ j = 1, \ldots, 3n$	$\nu = (p^{n-1} + \sum_{i=0}^{n-2} d_i p^i)\varepsilon_f$ $\dot{k}_i^1(\nu, t) = \alpha_{i+1}\nu H_0^{-1}(p)p^i\mu_f$, $i = 0, \ldots, n-2$ $\dot{k}_i^2(\nu, t) = \alpha_{n+i+1}\nu H_0^{-1}(p)p^i\rho_f$, $i = 0, \ldots, n-1$ $\dot{k}_i^3(\nu, t) = \alpha_{2n+i+1}\nu H_0^{-1}(p)p^i\theta_{Pf}$, $i = 0, \ldots, n-1$ $\alpha_j > 0;\ j = 1, \ldots, 3n-1$	For $c_i = d_i$, $\nu = \varepsilon$; $m = 0$ and $H_0(p) = C(p)$; $\dot{k}_i^2(\nu, t)$, $i = 0, \ldots, n-1$ is replaced by $\dot{k}_0^2(\nu, t) = \alpha_{n+1}\nu\rho$
Stability condition	$h(s) = \dfrac{s^{n-1} + \Sigma_{i=0}^{n-2} d_i s^i}{s^n + \Sigma_{i=0}^{n-1} a_i s^i}$ is strictly positive real.		

Observe that the μ given by Eq. (6.4-50) tends toward the structure of linear control given by Eq. (6.4-30) when the adaptation is completed and that the various coefficients k_i^1, k_i^2, k_i^3 become constant. Observe also that $\dot{k}_i^1(\nu, t)$, $\dot{k}_i^2(\nu, t)$, and $\dot{k}_i^3(\nu, t)$ given in Table 6.4-1 and ν given by Eq. (6.4-41), assure that μ_0 will satisfy the condition of Eq. (6.4-38) if θ_p, μ, and ρ are bounded.[1]

Once μ_0 is determined, one can obtain $\dot{k}(\nu, t)$ given in Table 6.4-1, and by integration one obtains $k(\nu, t)$. Note from Eq. (6.4-47) that if $\hat{b}_m$ is known, $k(\nu, t)$ can be made equal to $\hat{b}_m$ and the adaptation chain for $k(\nu, t)$ is no longer necessary.

Remark 1

For the case m = 0, one can choose $H_0(p) = C(p)$ given in Eq. (6.4-37), and also using Eq. (6.4-33) we shall finally obtain

$$\mu = \sum_{i=0}^{n-2} k_i^1(\nu, t)p^i\mu_f + k_0^2(\nu, t)\rho + \sum_{i=0}^{n-1} k_i^3(\nu, t)p^i\theta_{Pf} \tag{6.4-53}$$

and

$$\mu_0 = H_0^{-1}(p)\left\{\sum_{i=0}^{n-2}\sum_{j=0}^{n-2} H_0^j(p)[\dot{k}_i^1(\nu, t)p^{j+i}H_0^{-1}(p)\mu_f] + \sum_{j=0}^{n-2} H_0^j(p)[\dot{k}_0^2(\nu, t)p^jH_0^{-1}(p)\rho] + \sum_{i=0}^{n-1}\sum_{j=0}^{n-2} H_0^j(p)[\dot{k}_i^3(\nu, t)p^{j+i}H_0^{-1}(p)\theta_{Pf}]\right\} \tag{6.4-54}$$

Remark 2

Note that the transient signal μ_0 in Eq. (6.4-51) or (6.4-54) depends mainly on the first derivatives of $k_i^1(\nu, t)$, $k_i^2(\nu, t)$, $k_i^3(\nu, t)$. For low values of the adaptation gains α_i in Table 6.4-1 and a low level of the output generalized error, the derivatives of $k_i^1(\nu, t)$, $k_i^2(\nu, t)$, $k_i^3(\nu, t)$ can have low values. In such cases, the effect of the transient signal μ_0 upon the output generalized error is negligible compared to that of the signal μ, and consequently the generation of the signal μ_0 [and of course of $k(\nu, t)$] can be avoided.

---

[1]For $n - m \leq 2$ or for m = 0 the assumption on the boundness of θ_p and μ is not required.

Remark 3

The transient signal μ_0 applied through the gain $k(\nu, t)$ to the reference model can eventually be replaced by a signal directly added to the output generalized error (their sum forms the "augmented error" [25]). However, this design requires n additional integrators. This approach, together with the choice $c_i = d_i$, is considered in Ref. 25, and its extension to the multiinputs/multioutputs case is considered in Ref. 26.

Remark 4

The discrete-time versions of the design presented above and of the design considered in Ref. 25 can be found in Refs. 27 and 28 respectively.

6.5 IMPLEMENTATION ASPECTS FOR ADAPTIVE MODEL-FOLLOWING CONTROL SYSTEMS

In this section, we shall briefly discuss some details concerning implementation of AMFC systems. The problems to be considered are (1) rejection of the dc components, (2) saturation of the plant actuator, and (3) digital implementation.

Rejection of the dc Components

It is well known that the dynamic equations are usually written for variations of the controlled variables around specified steady-state values. Therefore, when implementing AMFC systems (as well as other MRAS's), this must be kept in mind. The problem of removing the dc components appears in this case when implementing both the reference model and the adaptation mechanism.

The first task is to reject the dc components, which can eventually appear as part of the measured generalized error. To simplify the presentation of the basic solutions, we shall consider the case of a single-input/single-output AMFC system. The extension to the case of a vector generalized error is straightforward.

Basically, one can distinguish two cases:

1. The dynamic input applied to the reference model is superimposed (via the feedforward gains) onto a steady-state

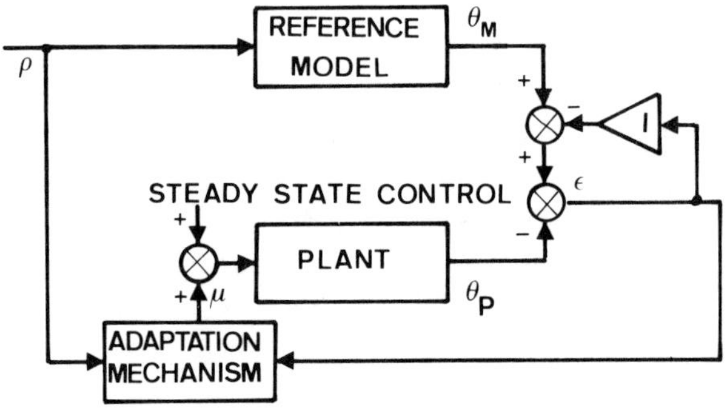

Fig. 6.5-1 Elimination of the dc component on the output generalized error in the case of a dynamic input superimposed on a steady-state plant control input.

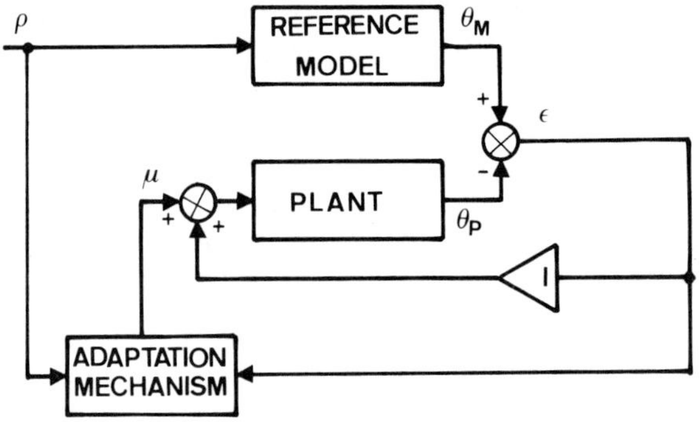

Fig. 6.5-2 Elimination of the dc component on the output generalized error provoked by the difference between the static and dynamic gain of the plant.

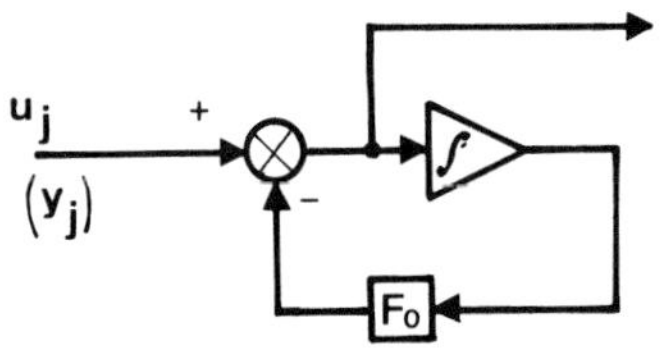

Fig. 6.5-3 Filter for the elimination of the dc component on the input and plant state components (washout filter).

plant control input. In this case, one can use, for example, the scheme indicated in Fig. 6.5-1. In Fig. 6.5-1, the integrator I will assure that $\varepsilon = 0$ in the steady state, independently of either the steady-state value of the plant control μ^0 or the steady-state value of the plant output θ_p^0.

2. The reference model defines the desired dynamic response as well as the desired steady-state output. In this case, since the static plant gain (defined as the ratio between the steady-state values of the output and input of the plant) in general differs from the dynamic gain (defined for variations of input and output around the steady-state values), one must use the scheme shown in Fig. 6.5-2. In Fig. 6.5-2, the integrator I acts only during the steady state, adding a steady-state control to the plant such that $\varepsilon = 0$.

The second problem concerns the implementation of the adaptation mechanism. Consider, for example, the implementation of the adaptation laws given in Table 6.3-1, which correspond to the case when the plant states are measurable. Referring to those equations, we must construct products of two variables. These products are of the form $v_i u_j$ or $v_i y_j$, where v_i is the ith component of the vector $\underline{v}$ related to the generalized state error vector $\underline{e}$ through the matrix D [see Eq. (6.3-15)] and u_j and y_j are the jth components of the input to the reference model and of the plant state, respectively. Since the dc components of the vector $\underline{e}$ have been removed, for the products $v_i y_j$ and $v_i u_j$ to have a sign sense, one must also remove the dc components from y_j and u_j. To do this, a high-band-pass filter of the type shown in Fig. 6.5-3 can be used. This filter must be designed so as to eliminate only the low frequencies (around zero). This type of filter, used for removing dc components, is also called a *washout* filter.

Saturation of the Plant Actuator

In the case of model-following control systems, if the time response of the reference model is much shorter than that of the plant, saturation of the plant actuator as well as of the adaptation mechanism can occur. The combined effect of the saturation of both the actuator and the adaptation mechanism can lead to the appearance of undamped nonlinear oscillations in the system. Therefore, these problems must be circumvented.

In practice, one can distinguish two basic situations: (1) Small variations of the output (or states) occur around a steady-state value. In this case, depending on the steady-state value of the plant control input and its saturation level, a compatible set of dynamics for the reference model can be chosen. These dynamics must be such that the model matching can be achieved without reaching the saturation level of the plant actuator. (2) Large variations of the output (or the states) occur, and we want to control only the final part of the transient. Such a situation appears, for example, when heating an oven. One likes to initially use the maximum heating power available and to control only the final part of the heating process in order to quickly and smoothly reach the desired temperature [27].

In such cases, during the time when the plant acutator is in saturation, the reference model must be *locked*, and the output (or

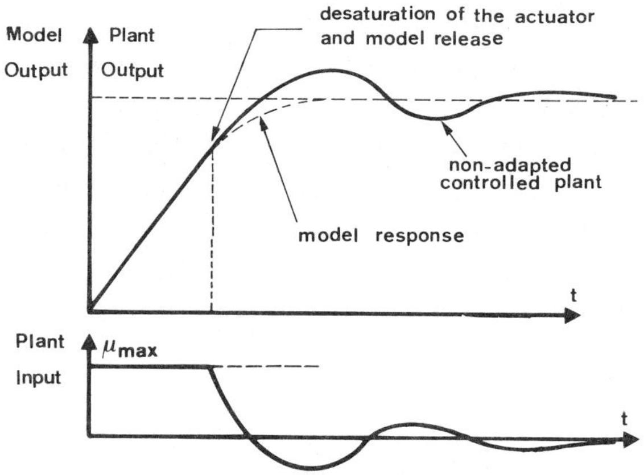

Fig. 6.5-4 Evolution of the plant and model outputs in a model-following control system with the model locked during the saturation of the plant actuator.

the states) of the model must follow those of the plant. The reference model will be released (it becomes again active) only when the plant actuator is no longer in saturation (due to the effect of the feedback). This is illustrated in Fig. 6.5-4, where the time response of the plant and of the reference model as well as the evolution of the plant control input are represented. After the actuator is no longer saturated and the reference model is activated, the adaptation mechanism will act, and the plant will follow the reference model.

Digital Implementation

For a high speed of adaptation to be achieved when digital computers are used for implementing AMFC systems, the *sampling period should be small not only with respect to the time response of the plant (or of the model) but with regard to the adaptation time*. Note that this adaptation time can be small if high adaptation gains are used and the input has a broad frequency spectrum.

6.6 CASE STUDY: AN ADAPTIVE SPEED CONTROLLER FOR DC ELECTRICAL DRIVES

A commonly used control scheme for dc electrical drives is shown in Fig. 6.6-1. This control scheme has both an internal control loop for the armature current and a speed control loop. The plant is formed by the power thyristor amplifier (the thyristor-controlled rectifier plus the firing control circuits), the dc motor, and the motor load.

The dynamic parameters of the plant are subject to variations (see Sec. 1.3-5). The parameter variations affect both the armature current control loop and the speed control loop [17, 29].

For the speed control loop in Fig. 6.6-1, the plant to be controlled comprises the armature current control loop and the dc motor. and the output is the motor speed. The plant can be characterized, for a given thyristor conduction time t_c (and for field current, moment of inertia, and friction, all constant), by the transfer function

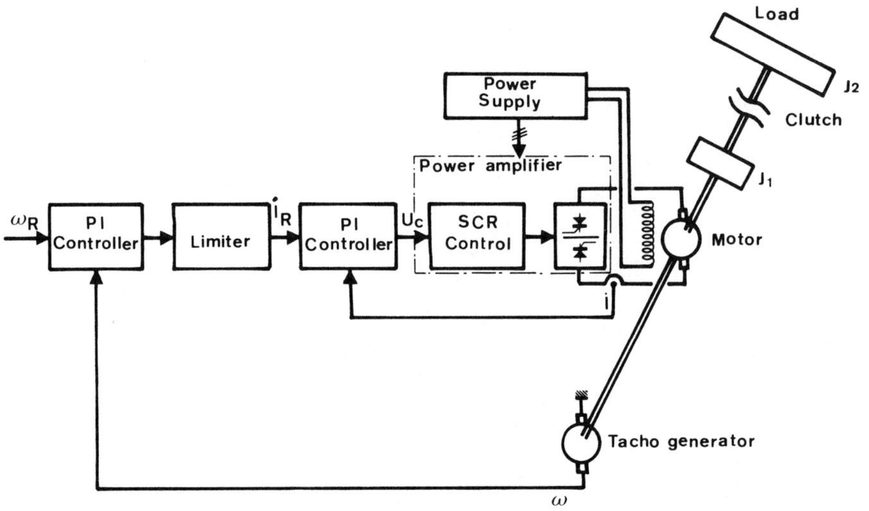

Fig. 6.6-1 Commonly used scheme for speed control of dc electrical drive. Motor speed (ω), armature current (i), and moment of inertia (j).

$$H_p(s) = \frac{Gg}{(1 + s\tau)(1 + \psi s)} = \frac{Gg}{\tau\psi} \frac{1}{(1/\tau\psi) + [(\tau + \psi)/\tau\psi]s + s^2} \tag{6.6-1}$$

where g is the gain of the tachogenerator (V/rad/sec), G = K/f, ψ = J/f, K is the motor torque constant (Nm/A), J is the moment of inertia seen at the motor axle (kg m^2), and f is the friction (Nm/rad/sec).

The objective of the speed control loop is to assure a particular speed time response independent of the variations of t_c, J, f, and K. Since the plant can be characterized for a given nominal working point by a second-order transfer function, it seems natural to define the desired response by that of a reference model characterized also by a second-order transfer function.

If the speed and the acceleration are the state variables used, the plant to be controlled will be described by

$$\dot{\underline{y}} = A_P\underline{y} + \underline{b}_P u_P \tag{6.6-2}$$

where

$$\underline{y} = \begin{bmatrix} y_1 \\ y_2 \end{bmatrix} = \begin{bmatrix} \omega \\ a \end{bmatrix} \tag{6.6-3}$$

$$A_P = \begin{bmatrix} 0 & 1 \\ -\frac{1}{\tau\psi} & -\frac{1}{\tau}(1 + \frac{\tau}{\psi}) \end{bmatrix}, \quad \underline{b}_P = \begin{bmatrix} 0 \\ \frac{A_0 Gg}{\tau\psi} \end{bmatrix} \tag{6.6-4}$$

ω is the motor speed and a is the acceleration and u_P is the reference setting of the armature current control loop.

We shall choose a reference model with a similar structure (and having a static gain equal to 1), described by

$$\dot{\underline{x}} = A_M\underline{x} + \underline{b}_M u_M \tag{6.6-5}$$

where

$$\underline{x} = \begin{bmatrix} x_1 \\ x_2 \end{bmatrix} = \begin{bmatrix} \omega_M \\ a_M \end{bmatrix} \tag{6.6-6}$$

$$A_M = \begin{bmatrix} 0 & 1 \\ -\alpha\frac{G_0 A_0 g}{\tau_0\psi_0} & -\frac{1}{\tau_0}(1 + \frac{\tau_0}{\psi_0}) \end{bmatrix}, \quad \underline{b}_M = \begin{bmatrix} 0 \\ \frac{\alpha G_0 A_0 g}{\tau_0\psi_0} \end{bmatrix} \tag{6.6-7}$$

ω_M is the desired speed, a_M is the desired acceleration and u_M is the reference speed. In Eq. (6.6-7), τ_0, ψ_0, and G_0 represent the nominal values of the plant dynamic parameters, and the parameter α is introduced in order to obtain the desired dynamics of the speed control loop.

If we define [as in Sec. 6.3, Eqs. (6.3-8) to (6.3-11)]

$$u_P = u_{P1} + u_{P2} \tag{6.6-8}$$

where u_{P1} is the linear control signal *computed for the nominal value of the dynamic parameters* and u_{P2} is the adaptation signal, we can then choose

$$u_{P1} = -K_P\underline{y} + K_u u_M \tag{6.6-9}$$

where

$$K_P = [k_{p1};\ 0], \quad K_U = k_U \tag{6.6-10}$$

and

$$u_{P2} = \Delta K_P(v,\ t)\underline{y} + \Delta K_U(v,\ t)u_M \tag{6.6-11}$$

where

$$\Delta K_P(v,\ t) = [\Delta k_{P1};\ \Delta k_{P2}]; \quad \Delta K_U = \Delta k_U \tag{6.6-12}$$

The introduction of a term $K_M\underline{x}$ in Eq. (6.6-9) is not necessary since the output of the plant can shift the coefficient $1/\tau\psi$ of the matrix A_P through the matrix K_P given by Eq. (6.6-10). Note that the plant and the reference model satisfy the perfect model-following conditions independently of the values of A_M, A_P, B_M, and B_P.

The adaptation mechanism contains a linear compensator given by Eq. (6.3-15) where $D = \underline{d}^T = [d_1,\ d_2]$.

For integral + proportional adaptation of $\Delta K_P(v,\ t)$ and $\Delta K_U(v,\ t)$ in Eq. (6.6-12), the adaptation laws of Table 6.3-1 (tracking) can be used. By choosing

$$\begin{gathered} F = f > 0, \quad F' = f' > 0, \quad G = \begin{bmatrix} 1 & 0 \\ 0 & \beta \end{bmatrix}, \quad \beta > 0 \\ M = m > 0, \quad M' = m' > 0, \quad N = n = 1 \end{gathered} \tag{6.6-13}$$

the adaptation signal $\underline{u}_{P2}$ in Eq. (6.6-11) will be given by

$$\begin{aligned} u_{P2} = {} & (\textstyle\int_0^t fvy_1\, d\tau)y_1 + f'vy_1^{\,2} + \beta[\textstyle\int_0^t fvy_2\, d\tau]y_2 \\ & + \beta f'vy_2^{\,2} + (\textstyle\int_0^t mvu_M\, d\tau\ u_M) + m'vu_M^{\,2} \end{aligned} \tag{6.6-14}$$

The AMFC system using the adaptation law will be asymptotically stable if for any $\underline{b}_p$ the transfer function

$$H(s) = \underline{d}^T(sI - A_M)^{-1}\underline{b}_p \tag{6.6-15}$$

is strictly positive real. However, any $\underline{b}_p$ can be expressed by

$$\underline{b}_p = \underline{b}_p^{\,0} r \tag{6.6-16}$$

where r is a positive constant [see Eqs. (6.6-4) and (6.6-7)]. Therefore in Eq. (6.6-15) b_p can be replaced by $b_p^{\,0}$.

In practice, we can distinguish situations which allow us to simplify the adaptation law of Eq. (6.6-14). These situations include cases where the variations of the term $(1/\tau)[1 + (\tau/\Psi)]$ of the matrix A_p are small (noninterrupted armature current and $\tau_0/\Psi_{min} < 0.25$; use of an adaptive current controller [29]). For these cases the gain β can be set equal to a very small value, since the performance of this adaptation chain will have a small effect upon the global system performance. *In practice, this allows us to use a simplified expression for* u_{P2} *obtained from Eq. (6.6-14) with* $\beta = 0$.

The AMFC system designed above can be used as part of a speed control scheme containing, in addition, a PI speed controller (it will represent the controlled plant), or it can directly play the role of an adaptive speed controller. In the first case, the PI controller is adjusted in accordance with the desired performance, assuming that the plant always behaves as the reference model (in this case the signal u_M is the output of the PI controller). In the second case, the adaptive speed control system has the detailed configuration given in Fig. 6.6-2, where an integrator has been added in order to assure null static error between the desired speed ω_M and actual speed ω (the integrator has no effect during transients). In this case the signal u_M is the reference speed.

For the scheme shown in Fig. 6.6-2, a state-variable filter is introduced in order to obtain filtered values of the measured speed and acceleration. Since the time constant of this filter is small compared to ψ (< 1:5), it can be neglected, and no such filters are

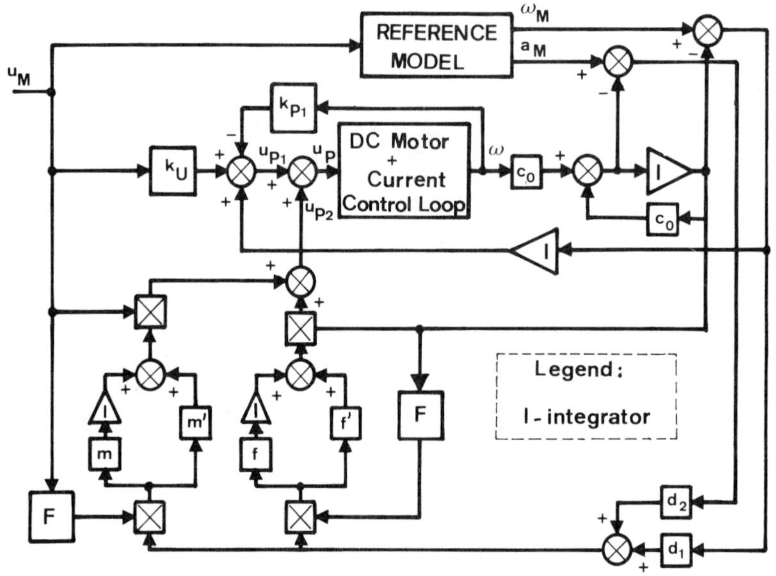

Fig. 6.6-2 Adaptive model-following speed control system of a dc electrical drive.

necessary in the model. In Fig. 6.6-2, the filters denoted by F serve to reject the dc components of u_M and ω.

The performance of the adaptive speed controller shown in Fig. 6.6-2, built in hybrid technology, is illustrated by its application to a 5.3-kW dc electrical drive (nominal speed, 1500 rpm), fed by a three-phase bidirectional controlled rectifier [17].[4]

We shall illustrate only the high speed of adaptation assured by the adaptive control scheme of Fig. 6.6-2 [see Figs. 6.6-3(a) and (b)]. At $t = 0$, the motor load has its maximal inertia. Until $t = t_0$ a PI speed controller (as in Fig. 6.6-1) adjusted for the low value of the load inertia is used (a ratio of 3.71 to 1 between maximum and minimum inertia). At $t = t_0$, the output signal of the PI controller

[4]The subsequent result has been provided by courtesy of ALSTHOM-Directions des Recherches, Département d'Automatique et d'Electronique Grenoble.

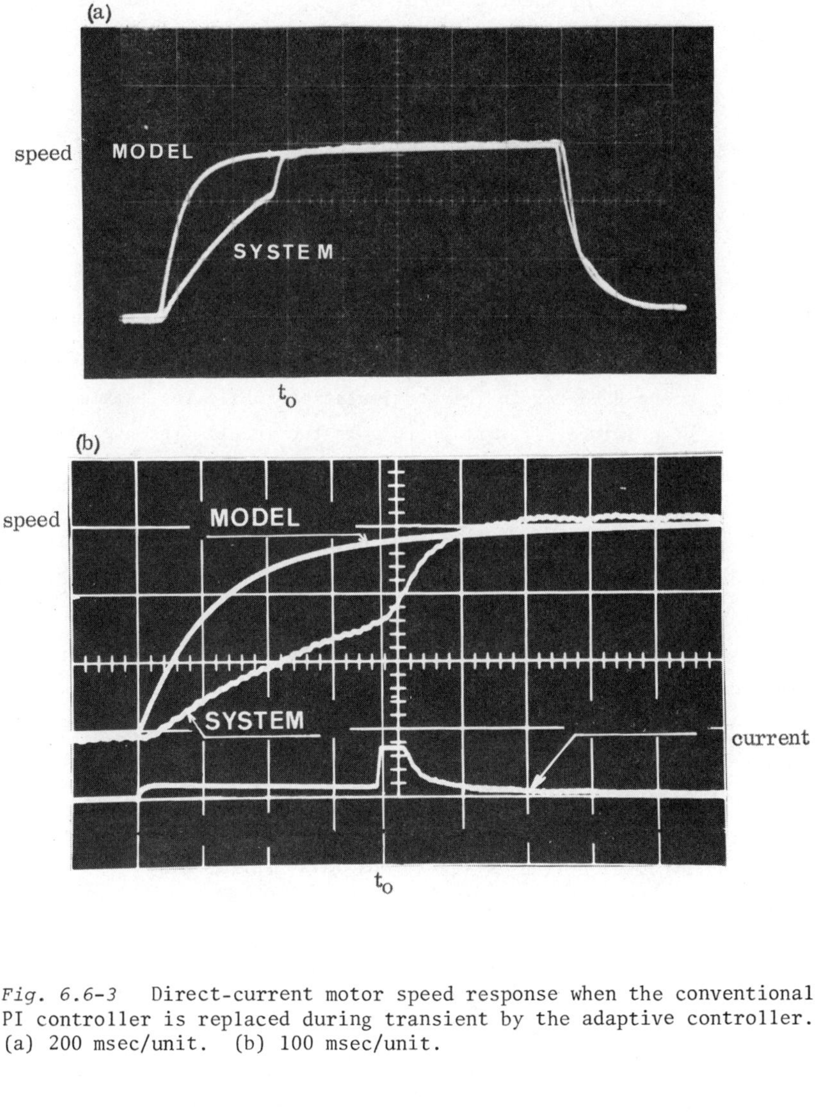

Fig. 6.6-3 Direct-current motor speed response when the conventional PI controller is replaced during transient by the adaptive controller. (a) 200 msec/unit. (b) 100 msec/unit.

is replaced by the signal u_P provided by the control scheme of Fig. 6.6-2. The speed of adaptation has been limited by the maximum value of the allowable armature current, as shown in Fig. 6.6-3(b). But even with this limitation, the slope of the adaptation process is comparable with the slope of the time response of the reference model. One can remark with regard to Fig. 6.6-3(a) that the next response to a step input variation is already acceptable.

6.7 CASE STUDY: AN AIRCRAFT LONGITUDINAL CONTROL PROBLEM

One of the possible applications of AMFC systems is in the design of active simulation facilities. For example, in aeronautics, the plant can be a plane used for in-flight simulation, while the estimated dynamics of a future aircraft represents the model. The use of LMFC techniques for the design of such in-flight simulators has been extensively considered in the past [1, 4, 8]. In the present case study, the example considered by Winsor and Roy [1] concerning the design of a desensitized model-following control system for an in-flight simulator is reformulated as the design of an AMFC system. The results obtained on a computer simulation for the two approaches are then compared.

The plant for this example represents the three-degree-of-freedom linearized perturbation longitudinal state equations of a conventional subsonic aircraft, a Convair C-131B, while the estimated dynamics of a large supersonic aircraft represents the model [1, 14].

The dynamics of the reference model is expressed by

$$\dot{\underline{x}} = A_M \underline{x} + B_M \underline{u}_M \tag{6.7-1}$$

where

$$\underline{x}_M^T = [\theta_M, q_M, \alpha_M, v_M], \quad \underline{u}_M^T = [\delta e_M, \delta t_M]$$

θ_M is the pitch attitude, q_M is the pitch rate, α_P is the angle of attack, v_M is the air speed, δe_M is the elevator command input, and δt_M is the throttle command input. The plant is described by

$$\dot{\underline{y}} = A_P \underline{y} + B_P \underline{u}_P \tag{6.7-2}$$

where

$$\underline{y}^T = [\theta_p, q_p, \alpha_p, v_p], \quad \underline{u}^T_p = [\delta e_p, \delta t_p, \delta z_p]$$

$$A_p = \begin{bmatrix} 0.0 & 1.0 & 0.0 & 0.0 \\ 1.401E\text{-}4 & (M_q + M_{\dot{\alpha}}) & -1.9513 & 0.0133 \\ -2.505E\text{-}4 & 1.0 & -1.3239 & -0.0238 \\ -0.561 & 0.0 & 0.3580 & -0.0279 \end{bmatrix}$$

$$B_p = \begin{bmatrix} 0.0 & 0.0 & 0.0 \\ -5.3307 & 6.447E\text{-}3 & -0.2669 \\ -0.16 & -1.155E\text{-}2 & -0.2511 \\ 0.0 & 0.106 & 0.0862 \end{bmatrix}$$

θ_p is the pitch attitude, q_p is the pitch rate, α_p is the angle of attack, v_p is the air speed, δe_p is the elevator command deflection, δt_p is the throttle control, and δz_p is the flap command deflection. The term a_{22} of the matrix A_p which represents the aircraft dynamic parameter $M_q + M_{\dot{\alpha}}$ has the nominal value -2.038, and it is assumed to vary from -0.558 to -3.558 (≃ 75% variation around the nominal value). The plant and the model satisfy Erzberger's conditions [Eqs. (6.2-17) and (6.2-18)], and, according to Winsor and Roy's results [1], it is possible to determine matrices K_U, K_p, and K_M for each value of $M_q + M_{\dot{\alpha}}$ in order to obtain perfect model following.

The Adaptive Control Structure

An AMFC configuration with signal-synthesis adaptation of the form given in Fig. 6.3-1(b) was adopted. The adaptation loop was superimposed on a perfect LMFC system designed for $M_q + M_{\dot{\alpha}} = -2.038$ [1]. Since the matrix A_M is "ill-conditioned," in order to compute the matrix D the Lyapunov equation in Table 6.3-1 has been solved using Jameson's method [19] (see also Sec. 6.3).

Simulation Results [14]

We shall show only some results illustrating the performance of the AMFC system in the tracking regime. (For more details, see Ref. 14.)

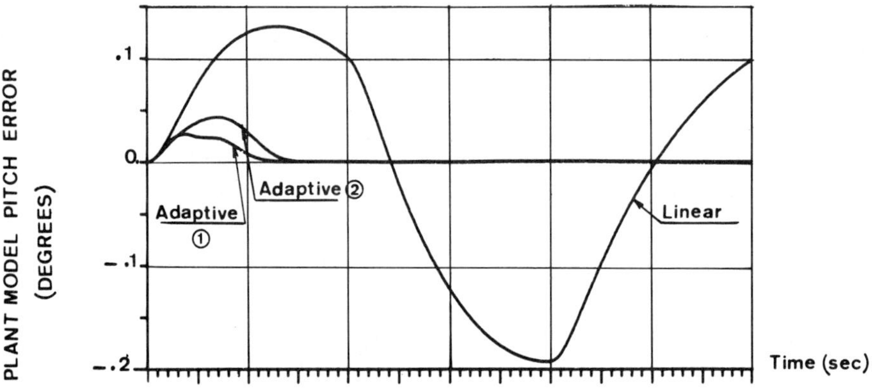

(a)

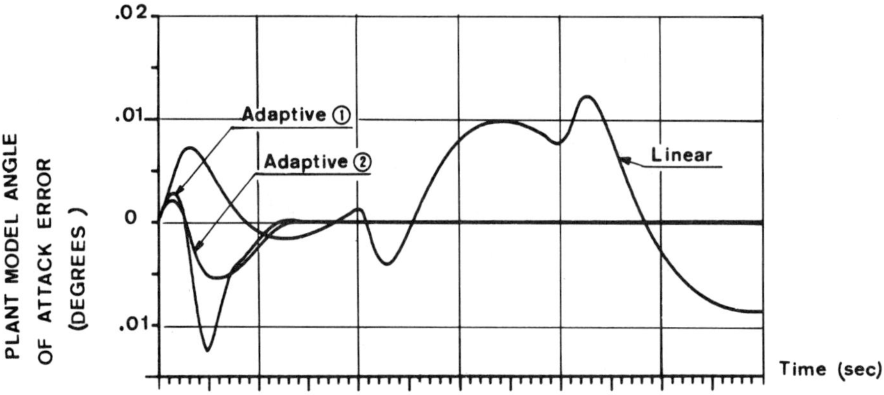

(b)

Fig. 6.7-1 Plant model state error response in tracking regime for $M_q + M_{\dot{\alpha}} = -3.558$. (a), (b), (c) Evolution of three components of the state error vector. (d) Evolution of the command inputs.

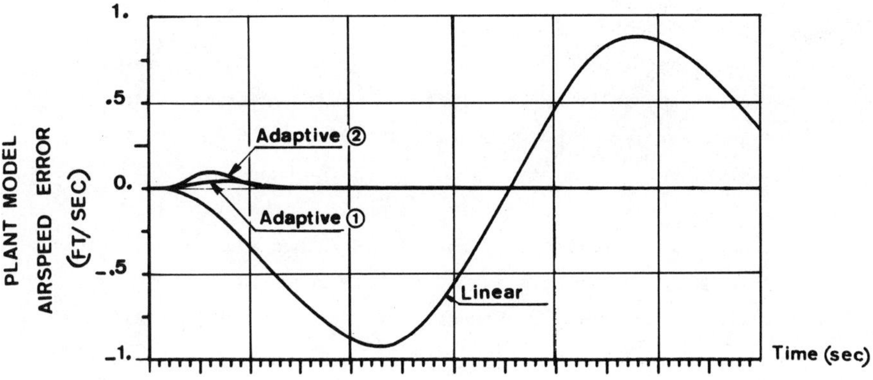

(c)

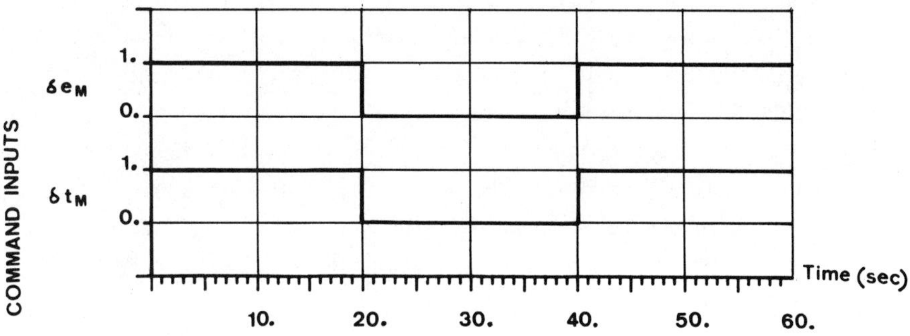

(d)

Fig. 6.7-1 (Continued)

Figure 6.7-1 illustrates the evolution of three components of the generalized state error vector for $M_q + M_{\dot{\alpha}} = -3.558$ ($\simeq$ 75% variation below the nominal value). The curves marked *Adaptive 1* represent the evolution of the error when the computation of the linear compensator D is done using a diagonal matrix $Q_1 = \text{diag}[2.5, 2.5, 2.5, 2.5]$ in the Lyapunov equation (Table 6.3-1). One can see that the performance results are better than in the linear case except for the *angle of attack* during the first 10 sec. Of course, the choice of another weighting matrix Q in the Lyapunov equation allows us to improve the response of this component of the state error vector. This is illustrated by the curves marked *Adaptive 2*, which are obtained if the matrix Q is modified in order to augment the contribution of the angle of attack state error in the generation of the vector $\underline{v}$. ($Q_2 = \text{diag}[1, 1, 7, 1]$. Note that the traces of Q_1 and Q_2 are equal.)

From Fig. 6.7-1, one also concludes that the adaptation process is almost achieved in about 15 sec. This means, for example, that for step variations of the parameter $M_q + M_{\dot{\alpha}}$ occurring with a period longer than 15 sec, we can consider that the design hypotheses are fulfilled. Of course, if instead of step variations other smoother variations are considered, the duration of the adaptation process will be shorter (for the same type of input $\underline{u}_M$). In fact, as the theoretical analysis given at the end of Sec. 6.3-2 has shown, the adaptive control system also performs well when the plant parameters are continuously time-varying. The simulation results have confirmed this [14].

6.8 CONCLUDING REMARKS

We wish to emphasize the following basic ideas from this chapter:

1. Adaptive model-following control systems are an extension of linear model-following control systems for the case where the parameters of the plant are poorly known or vary during operation.

2. The AMFC problem can be formulated as an MRAS design for the case where a solution for perfect model following exists.
3. The conditions for the existence of a solution for perfect model following are essentially related to the structure of the reference model and of the plant to be controlled. The reference model should be chosen such that perfect model following can be achieved for any possible values of the plant parameters.
4. In the case of the control of plants described by state equations (state tracking), information about the range of variation of the plant input matrix B_p is required for designing the adaptation mechanism. Such information is not necessary in the case of description by input/output relations.
5. Adaptive state observers, which will be discussed in Chap. 8, can be used for the implementation of AMFC systems which track the model state when the states of the plant are not directly accessible.
6. The design method for plants described in terms of input/output relations (presented in Sec. 6.4) can be applied only if for a certain working point the zeros of the plant transfer function are in the left half plane.
7. In the case of access to the states of the plant, the number of adaptation chains is equal to the number of parameters in the feedforward gain matrix and plant feedback gain matrix, which must be adjusted. In the case of a single-input plant, this corresponds to a maximum of $n + 1$ adaptation chains (n is the dimension of the state vector). In the case of single-input/single-output plants described in terms of input/output relations, the number of adaptation chains is in general equal to $3n$ (where n is the degree of the output differential equation).
8. The use of integral + proportional (or integral + relay) adaptation in the case of a state-space description has a

beneficial effect upon the convergence of the generalized state error vector in comparison to the use of integral adaptation.

9. The parallel AMFC systems are essentially used for tracking the reference model in the presence of an external input. The series-parallel AMFC systems are essentially used for state (or output) regulation.
10. The implementation of AMFC systems must take into account the problems related to the rejection of the dc components and the effect of the eventual saturation of the plant actuator at the beginning of the transients.
11. In the case of digital implementation, the sampling period should be small in comparison with the adaptation time, which can be smaller than the smallest significant time constant of the plant if the input is rich.

PROBLEMS

6.1 [30] In the case of a dc drive where the effect of the current control loop can be neglected but where the effect of the bearing friction is important, the dynamic behavior of the dc drive can be described by

$$\dot{\omega} + \frac{1}{\psi}\omega = Gu_p + \hat{f}(\omega)$$

$$\psi = \frac{J}{f} \qquad G = \frac{K}{f}$$

where ω is the motor speed, J is the moment of inertia, f is the viscous friction, u_p is the input voltage to the power amplifier which feeds the dc motor, K is the gain of the plant (dc motor + power amplifier), and $\hat{f}(\omega)$ denotes the effect of the bearing friction. Except for the region around zero speed, $\hat{f}(\omega)$ is almost constant, but its sign depends on the sign of ω (i.e., the sense of rotation). Therefore one can approximate $\hat{f}(\omega)$ by $\hat{f}(\omega) = \alpha \operatorname{sgn} \omega$, where α is unknown. In addition one assumes that during operation J, f, and K are subject to variations.

Because of the form of $\hat{f}(\omega)$, one chooses the following control law: $u_P = u_{P1} + u_{P2}$, where u_{P1} is the linear control signal, defined as $u_{P1} = k_U u_M - k_P \omega$, and u_{P2} is the adaptation signal, defined as $u_{P2} = \Delta k_U(e, t)u_M + \Delta k_P(e, t)\omega + k(e. t) \operatorname{sgn} \omega$. Design an AMFC system with integral + proportional adaptation such that the dc drive behaves like the following reference model:

$$\dot{\omega}_M + \frac{1}{\psi_0} \omega_M = G_0 u_M$$

where u_M is the output of a speed controller adjusted for the above model. (The generalized error is defined as $e = \omega_M - \omega$.)

6.2 [31] The dynamics relating the input to the steering gear and the course angular velocity of a ship can be expressed in state-space form as

$$\dot{y}_1 = y_2$$
$$\dot{y}_2 = ay_2 - K(b_1 y_1 + b_2 y_1^3) + GKu_P$$

where ψ is the course of the ship, $y_1 = \dot{\psi}$ is the course angular velocity, and $u_P = \delta_g$ is the input to the steering gear. In the above equations, the parameters b_1 and b_2 are subject to variations during ship operation.

(a) Consider that b_2 is known and constant. Design an AMFC system, where the desired behavior of the ship is expressed by

$$\dot{x}_1 = x_2$$
$$\dot{x}_2 = ax_2 - K(b_{1M} x_1 + b_{2M} x_1^3) + GKu_M$$

where $u_M = \delta$ is the output of the course controller, $x_1 = \dot{\psi}_M$ is the desired course angular velocity, $a = 3.4$, $K = 0.05$, $b_{1M} = 1.24$, and $b_{2M} = 1.06$, by neglecting the effect of x_1^3 and y_1^3 and assuming that y_1 and y_2 are measurable.

(b) Consider that b_1 and b_2 are subject to variations. Design an AMFC system where the desired behavior of the ship is expressed by a combined parallel and series-parallel reference model described by

$$\dot{x}_1 = x_2$$
$$\dot{x}_2 = ax_2 - K(b_{1M}x_1 + b_{2M}y_1^3) + GKu_M$$

(c) Does the control loop of the ship course need an additional adaptation mechanism when b_1 in part (a) or b_1 and b_2 in part (b) vary during operation?

6.3 Design a parallel AMFC system when the adaptation signal in Eq. (6.3-11) has the form $\underline{u}_{P2} = \Delta K_M(\underline{e}, t)\underline{x} + \Delta K_U(\underline{e}, t)\underline{u}_M$ or when the linear control signal in Eq. (6.3-10) has the form $\underline{u}_P = K\underline{e} + K_M\underline{x} + K_U\underline{u}_M$. Compare these designs with those given in Sec. 6.3-2.

6.4 Model-following control [32] can also be used for decoupling control if the reference model is decoupled and perfect model-following conditions are satisfied. Consider a reference model

$$\dot{\underline{x}} = A_M\underline{x} + B_M\underline{u}_M$$

where

$$A_M = \begin{bmatrix} 0.0 & 1.0 & 0.0 & 0.0 \\ -0.04 & -0.3 & 0.0 & 0.0 \\ 0.0 & 0.0 & -0.4 & 0.0 \\ 0.0 & 0.0 & 0.0 & -0.05 \end{bmatrix}, \quad B_M = \begin{bmatrix} 0.0 & 0.0 \\ 0.1 & 0.0 \\ 0.4 & 0.0 \\ 0.0 & 5.0 \end{bmatrix}$$

and $\underline{x}$ and $\underline{u}_M$ have the same significance as in Eq. (6.7-1). Consider also the controlled plant given in Eq. (6.7-2).

(a) Show that the above reference model corresponds to a decoupled system, the first input acting on the first three state variables and the second input acting on the last state variable.

(b) Check if the perfect model-following conditions are satisfied.

6.5 Show that the adaptation algorithms given in Table 6.3-2 for discrete-time parallel and series parallel AMFC systems lead to asymptotically hyperstable equivalent feedback systems.

6.6 Give the form of the adaptation algorithms for the AMFC systems given in Table 6.3-2 when one uses adaptation algorithms with decreasing adaptation gain. (Hint: See Secs. 5.3 and 5.4.)

REFERENCES

1. C. A. Winsor and R. J. Roy, The application of specific optimal control to the design of desensitized model following control systems, IEEE Trans. Autom. Control, *AC-15*, 326-333 (1970).
2. R. B. Newell and D. G. Fisher, Experimental evaluation of optimal, multivariable regulatory controllers with model following capabilities, Automatica, *8*, 247-262 (1972).
3. H. Kaufman and P. Berry, Development of a digital adaptive optimal linear regulator flight controller, in *Proceedings of the Sixth IFAC World Congress*, Part I-D, Boston, 1975, pp. 58.5.1-58.5.8. Edited by International Federation of Automatic Control. Distributed by Instruments Society of America, Pittsburgh, 1975.
4. J. S. Tyler, The characteristics of model following systems as synthesized by optimal control, IEEE Trans. Autom. Control, *AC-9*, 485-498 (1964).
5. S. J. Asseo, Application of optimal control to perfect model following, in *Proceedings of the JACC*, Ann Arbor, 1968, pp. 1056-1070.
6. C. Foulard and P. Zoydo-Crespo, Commande modale et commande découplée multidimensionnelle par des techniques de poursuite d'un modèle de référence, in *Proceedings of the Fourth IFAC/IFIP International Conference on Digital Computer Applications to Process Control*, Vol. I, Zurich, 1974, pp. 88-100. Springer-Verlag, New York, 1974.
7. R. T. Curran, Equi-controllability and the model following problem, Technical Report No. 6303-2, Center for Systems Research, Stanford University, Stanford, Calif., July 1971.
8. H. Erzberger, Analysis and design of model following systems by state space techniques, in *Proceedings of the Joint Automatic Control Conference*, Ann Arbor, 1968, pp. 572-581.
9. Y. T. Chen, Perfect model following with a real model, in *Proceedings of the JACC*, paper 10-5, 1973.

10. A. S. Morse, Structure and design of linear model following control systems, IEEE Trans. Autom. Control, *AC-18*, 346-354 (1973).
11. J. M. Dion and I. D. Landau, Aperçu des méthodes algébriques utilisées pour l'étude de la commande des systèmes linèaires, Part 1, Rev. RAIRO Jaune, No. 2, 125-160 (1977).
12. W. A. Wolovich, The use of state feedback for exact model matching, SIAM J. Control, *10*, no. 3, 512-523 (1972).
13. S. H. Wang and C. A. Desoer, The exact model matching of linear multivariable systems, IEEE Trans. Autom. Control, *AC-17*, 347-349 (1972).
14. I. D. Landau and B. Courtiol, Design of multivariable adaptive model following control systems, Automatica, *10*, 483-494 (1974).
15. H. Schneider and G. P. Barner, *Matrices and Linear Algebra*, Holt, Rinehart and Winston, New York, 1968.
16. D. G. Luenberger, Canonical forms for linear multivariable systems, IEEE Trans. Autom. Control, *AC-12*, 290-293 (1967).
17. B. Courtiol and I. D. Landau, High speed adaptation system for controlled electrical drives, Automatica, *11*, 119-127 (1975).
18. I. D. Landau, Sur une méthode de synthèse des systèmes adaptatifs avec modèle utilisés pour la commande et l'identification d'une classe de procédés physiques, These de Doctorat és Sciences, No. CNRS:AO 8495, University of Gernoble, Grenoble, 1973.
19. S. Barnett and C. Storey, *Matrix Methods in Stability Theory,* Barnes & Noble, New York, 1970.
20. V. M. Popov, Cîteva noţiuni generale de stabilitate de interes special în teoria sistemelor automate, Stud. Cercet. Energ. Electroteh., *20*, 693-711 (1970).
21. R. E. Kalman and J. E. Bertram, Control system analysis and design via the second method of Lyapunov, Trans. ASME, J. Basic Eng., 371-393 (June 1960).
22. E. Irving, J. P. Barret, C. Charcossey, and J. P. Monville, Improving Power Network Stability and Unit Stress with Adaptive Generator Control, Automatica, *15*, no. 1 (1979).

23. J. M. Dion, Etude des systèmes adaptatifs à amortissement constant, Thèse 3me Cycle, Institut National Polytechnique de Grenoble, Grenoble, March 1977.

24. G. Bethoux and B. Courtiol, A hyperstable discrete model reference adaptive control systems, in *Proceedings of the Third IFAC Symposium on Sensitivity, Adaptivity and Optimality,* Ischia, 1973, pp. 282-289. (Instruments Society of America, Pittsburgh, 1973.)

25. R. V. Monopoli, Model reference adaptive control with an augmented error signal, IEEE Trans. Autom. Control, *AC-19*, 474-484 (1974).

26. R. V. Monopoli and C. C. Hsing, Parameter adaptive control of multivariable systems, Int. J. Control, *22*, 313-327 (1975).

27. R. Benejean, La commande adaptative à modèle de reference evolutif (application à la commande sans accés aux variables d'etat), These 3me Cycle, Institut National Polytechnique de Grenoble, Grenoble, March 1977.

28. T. Ionescu and R. Monopoli, Discrete model reference adaptive control with an augmented error signal, Automatica, *13*, 507-518 (1977).

29. B. Courtiol, Applying model reference adaptive techniques for the control of electromechanical systems, in *Proceedings of the Sixth International Federation of Automatic Control World Congress,* Part ID, Boston, 1975, pp. 58.2.1-58.2.9. Distributed by Instruments Society of America, Pittsburgh, 1975.

30. J. W. Gilbart and G. C. Winston, Adaptive compensation for an optical tracking telescope, Automatica, *10*, 125-131 (1974).

31. J. van Amerongen and A. J. Udink ten Cate, Model reference adaptive autopilots for ships, Automatica, *11*, 441-450 (1975).

32. B. Courtiol, Etude et application de la commande adaptative avec modèle, Thèse de Docteur Ingénieur, No. CNRS:AO 12663, Institut Polytechnique de Grenoble, Grenoble, July 1976.

Chapter Seven

PARAMETRIC IDENTIFICATION USING MODEL REFERENCE ADAPTIVE SYSTEMS

7.1 INTRODUCTION

Identification of a dynamic process contains four basic steps [1, 2]:

1. *Structural identification*, which allows us to characterize the structure of the mathematical model of the process to be identified
2. *Input/output data acquisition*
3. *Parametric identification*, which allows us to determine (more exactly, to estimate) the parameters of the mathematical model of the process
4. *Validation* of the identified model

In this chapter, we shall focus our attention on the third aspect of identification, the *parametric identification*, and in particular on two aspects of parametric identification:

1. *Recursive parameter identification* (i.e., one likes to obtain an estimation of the parameters, as the process develops, without being obliged to use all the past input/output data at each step)
2. *Parameter tracking* (i.e., one likes to track the values of the parameters of a process which can vary during operation)

Note from the beginning that in order to obtain good results in recursive identification and parameter tracking, a certain degree

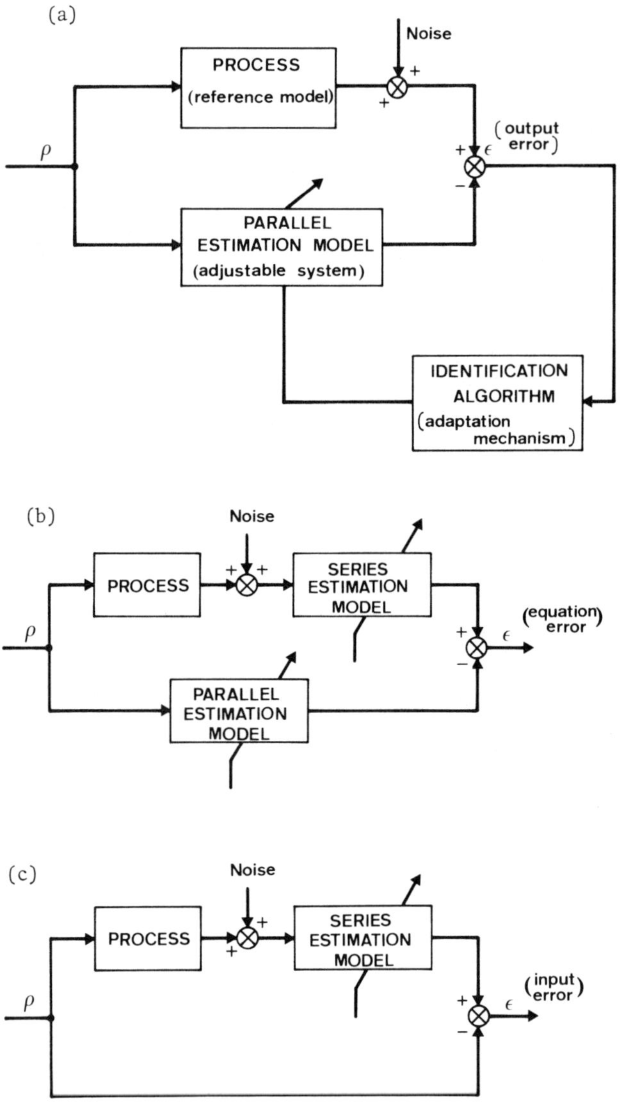

Fig. 7.1-1 Three basic identification structures and their correspondence with MRAS's. (a) Output error method (parallel MRAS's). (b) Equation error method (series-parallel MRAS's). (c) Input error method (series MRAS's).

of structural identification must be a priori achieved. This can be done either from the phenomenological analysis of the process to be identified or from an off-line identification based on a previous set of input/output data.

Recursive identification of the dynamic parameters of a process (on-line identification) as well as the tracking of the process parameters when they are time-varying (real-time identification) can both be formulated as a model reference adaptive problem [3, 4]. The *process to be identified* represents the *reference model*. The *adjustable system* is constituted by an *adjustable model* having the *structure of the mathematical model of the process* (also called the *estimation model*) whose parameters are driven by an *adaptation mechanism* (which implements an identification algorithm).

However, depending on the configuration of the estimation model, the signals which are fed in, and how the error between the outputs of the process and the estimation model is obtained, one can distinguish three basic configurations, called in the field of identification the *output error method*, the *equation error method*, and the *input error method*. They are illustrated schematically in Fig. 7.1-1 [2-5]. These configurations correspond to the three basic structures of model reference adaptive systems (MRAS's), and the correspondence is indicated in Table 7.1-1.

An MRAS used for identification plays the role of an *identifier* which can be of *parallel*, *series parallel*, or *series* type. In this chapter, to designate them, we shall use the shortened terms indicated in the right-hand column of Table 7.1-1. Note that all the

Table 7.1-1 Correspondence Between Identification Structures and Model Reference Adaptive System Structures

Identification structure	MRAS	Terminology used in Chap. 7
Output error method	Parallel	Parallel identifier
Equation error method	Series-parallel	Series-parallel identifier
Input error method	Series	Series identifier

recursive-type identification algorithms can be cast, independently of the considerations used for their derivation, into one of the block diagrams shown in Fig. 7.1-1 and that their implementation leads to an MRAS structure.

The equation error method (and its counterpart, the series-parallel MRAS) is the most used configuration for recursive identification using algorithms derived from statistical considerations. It has the attractiveness of a direct link between the parameter error and the measured output error (and leads to the simplest identification algorithm: least squares). However (as will be shown in Secs. 7.2 and 7.3), this configuration has an important drawback: The measurement noise leads normally to an erroneous estimation of the process parameters (the estimated parameters are biased).

From statistical considerations, recursive identification algorithms for equation error configurations assuring in most of the cases unbiased parameter estimates have been derived [2, 3]. However, the convergence of the identification with regard to the initial parameter error is often difficult to prove. In general, these algorithms must already have a sufficiently accurate initialization of the estimation model so that the convergence be assured.

On the other hand, MRAS's which are designed from stability considerations, when used for identification, allow us to specify the conditions of the convergence with respect to the initial parameter error, but the performance in the presence of measurement noise needs to be examined separately. As will be shown in Sec. 7.3, asymptotically stable discrete-time parallel MRAS's with decreasing adaptation gain can assure an unbiased parameter estimation in the presence of noise.

Note also that the methodology based on the use of hyperstability and positivity concepts for designing MRAS's can also be applied to the analysis of the stability properties of the identifiers derived from statistical considerations [5, 6].

For all the identification algorithms, the convergence of the identified parameters to the true values depends also on the input characteristics. This is also true when using MRAS's for identification.

The various structures of MRAS's cited in Table 7.1-1 have been presented in Chap. 2, and their designs have been discussed in detail in Chaps. 4 and 5. In this chapter, we shall discuss them from the identification point of view. Within the three basic structures, other MRAS configurations exist which can be used for simultaneous adaptive state observation and parameter identification. These configurations will be discussed in Chap. 8.

Historically, continuous-time MRAS's with constant adaptation gain have first been used for identification and parameter tracking. They still have their usefulness for real-time identification of high-speed dynamic processes such as those encountered in the control of electrical variables. Continuous-time identifiers will be discussed in Sec. 7.2. The problem of input characteristics assuring the parameter convergence and the qualitative effect of measurement noise upon the various configuration will be emphasized.

Section 7.3 is devoted to the presentation of discrete-time identifiers used for recursive identification of linear time-invariant processes and for parameter tracking. A detailed analytical analysis of the effect of the measurement noise upon the various types of identifiers is presented.

In Sec. 7.4 an application of continuous-time identifiers to the real-time identification and adaptive control of an dc/ac converter is presented.

A comparison of recursive discrete-time identifiers derived from MRAS's with several other recursive identification techniques is presented in Sec. 7.5 using simulated and real data.

7.2 CONTINUOUS-TIME IDENTIFIERS USING MODEL REFERENCE ADAPTIVE TECHNIQUES

7.2-1 Introduction

As mentioned in Sec. 7.1, the identification of the values of the parameters of a dynamic process can be formulated as an MRAS problem. The *reference model* will be constituted by the *dynamic process* (with unknown parameters) and the *adjustable system* will have the structure of the *mathematical model of the process*. If the structure of the mathematical model has been well chosen and the identifier performs well, the values of the parameters of the adjustable model will converge to significant ones, allowing us to characterize the input/output behavior of the dynamic process. Since the *adjustable system* will provide an estimation of the parameters of the process mathematical model, it will be called an *estimation model*.

In this section, we shall emphasize how the MRAS designs developed in Chap. 4 can be used for identification. Two basic types of design problem were discussed in Chap. 4: (1) when the states of the reference model and of the adjustable system are available and (2) when only input and output measurements are available.

The designs developed in Chap. 4 for the case where the states of the reference model and of the adjustable system are available have in practice a limited direct use for identification. However, they will be briefly discussed next because it allows us to highlight such important basic aspects common to all identifiers derived from MRAS's as

Conditions on the input characteristics assuring parameter convergence

Initialization of the estimation model

Qualitative effect of the measurement noise

Advantages and disadvantages of the basic configurations (i.e., parallel versus series-parallel)

Among the designs developed in Chap. 4 for the case where only input and output measurements are available, we shall discuss only those leading to parameter identifiers. The use of the designs developed in Secs. 4.2-3 and 4.2-4 which allow us to make simultaneous parameter identification and adaptive state observation will be discussed in Chap. 8.

7.2-2 Continuous-Time Identifiers for Processes with Accessible States

We shall first reformulate the identification problem in terms of the MRAS design problem as considered in Sec. 4.1-1. For the remainder of this chapter, we shall systematically use the changes to the formulations given in Chap. 4 that are indicated in Table 7.2-1.

Parallel Identifier (Output Error Method)

Consider first a parallel MRAS with an integral adaptation law (Sec. 4.1-1). The corresponding identifier will be described by the following [7]:

The process to be identified:

$$\dot{\underline{x}} = A_p\underline{x} + B_p\underline{u}, \quad \underline{x}(0) = \underline{x}_0 \tag{7.2-1}$$

The parallel estimation model:

$$\dot{\underline{y}} = A_S(\underline{v}, t)\underline{y} + B_S(\underline{v}, t)\underline{u} \quad [\underline{y}(0) = \underline{y}_0, A_S(0) = A_{S0}, B_S(0) = B_{S0}] \tag{7.2-2}$$

The state generalized error (output error):

$$\underline{e} = \underline{x} - \underline{y} \tag{7.2-3}$$

Table 7.2-1 Comparison of the Terminology and Notations of Chaps. 4 and 7

Terminology and notations used in Chap. 4	Terminology and notations used in Chap. 7
Reference model	Process to be identified
Model parameters: A_M, B_M	Process parameters: A_P, B_P
Model output: θ_M	Process output: θ_P
Adjustable system	Estimation model
Adaptation law (algorithm)	Identification law (algorithm)

The identification law (integral):

$$\underline{v} = D\underline{e} \tag{7.2-4}$$

$$A_S(\underline{v}, t) = \int_0^t F_A\underline{v}(G_A\underline{y})^T \, d\tau + A_S(0) \tag{7.2-5}$$

$$B_S(\underline{v}, t) = \int_0^t F_B\underline{v}(G_B\underline{u})^T \, d\tau + B_S(0) \tag{7.2-6}$$

In Eqs. (7.2-1) to (7.2-3), $\underline{x}$ is the process state vector (n-dimensional), $\underline{y}$ is the state vector of the estimation model (n dimensional), and $\underline{u}$ is the process input vector (m-dimensional) belonging to the class of piecewise continuous vector functions. A_P and B_P are constant matrices of appropriate dimension, to be identified, and $A_S(\underline{v}, t)$, $B_S(\underline{v}, t)$ are the adjustable matrices of the estimation model which will provide an estimation of the matrices A_P and B_P. In Eqs. (7.2-4) to (7.2-6), D is a positive definite matrix satisfying certain conditions (to follow), and F_A, G_A, F_B, and G_B are arbitrary positive definite matrices.

The identifier will assure that $\lim_{t\to\infty} \underline{e}(t) = \underline{0}$ for any initial conditions $\underline{x}(0)$, $\underline{y}(0)$, $A_P - A_S(0)$, and $B_P - B_S(0)$ and for all piecewise continuous input vector functions $\underline{u}$ if the matrix D is such that the transfer matrix

$$H(s) = D(sI - A_P)^{-1} \tag{7.2-7}$$

is strictly positive real. The condition of Eq. (7.2-7) will be satisfied if the matrix D satisfies the Lyapunov equation

$$DA_p + A_p^T D = -Q \tag{7.2-8}$$

for any arbitrary positive definite matrix Q (see Appendix B, Lemma B.2-3).

Conditions for Parameter Convergence. If the matrix D satisfies Eq. (7.2-8), the identifier of Eqs. (7.2-1) to (7.2-6) will assure that $\lim_{t\to\infty} \underline{e}(t) = 0$. *However, for identification purposes, we are interested in finding the additional conditions allowing us to obtain*

$$\lim_{t\to\infty} A_S(\underline{v}, t) = A_P, \quad \lim_{t\to\infty} B_S(\underline{v}, t) = B_P \tag{7.2-9}$$

Subtracting Eq. (7.2-2) from Eq. (7.2-1) and taking the limit of this expression when t tends toward ∞ ($t \to \infty$), one obtains

$$[A_P - \lim_{t\to\infty} A_S(\underline{v}, t)]\underline{y} + [B_P - \lim_{t\to\infty} B_S(\underline{v}, t)]\underline{u} = 0 \qquad (7.2\text{-}10)$$

since $\underline{e} = \underline{0}$ is a globally asymptotically stable equilibrium point for the considered identifier. Because we are only interested in the limit behavior of the above expression, taking into account that $\lim_{t\to\infty} \underline{e}(t) = \lim_{t\to\infty}(\underline{x} - \underline{y}) = 0$, one can replace $\underline{y}$ by $\underline{x}$ and Eq. (7.2-10) can be rewritten as

$$[A_P - \lim_{t\to\infty} A_S(\underline{v}, t)]\underline{x} + [B_P - \lim_{t\to\infty} B_S(\underline{v}, t)]\underline{u} = 0 \qquad (7.2\text{-}11)$$

If $\underline{u}$ and $\underline{x}$ are linearly independent, Eq. (7.2-11) implies that Eq. (7.2-9) holds. Therefore, we must examine under what conditions $\underline{u}$ and $\underline{x}$ can be made linearly independent [8].[1]

From Eq. (7.2-1), using the differential operator $p = d/dt$, one obtains the following expression for $\underline{x}$:

$$\underline{x} = (pI - A_P)^{-1}B_P\underline{u} \qquad (7.2\text{-}12)$$

Introducing the notations

$$\Delta A = A_P - \lim_{t\to\infty} A_S(\underline{v}, t) \qquad (7.2\text{-}13)$$

and

$$\Delta B = B_P - \lim_{t\to\infty} B_S(\underline{v}, t) \qquad (7.2\text{-}14)$$

one obtains, from Eq. (7.2-11) in the steady state, the following identity:

$$[\Delta A(pI - A_P)^{-1}B_P + \Delta B]\underline{u} \equiv 0 \qquad (7.2\text{-}15)$$

The term $(pI - A_P)^{-1}$ in the last identity can be expressed as [8]

$$(pI - A_P)^{-1} = \frac{1}{\alpha(p)} \sum_{i=1}^{n} \alpha_i(p)A_P^{\,i-1} \qquad (7.2\text{-}16)$$

where $\alpha(p)$ is the characteristic polynomial of A_P,

$$\alpha(p) = \det(pI - A_P) = p^n + a_{n-1}p^{n-1} + \cdots + a_1p + a_0 \qquad (7.2\text{-}17)$$

[1]The following proof closely follows Ref. 8.

and the polynomials $\alpha_i(p)$ are defined as

$$\alpha_i(p) = \begin{cases} \alpha(p), & i=0 \\ p\alpha_{i+1}(p) + a_i, & i = 1, 2, \ldots, n-1 \\ 1, & \end{cases} \qquad (7.2\text{-}18)$$

Therefore, Eq. (7.2-15) can be rewritten as

$$\left[\Delta A \sum_{i=1}^{n} \alpha_i(p) A_p^{\,i-1} B_p + \Delta B \alpha(p)\right] \underline{u} \equiv 0 \qquad (7.2\text{-}19)$$

Using the expressions given by Eqs. (7.2-17) and (7.2-18) for $\alpha_i(p)$ and $\alpha(p)$ and rearranging the terms of the left-hand side of Eq. (7.2-19) according to the power of p, one obtains

$$\begin{aligned}[\Delta B p^n &+ (\Delta B a_{n-1} + \Delta A B_p)p^{n-1} + (\Delta B a_{n-2} + \Delta A B_p a_{n-1} \\ &+ \Delta A A_p B_p)p^{n-2} + \cdots + (\Delta B a_0 + \Delta A B_p a_1 + \Delta A A_p B_p a_2 \\ &+ \cdots + \Delta A A_p^{\,n-2} B_p a_{n-1} + \Delta A A_p^{\,n-1} B_p)]\underline{u} = 0 \end{aligned} \qquad (7.2\text{-}20)$$

Consider now that the components of $\underline{u}$, $u_1, u_2, \ldots, u_m$ are linearly independent piecewise continuous frunctions. By using the notations

$$\Delta B = \{\delta b_{ij}\},\ \Delta A = \{\delta a_{ij}\},\ \Delta A B_p = \{\delta a_{ij}^{\,1}\},\ \ldots,$$

$$\Delta A A_p^{\,k} B_p = \{\delta a_{ij}^{\,k+1}\} \qquad (7.2\text{-}21)$$

where $k = 0, 1, 2, \ldots, n-1$, Eq. (7.2-20) is then reduced to $m \times n$ equations of the form

$$\begin{aligned}[\delta b_{ij} p^n &+ (\delta b_{ij} a_{n-1} + \delta a_{ij}^{\,1})p^{n-1} + \cdots + (\delta b_{ij} a_0 + \delta a_{ij}^{\,1} a_1 \\ &+ \cdots + \delta a_{ij}^{\,n-1} a_{n-1} + \delta a_{ij}^{\,n})]u_j = 0 \\ &\text{for all } i = 1, \ldots, n,\ j = 1, \ldots, m \end{aligned} \qquad (7.2\text{-}22)$$

Equation (7.2-22) implies that $\delta b_{ij} = 0$, $\delta a_{ij}^{\,k} = 0$ for $k = 1, \ldots, n-1$ only if u_j is not a solution of the differential equation (7.2-22).

Now, consider the class of input vector functions characterized by the fact that each component u_j of the input vector function is a sum of q sinusoidal signals with distinct frequencies. On the other hand, depending on whether n is even or odd, the solution of the differential equation (7.2-22) can have at most n/2 or (n - 1)/2 sinusoidal components, respectively. Therefore, if each u_j contains more than n/2 sinusoidal signals, u_j cannot be a solution of Eq. (7.2-22). One concludes that if q (the number of sinusoidal signals in each u_j) is the nearest integer allowing us to satisfy the condition

$$q \geq \frac{n + 1}{2} \tag{7.2-23}$$

and all the components u_j of the vector $\underline{u}$ are linearly independent, then

$$\begin{cases} \delta b_{ij} = 0 \\ \delta a_{ij}^{k} = 0, \quad k = 1, 2, \ldots, n - 1 \text{ for all } i = 1, \ldots, n, \\ \qquad\qquad j = 1, \ldots, m \end{cases} \tag{7.2-24}$$

Equation (7.2-24) implies that

$$\Delta B = 0 \tag{7.2-25}$$

and that

$$\Delta AB_p = 0, \ \Delta AA_pB_p = 0, \ \ldots, \ \Delta AA_p^{n-1}B_p = 0 \tag{7.2-26}$$

But Eq. (7.2-26) can also be written as

$$\Delta A[B_p, A_pB_p, \ldots, A^{n-1}B_p] = \Delta A \langle A_p | B_p \rangle = 0 \tag{7.2-27}$$

where $\langle A_p | B_p \rangle = [B_p, A_pB_p, \ldots, A^{n-1}B_p]$ is the controllability matrix of the process to be identified. If the process to be identified is completely controllable, then Eq. (7.2-27) implies that $\Delta A = 0$. Therefore, one obtains the results stated without proof in Sec. 4.1-1, which in the case of identification become the following.

THEOREM 7.2-1. Given a globally asymptotically stable parallel identifier described by Eqs. (7.2-1) to (7.2-6) with each of the components of the input vector $\underline{u}$ formed by a sum of sinusoidal signals of distinct frequencies, one obtains

$$\lim_{t\to\infty} A_S(\underline{v}, t) = A_p, \quad \lim_{t\to\infty} B_S(\underline{v}, t) = B_p$$

if

1. The plant to be identified is completely controllable
2. The components of the vector $\underline{u}$ are linearly independent
3. Each component of the vector $\underline{u}$ contains at least $(n + 1)/2$ distinct frequencies.

Theorem 7.2-1 has an important consequence for the practice of identification: To obtain an accurate identification we must use input signals with a broad frequency spectrum containing enough components with significant energy in accordance with the number of parameters characterizing the process to be identified.

Since a random signal over a finite interval can be decomposed into an infinite sum of sinusoidal signals of distinct frequencies, Theorem 7.2-1 also holds when $\underline{u}$ is a vector with linear independent random components. A direct proof for the case of a single-input/single-output identifier with a random input can be found in Ref. 9.

A similar result is obtained if instead of an integral adaptation law one uses an integral + proportional identification law. However, in general this kind of identification law will accelerate the convergence of the generalized error but will slow down the rate of convergence of the parameters. The integral + proportional law (with positive proportional adaptation gains) is interesting only for the first stages of the identification if a large initial generalized error occurs [10, 11].

The Effect of Measurement Noise. Another important point to be clarified when one uses parallel MRAS's for identification is the effect of the measurement noise upon the estimated parameters. Next we shall present a qualitative analysis (a detailed analysis will be presented in Sec. 7.3 for discrete-time recursive identifiers).

To simplify the analysis of the measurement noise effect, let us consider as an example a first-order process to be identified (b_p is supposed known):

$$\dot{x} = -a_p x + b_p u, \quad a_p > 0 \tag{7.2-28}$$

The parallel estimation model will be described by

$$\dot{y} = -a_S(e, t)y + b_P u \tag{7.2-29}$$

The generalized error (output error) will be

$$e = x - y \tag{7.2-30}$$

and $a_S(e, t)$ will be given by the integral identification law:

$$-a_S(e, t) = \int_0^t fey\, d\tau - a_S(0) \tag{7.2-31}$$

where f is a positive gain. The generalized error e can be directly used in Eq. (7.2-31) since the transfer function

$$h(s) = (sI - A_P)^{-1} = \frac{1}{s + a} \tag{7.2-32}$$

is strictly positive real.

Suppose now that x is disturbed by a zero mean noise β of finite variance; therefore the measured $\hat{x}$ used for identification will be

$$\hat{x} = x + \beta \tag{7.2-33}$$

and Eq. (7.2-31) becomes

$$\begin{aligned} -a_S(e, t) &= \int_0^t f(e + \beta)y\, d\tau - a_S(0) \\ &= \int_0^t fey\, d\tau - a_S(0) + \int_0^t f\beta y\, d\tau, \quad f > 0 \end{aligned} \tag{7.2-34}$$

The third term in the second line on the right in Eq. (7.2-34) can cause an error in the value obtained for a_S. We can distinguish two basic situations:

Case 1: The variable y is slowly varying as compared to β. Over a certain interval, one can assume that y is constant. Since β is a zero mean signal, one observes that with a low gain f the effect of the third term on the right in Eq. (7.2-34) is negligible.

Case 2: The adaptation gain f is high. Even if y is slowly time-varying compared to β, the parameter $a_S(e, t)$ will have a component correlated with β. Therefore, y will also be correlated with β [by Eq. (7.2-29)]. In this case, the effect of the third term on the right in Eq. (7.2-34) can no longer be neglected and will cause two phenomena:

1. An error in the mean value of the estimated parameter; the mean value of the estimated parameter is *biased*.
2. The estimated parameter will oscillate around the biased mean value.

Therefore, *in the presence of measurement noise the adaptation gain must be small in order to identify the values of the process parameters with good precision. As a counterpart, the identification process will be slow.* These conclusions have been confirmed by extensive experimental observations [10, 12].

The Computation of the Matrix D (the strictly positive real condition). Another problem related to the use of the identifier of Eqs. (7.2-1) to (7.2-6) is the condition imposed on matrix D to be such that H(s) given by Eq. (7.2-7) is a strictly positive real transfer matrix. In fact, the matrix A_p being unknown, D cannot be directly computed by using Eq. (7.2-8).

This constraint can be partially circumvented by using the observation that for a given D there is a region in the parameter space for which Eq. (7.2-8) is satisfied, Q being any positive definite matrix.

Therefore, if D is computed using the initial value of the estimated matrix $A_S(0)$, then there is a region in the parameter space around the values of $A_S(0)$ for which Eq. (7.2-8) is satisfied and for which H(s) is a strictly positive real transfer matrix. For lower-dimensional systems, this region can be estimated. To be sure that the identifier is asymptotically stable, one needs a good initialization of the matrix A_S of the estimation model [$A_S(0)$ is not too far from A_p]. This can be obtained from a prior off-line identification which gives an idea about the range of A_S or by on-line identification with less precise methods which do not require a good initialization.

Note also that the *condition on H(s) given by Eq. (7.2-7) is not a necessary condition (it is only a sufficient condition).* This means that violation of the strictly positive real condition on H(s) does not necessarily imply instability of the identifier.

We conclude the analysis of the parallel (output error) identifier by restating two of its principal characteristics:

1. The convergence of the identifier for a given matrix D can be rigorously guaranteed only for processes whose matrix A_P satisfies the condition that $D(sI - A_P)^{-1}$ is a strictly positive real transfer matrix. For high-order processes this requires a good initialization of the estimation model.
2. In the presence of measurement noise, good precision of identification can be obtained by reducing the adaptation gains, with the disadvantage of slower convergence.

Series-Parallel Identifier (Equation Error)

Using the design presented in Sec. 4.1-2 (case a) and making the same substitutions as for the parallel identifier, one obtains the series-parallel identifier with an integral identification described by the following [13, 14]:

The process to be identified:

$$\dot{\underline{x}} = A_P\underline{x} + B_P\underline{u} \tag{7.2-35}$$

The series-parallel estimation model:

$$\dot{\underline{y}} = A_S(\underline{v}, t)\underline{x} + B_S(\underline{v}, t)\underline{u} - K\underline{e} \tag{7.2-36}$$

The state generalized error (equation error):

$$\underline{e} = \underline{x} - \underline{y} \tag{7.2-37}$$

The identification law (integral):

$$\underline{v} = D\underline{e} \tag{7.2-38}$$

$$A_S(\underline{v}, t) = \int_0^t F_A\underline{v}(G_A\underline{x})^T\, d\tau + A_S(0) \tag{7.2-39}$$

$$B_S(\underline{v}, t) = \int_0^t F_B\underline{v}(G_B\underline{u})^T\, d\tau + B_S(0) \tag{7.2-40}$$

In these equations, A_P, B_P, $\underline{x}$, $\underline{y}$, $\underline{u}$, D, F_A, F_B, G_A, and G_B have the same significance as in the parallel case [Eqs. (7.2-1) to (7.2-6)] and K is a Hurwitz matrix [(n × n)-dimensional].

The series-parallel identifier will assure that $\lim_{t\to\infty} \underline{e}(t) = \underline{0}$ for any initial conditions $\underline{x}(0)$, $\underline{y}(0)$, $A_p - A_S(0)$, and $B_p - B_S(0)$ and for all piecewise continuous input vector functions $\underline{u}$ if the matrix D is such that the transfer matrix

$$H(s) = D(sI - K)^{-1} \tag{7.2-41}$$

is strictly positive real. The matrix D satisfying the above condition can be obtained by solving the Lyapunov equation (see Appendix B, Lemma B.2-3)

$$DK + K^T D = -Q \tag{7.2-42}$$

for any arbitrary positive definite matrix Q.

Note immediately that in this case we do not need any knowledge about A_p in order to compute the matrix D. The matrix D depends only on the matrices K and Q chosen by the designer. *This identifier can be used when no prior information about the value of* A_p *is available* (which is not the case for the parallel identifier).

The conditions of parameter convergence when the components of the input vector $\underline{u}$ are a sum of sinusoidal signals are the same as those of Theorem 7.2-1, the problem being similar [8].

The Effect of Measurement Noise. It remains for us to examine qualitatively the effect of measurement noise (a detailed analysis will be presented in Sec. 7.3 for discrete-time recursive identifiers). Again, we consider a first-order example. The process to be identified (b_p is supposed known) is described by

$$\dot{x} = -a_p x + b_p u, \quad a_p > 0 \tag{7.2-43}$$

The series-parallel estimation model is described by

$$\dot{y} = -a_S(e, t)x + b_p u + ke, \quad k > 0 \tag{7.2-44}$$

The generalized error (equation error) will be given by

$$e = x - y \tag{7.2-45}$$

and $a_S(e, t)$ will be given by the following integral identification law:

$$-a_S(e, t) = \int_0^t fex\, d\tau - a_S(0) \tag{7.2-46}$$

where f is a positive gain. The generalized error e can be directly used in Eq. (7.2-46) since the transfer function of Eq. (7.2-41),

$$h(s) = \frac{1}{s + k} \tag{7.2-47}$$

is strictly positive real. If x is disturbed by a zero mean measurement noise β of finite variance, for identification we shall use the measured value

$$\hat{x} = x + \beta \tag{7.2-48}$$

and Eq. (7.2-46) becomes

$$\begin{aligned} -a_S(e, t) &= \int_0^t f(e + \beta)(x + \beta)\, d\tau - a_S(0) \\ &= \int_0^t fex\, d\tau - a_S(0) + \int_0^t f\beta(x + e)\, d\tau + \int_0^t f\beta^2\, d\tau \end{aligned} \tag{7.2-49}$$

One can see immediately that even for a small gain f the last integral in Eq. (7.2-49) will introduce an error because even if β is a zero mean disturbance, β^2 is always greater than zero. This will cause a *bias* on the identified parameters.

Therefore, the *series-parallel identifier, despite the advantage of not requiring any initial estimation of* A_P, *has the disadvantage of introducing a bias on the estimated parameters in the presence of measurement noise even for small adaptation gains.*

From the practical point of view, series-parallel identifiers can be used as a first step of an identification procedure, allowing us to obtain only a biased estimation of the process parameters in the presence of zero mean measurement noise. These first parameter estimations can be used to implement a parallel identifier with low adaptation gain which will lead to an accurate estimation of the process parameters.

7.2-3 Continuous-Time Identifiers for Single-Input/Single-Output Processes (with Inaccessible States)

All the designs presented in Sec. 4.2 for MRAS's using only input and output measurements can be used to implement identifiers for single-input/single-output dynamic processes with

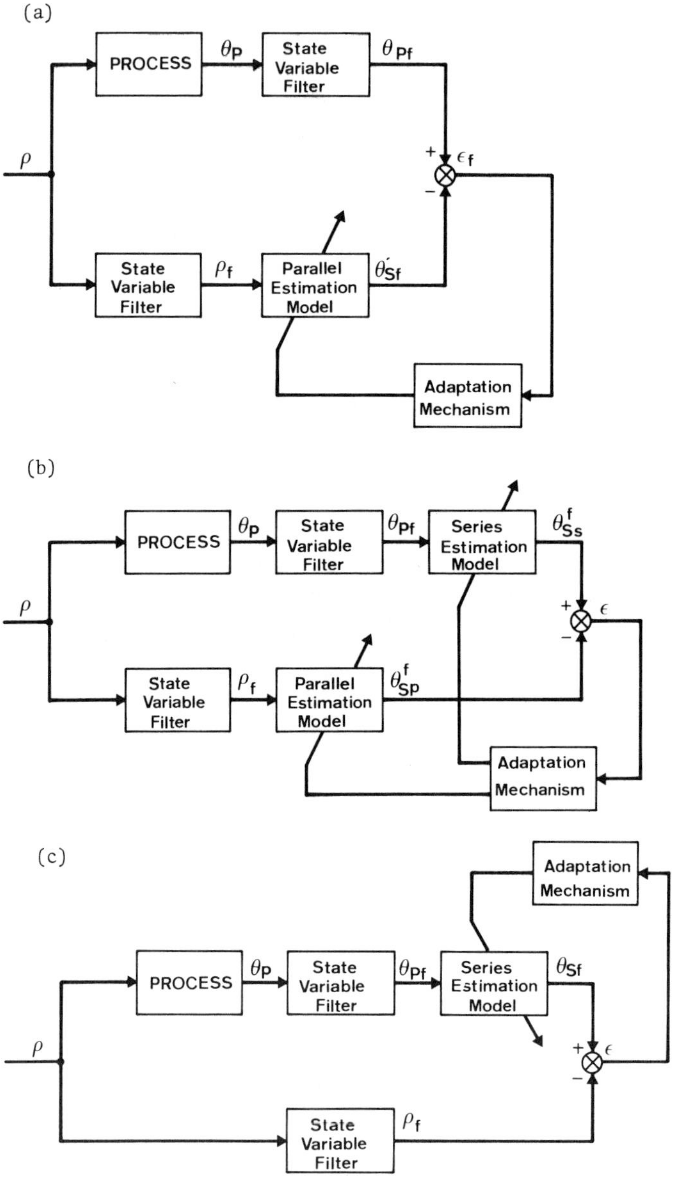

Fig. 7.2-1 Identifiers for single-input/single-output systems using state-variable filters. (a) Parallel identifier. (b) Series-parallel identifier. (c) Series identifier.

inaccessible states. However, as mentioned in Sec. 7.2-1, in the remaining of this section we shall not discuss the use of the designs given in Secs. 4.2-3 and 4.2-4 which can be used for simultaneous parameter identification and state observation (the use of these designs will be discussed in Chap. 8).

The design presented in Sec. 4.2-2 for parallel MRAS's which utilizes state-variable filters and its extensions for series-parallel and series MRAS's can be used for the implementation of the three basic identifier configurations derived from the output error method, equation error method, and input error method, respectively.

From the design method presented in Sec. 4.2-5, one also can derive a parallel-type identifier as well as a series-parallel identifier with the observation that in this case the estimation model is not a minimal realization and the parameters of the transfer functions cannot be directly obtained.

Identifiers Using State-Variable Filters

The structures of the various identifiers and the corresponding integral identification laws for those derived from the design given in Sec. 4.2-2 are summarized in Table 7.2-2. Block diagrams of these identifiers are shown in Figs. 7.2-1(a), (b), and (c). Note that the series identifier (last column) is identical to the series-parallel identifier if $b_0 = 1$ and $b_i = 0$ for $i \geq 1$. Note also that for the series identifier the linear compensator which forms the signal ν is not necessary if $b_i = 0$ for $i \geq 2$.

The considerations concerning the stability condition for parallel identifiers and series-parallel identifiers discussed in Sec. 7.2 are applicable here. Nevertheless, note that in this case the series-parallel identifier does not need a linear compensator. The considerations concerning the stability conditions for parallel identifiers are also applicable to the series identifiers if $b_i \neq 0$ for $i \geq 2$.

Table 7.2-2 Identifiers for S.I.S.O. Systems Using State-Variable Filters.

	Parallel identifier (output error method)	Series-parallel identifier (equation error method)	Series identifier (input error method)
The process to be identified*	$\left(\sum_{i=0}^{n} a_i p^i\right)\theta_P = \left(\sum_{i=0}^{m} b_i p^i\right)\rho$	$\left(\sum_{i=0}^{n} a_i p^i\right)\theta_P = \left(\sum_{i=0}^{m} b_i p^i\right)\rho$	$\left(\sum_{i=0}^{n} a_i p^i\right)\theta_P = \left(\sum_{i=0}^{m} b_i p^i\right)\rho$
The state-variable filter acting on the process output	$\left(\sum_{i=0}^{n-1} c_i p^i\right)\theta_{Pf} = \theta_P$	$\left(\sum_{i=0}^{n} c_i p^i\right)\theta_{Pf} = \theta_P$	$\left(\sum_{i=0}^{n} c_i p^i\right)\theta_{Pf} = \theta_P$
The state-variable filter acting on the process input	$\left(\sum_{i=0}^{n-1} c_i p^i\right)\rho_f = \rho$	$\left(\sum_{i=0}^{n} c_i p^i\right)\rho_f = \rho$	$\left(\sum_{i=0}^{n} c_i p^i\right)\rho_f = \rho$
The parallel estimation model	$\sum_{i=0}^{n} \hat{a}_i(\nu, t)p^i\theta_S' = \sum_{i=0}^{m} \hat{b}_i(\nu, t)p^i\rho_f$	$\theta_{Sp}^{\ f} = \sum_{i=0}^{m} \hat{b}_i(\varepsilon, t)p^i\rho_f$	
The series estimation model		$\theta_{Ss}^{\ f} = \sum_{i=0}^{n} \hat{a}_i(\varepsilon, t)p^i\theta_{Pf}$	$\sum_{i=0}^{m} \hat{b}_i(\nu, t)p^i\theta_{Sf} = \sum_{i=0}^{n} \hat{a}_i(\nu, t)p^i\theta_{Pf}$ $\cdot\ \hat{a}_i(\nu, t)p^i\theta_{Pf}$
The filtered generalized error	$\varepsilon_f = \theta_{Pf} - \theta_S'$	$\varepsilon_f = \theta_{Ss}^{\ f} - \theta_{Sp}^{\ f}$	$\varepsilon_f = \theta_{Sf} - \rho_f$

The identification law (integral)	$\nu = \left(\sum_{i=0}^{n-1} d_i p^i\right)\varepsilon_f$ $\hat{a}_i(\nu, t) = -\int_0^t k_{ai}\nu p^i \theta_S' \, d\tau + \hat{a}_i(0), \quad k_{ai} > 0$ $\hat{b}_i(\nu, t) = \int_0^t k_{bi}\nu p^i \rho_f \, d\tau + \hat{b}_i(0), \quad k_{bi} > 0$	$\hat{a}_i(\varepsilon, t) = -\int_0^t k_{ai}\varepsilon_f p^i \theta_{Pf} \, d\tau + \hat{a}_i(0), \quad k_{ai} > 0$ $\hat{b}_i(\varepsilon, t) = \int_0^t k_{bi}\varepsilon_f p^i \rho_f \, d\tau + \hat{b}_i(0), \quad k_{bi} > 0$	$\nu = \left(\sum_{i=0}^{m-1} d_i p^i\right)\varepsilon_f$ $\hat{a}_i(\nu, t) = -\int_0^t k_{ai}\nu p^i \theta_{Pf} \, d\tau + \hat{a}_i(0), \quad k_{ai} > 0$ $\hat{b}_i(\nu, t) = \int_0^t k_{bi}\nu p^i \theta_{Sf} \, d\tau + \hat{b}_i(0), \quad k_{bi} > 0$
Stability condition	$h(s) = \dfrac{\sum_{i=0}^{n-1} d_i s^i}{\sum_{i=0}^{n} a_i s^i}$ must be strictly positive real.	None	$h(s) = \dfrac{\sum_{i=0}^{m-1} d_i s^i}{\sum_{i=0}^{m} b_i s^i}$ must be strictly positive real.

*It is assumed that one of the parameters: a_0, a_n, or b_0 is equal to 1 and that the corresponding adjustable parameter is directly set equal to 1.

The effect of the measurement noise upon the performances of the parallel identifier and series-parallel identifier is qualitatively the same as discussed in Sec. 7.2-2. However, the state-variable filters can have a beneficial effect upon reducing the influence of the measurement noise if the noise spectrum is principally located outside the band pass of the state-variable filters. Note that by reducing the band pass of the state-variable filters the identification process becomes slower [10, 15]. The series identifier does not offer any advantages concerning the effect of the output measurement noise in comparison with the series-parallel identifier. The series identifier can be advantageous when either a noise is superimposed on the process input or the measured input is disturbed by a noise. In this case, the bias caused by the imperfect measurement of the effective process input (both situations correspond, in fact, to this one) can be reduced since in the identification law the measured input appears only through the generalized error.

The considerations concerning the use of integral + proportional identification laws presented in Sec. 7.2-2 are also applicable for the identifiers given in Table 7.2-2 [10].

The conditions upon the input characteristics assuring the convergence of the estimated parameters to the true values of those of the process mathematical model can be obtained along the same line as in Sec. 7.2-2 [16]. For the parallel identifier one has the following result (for $a_n = 1$).

THEOREM 7.2-2. Given a single-input/single-output globally asymptotically stable parallel identifier (Table 7.2-2) with an input $\rho(t)$ formed by a sum of sinusoidal signals of distinct frequencies, one has

$$\lim_{t\to\infty} \hat{a}_i(\nu, t) = a_i, \quad i = 0, 1, \ldots, n - 1$$

$$\lim_{t\to\infty} \hat{b}_i(\nu, t) = b_i, \quad i = 0, 1, \ldots, m$$

if

1. The plant to be identified is characterized by an irreducible transfer function with the degree of the denominator equal to n and the degree of the numerator equal to m
2. The filtered input ρ_f contains at least $(n + m + 1)/2$ distinct frequencies.

Notes:

1. The conditions of the Theorem 7.2-2 remain unchanged if instead of $a_n = 1$ one has either $a_0 = 1$ or $b_0 = 1$.
2. Similar results are obtained for the case of series-parallel and series identifiers [16].
3. Since condition 2 of Theorem 7.2-2 concerns ρ_f, the sinusoidal components of ρ should not be far outside the band pass of the state-variable filter acting on the input.

7.3 RECURSIVE IDENTIFIERS FOR DISCRETE-TIME PROCESSES

7.3-1 Introduction

In using recursive identifiers derived from discrete-time MRAS's, one can distinguish two basic situations.

Case 1: Frequent changes in the values of the parameters of the mathematical model of the process to be identified occur. The identifiers are supposed to be able to track the values of these parameters when such changes occur.

Case 2: Recursive identification of linear time-invariant discrete-time processes is desired. In this case, the unknown parameters of the mathematical model are supposed to be constant over a long time period.

Case 1 identification can be solved satisfactorily in most of the cases using discrete-time adaptation algorithms of class A (see Sec. 5.3), which have constant adaptation gains.

Case 2 identification can be solved using either class A adaptation algorithms or class B adaptation algorithms, which have decreasing adaptation gains (see Sec. 5.3) since both have a recursive character.

However, if a parallel configuration (the output error method) is used, the algorithms of class B can lead to an unbiased parameter estimation in the presence of measurement noise. This is not the case for the algorithms of class A, where in general a compromise between precision and speed of convergence must be made. Therefore for case 2 identification, algorithms of class B are more suitable.

Note also that the algorithms of class B (with decreasing adaptation gain) can eventually be used for case 1 identification if the parameter changes are not too frequent. In this case, one can detect, by examining the generalized error, a parameter change, and we reinitialize the adaptation gains at a high value.

We shall first present the recursive identifiers for case 1 (with constant adaptation gain) and case 2 (with decreasing adaptation gain) derived from the designs given in Sec. 5.3 for single-input/single-output MRAS's. After this presentation, we shall focus our attention on the effect of measurement noise upon the various identifiers. From this analysis we shall see that a *parallel identifier with decreasing adaptation gain allows us to obtain unbiased parameter estimation in the presence of a zero mean measurement noise and that the convergence of the estimated parameters to the right values occurs with probability 1.* A detailed evaluation of the recursive identifiers for case 2 (with decreasing adaptation gains) and a comparison with other recursive identification techniques can be found in the case study presented in Sec. 7.5.

7.3-2 *Basic Recursive Identifiers*

The procedure used in Sec. 7.2 for transposing the design of continuous-time MRAS's in order to construct identifiers is applicable in the discrete-time case. Almost the same notations and terminology indicated in Table 7.2-1 will be used. We restrict our presentation in this section to the case of recursive identifiers for single-input/single-output processes with integral + proportional identification algorithms and to the most used configurations: the parallel identifier (the output error method) and the series-parallel identifier (the equation error method). The identifiers to be presented are derived from the designs given in Sec. 5.3.

The presentation of integral + proportional identification algorithms in the case of discrete-time identifiers is interesting because, as was shown experimentally (see Sec. 7.5), the proportional term can eventually have a beneficial effect upon the parameters convergence. This can be obtained if the proportional adaptation gain is *negative* but satisfying a certain condition with respect to the value of the integral adaptation gain.

The structures of the identifiers and the algorithms for case 1 and case 2 identification are summarized in Table 7.3-1.

Let us make several comments concerning the use of the recursive identifiers specified in Table 7.3-1:

1. If $G_k' = 0$ or $G' = 0$, one obtains recursive identifiers with integral identification algorithms.
2. The series-parallel identifier with integral identification law and decreasing adaptation gain (case 2) corresponds to the classical recursive least-squares algorithm for identification using the equation error method.
3. The series-parallel identifier does not require any prior knowledge about the plant parameters, but, as will be shown next, it leads to poor results in the presence of measurement noise (the estimated parameters are highly biased).
4. The term $\Sigma_{i=1}^{n} d_i \varepsilon_{k-i}$ which is added to the expression of the generalized error (output error) in the case of the parallel identifier allows us to specify the conditions assuring the convergence of the identification algorithm (i.e., the asymptotic stability of the identifier). This condition is a sufficient one, guaranteeing the convergence

Table 7.3-1 Recursive Identifiers [$\theta_p(k)$ = Process Output, $\theta_S(k)$ = Estimation Model Output, $\rho(k)$ = Process Input]

	Parallel identifier (output error method)	Series-parallel identifier (equation error method)
The process to be identified	$\theta_p(k) = \underline{p}^T \underline{x}_{k-1}$	
	$\underline{p}^T = [a_1, \ldots, a_n, b_0, \ldots, b_m]$,	$\underline{x}_{k-1}^T = [\theta_p(k-1), \ldots, \theta_p(k-n), \rho(k), \ldots, \rho(k-m)]$
The estimation model	$\theta_S(k) = \hat{\underline{p}}^T(k)\underline{y}_{k-1} = [\hat{\underline{p}}^I(k) + \hat{\underline{p}}^P(k)]^T \underline{y}_{k-1}$	$\theta_S(k) = \hat{\underline{p}}^T(k)\underline{x}_{k-1} = [\hat{\underline{p}}^I(k) + \hat{\underline{p}}^P(k)]^T \underline{x}_{k-1}$
	$\theta_S^0(k) = [\hat{\underline{p}}^I(k-1)]^T \underline{y}_{k-1}$	$\theta_S^0(k) = [\hat{p}^I(k-1)]^T \underline{x}_{k-1}$
	$\hat{\underline{p}}^T(k) = [\hat{a}_1(k), \ldots, \hat{a}_n(k), \hat{b}_0(k), \ldots, \hat{b}_m(k)]$	$\hat{\underline{p}}^T(k) = [\hat{a}_1(k), \ldots, \hat{a}_n(k), \hat{b}_0(k), \ldots, \hat{b}_m(k)]$
	$\underline{y}^T_{k-1} = [\theta_S(k-1), \ldots, \theta_S(k-n), \rho(k), \ldots, \rho(k-m)]$	
Generalized error	$\varepsilon_k^0 = \theta_p(k) - \theta_S^0(k) = \theta_p(k) - [\hat{\underline{p}}^I(k-1)]^T \underline{y}_{k-1}$	$\varepsilon_k^0 = \theta_p(k) - \theta_S^0(k) = \theta_p(k) - [\hat{p}^I(k-1)]^T \underline{x}_{k-1}$
	$\varepsilon_k = \theta_p(k) - \theta_S(k)$	$\varepsilon_k = \theta_p(k) - \theta_S(k)$
Class A integral + proportional identification algorithm (constant adaptation gain)	$\nu_k^0 = \varepsilon_k^0 + \sum_{i=1}^{n} d_i \varepsilon_{k-1}$	
	$\hat{\underline{p}}^I(k) = \hat{\underline{p}}^I(k-1) + \dfrac{G\underline{y}_{k-1}}{1 + \underline{y}_{k-1}^T(G + G')\underline{y}_{k-1}} \nu_k^0$	$\hat{\underline{p}}^I(k) = \hat{\underline{p}}^I(k-1) + \dfrac{G\underline{x}_{k-1}}{1 + \underline{x}_{k-1}^T(G + G')\underline{x}_{k-1}} \varepsilon_k^0$
	$\hat{\underline{p}}^P(k) = \dfrac{G'\underline{y}_{k-1}}{1 + \underline{y}_{k-1}^T(G + G')\underline{y}_{k-1}} \nu_k^0$	$\hat{\underline{p}}^P(k) = \dfrac{G'\underline{x}_{k-1}}{1 + \underline{x}_{k-1}^T(G + G')\underline{x}_{k-1}} \varepsilon_k^0$
	$\hat{\underline{p}}(k) = \underline{p}^I(k) + \underline{p}^P(k)$	$\hat{\underline{p}}(k) = \hat{\underline{p}}^I(k) + \hat{\underline{p}}^P(k)$

	$G > 0,\ G' + \frac{1}{2}G \geq 0$	$G > 0,\ G' + \frac{1}{2}G \geq 0$
Class B integral + proportional identification algorithm (decreasing adaptation gain)	$\nu_k^0 = \varepsilon_k^0 + \sum_{i=1}^{n} d_i \varepsilon_{k-i}$ $\hat{\underline{p}}^I(k) = \hat{\underline{p}}^I(k-1) + \frac{G_{k-1}\underline{y}_{k-1}}{1 + \underline{y}_{k-1}^T(G_{k-1} + G'_{k-1})\underline{y}_{k-1}} \nu_k^0$ $\hat{\underline{p}}^P(k) = \frac{G'_{k-1}\underline{y}_{k-1}}{1 + \underline{y}_{k-1}^T(G_{k-1} + G'_{k-1})\underline{y}_{k-1}} \nu_k^0$ $\hat{\underline{p}}(k) = \hat{p}^I(k) + \hat{p}^P(k)$ $G_k = G_{k-1} - \frac{1}{\lambda} \frac{G_{k-1}\underline{y}_{k-1}\underline{y}_{k-1}^T G_{k-1}}{1 + (1/\lambda)\underline{y}_{k-1}^T G_{k-1}\underline{y}_{k-1}}$ $G_0 > 0,\ \lambda > 0.5$ $G'_k = \alpha G_k;\ \alpha \geq -0.5$	$\hat{\underline{p}}^I(k) = \hat{\underline{p}}^I(k-1) + \frac{G_{k-1}\underline{x}_{k-1}}{1 + \underline{x}_{k-1}^T(G_{k-1} + G'_{k-1})\underline{x}_{k-1}} \nu_k^0$ $\hat{\underline{p}}^P(k) = \frac{G'_{k-1}\underline{x}_{k-1}}{1 + \underline{x}_{k-1}^T(G_{k-1} + G'_{k-1})\underline{x}_{k-1}} \varepsilon_k^0$ $\hat{\underline{p}}(k) = \hat{\underline{p}}^I(k) + \hat{\underline{p}}^P(k)$ $G_k = G_{k-1} - \frac{1}{\lambda} \frac{G_{k-1}\underline{x}_{k-1}\underline{x}_{k-1}^T G_{k-1}}{1 + (1/\lambda)\underline{x}_{k-1}^T G_{k-1}\underline{x}_{k-1}}$ $G_0 > 0;\ \lambda > 0.5$ $G'_k = \alpha G_k;\ \alpha \geq -0.5$
Stability conditions	Class A $h(z) = \frac{1 + \Sigma_{i=1}^{n} d_i z^{-i}}{1 - \Sigma_{i=1}^{n} a_i z^{-i}}$ must be strictly positive real. Class B $h'(z) = h(z) - \frac{1}{2\lambda}$ must be strictly positive real.	None

Table 7.3-2 Parallel Recursive Identifier with Extended Adjustable Parameter Vector

The process to be identified	$\theta_p(k) = \underline{p}^T\underline{x}_{k-1}$
	$\underline{p}^T = [a_1, a_2, \ldots, a_n, b_0, \ldots, b_m]$
	$\underline{x}_{k-1}^T = [\theta_p(k-1), \ldots, \theta_p(k-n), \rho(k), \ldots, \rho(k-m)]$
The estimation model	$\theta_S(k) = \hat{\underline{p}}^T(k)\underline{y}_{k-1} = [\hat{\underline{p}}^I(k) + \hat{\underline{p}}^P(k)]^T\underline{y}_{k-1}$
	$\theta_S^0(k) = [\hat{\underline{p}}^I(k-1)]^T\underline{y}_{k-1}$
	$\hat{\underline{p}}^T(k) = [\hat{a}_1(k), \ldots, \hat{a}_n(k), \hat{b}_0(k), \ldots, \hat{b}_m(k)]$
	$\underline{y}_{k-1}^T = [\theta_S(k-1) \ldots, \theta_S(k-n), \rho(k), \ldots, \rho(k-m)]$
Generalized error	$\varepsilon_k^0 = \theta_p(k) - \theta_S^0(k) = \theta_p(k) - [\hat{\underline{p}}^I(k-1)]^T\underline{y}_{k-1}$
	$\varepsilon_k = \theta_p(k) - \theta_S(k)$
Extended parameter and observation vectors	$[\hat{\underline{p}}_e(k)]^T = [\hat{\underline{p}}^T(k), -\hat{\underline{d}}^T(k)]$
	$\tilde{\underline{y}}_{k-1}^T = [\underline{y}_{k-1}^T, \underline{e}_{k-1}^T]$
	$\hat{\underline{d}}^T(k) = [\hat{d}_1(k), \ldots, \hat{d}_n(k)]$
	$\underline{e}_{k-1}^T = [\varepsilon_{k-1}, \ldots, \varepsilon_{k-n}]$

Class A integral + proportional identification algorithm (constant adaptation gain)	$\nu_k^0 = \varepsilon_k^0 + [\hat{\underline{d}}^I(k-1)]^T \underline{e}_{k-1}$ $= \theta_p(k) - [\hat{\underline{p}}_e^I(k-1)]^T \tilde{\underline{y}}_{k-1}$ $\hat{\underline{p}}_e^I(k) = \hat{\underline{p}}_e^I(k-1) + \dfrac{\tilde{G}\tilde{\underline{y}}_{k-1}}{1 + \tilde{\underline{y}}_{k-1}^T(\tilde{G} + \tilde{G}')\tilde{\underline{y}}_{k-1}} \nu_k^0$ $\hat{\underline{p}}_e^P(k) = \dfrac{\tilde{G}'\tilde{\underline{y}}_{k-1}}{1 + \tilde{\underline{y}}_{k-1}^T(\tilde{G} + \tilde{G}')\tilde{\underline{y}}_{k-1}} \nu_k^0$ $\hat{\underline{p}}_e(k) = \hat{\underline{p}}_e^I(k) + \underline{p}_e^P(k)$ $\tilde{G} > 0,\ \tilde{G}' + \frac{1}{2}\tilde{G} \geq 0$
Class B integral + proportional identification algorithm decreasing adaptation gain)	$\nu_k^0 = \theta_p(k) - [\hat{\underline{p}}_e^I(k-1)]^T \tilde{\underline{y}}_{k-1}$ $\hat{\underline{p}}_e^I(k) = \hat{\underline{p}}_e^I(k-1) + \dfrac{\tilde{G}_{k-1}\tilde{\underline{y}}_{k-1}}{1 + \tilde{\underline{y}}_{k-1}^T(\tilde{G}_{k-1} + \tilde{G}'_{k-1})\tilde{\underline{y}}_{k-1}} \nu_k^0$ $\hat{\underline{p}}_e^P(k) = \dfrac{\tilde{G}'_{k-1}\tilde{\underline{y}}_{k-1}}{1 + \tilde{\underline{y}}_{k-1}^T(\tilde{G}_{k-1} + \tilde{G}'_{k-1})\tilde{\underline{y}}_{k-1}} \nu_k^0$ $\hat{\underline{p}}_e(k) = \hat{\underline{p}}_e^I(k) + \hat{\underline{p}}_e^P(k)$ $\tilde{G}_k = \tilde{G}_{k-1} - \dfrac{1}{\lambda} \dfrac{\tilde{G}_{k-1}\tilde{\underline{y}}_{k-1}\tilde{\underline{y}}_{k-1}^T\tilde{G}_{k-1}}{1 + (1/\lambda)\tilde{\underline{y}}_{k-1}^T\tilde{G}_{k-1}\tilde{\underline{y}}_{k-1}}$ $\tilde{G}_0 > 0;\ \lambda > 0.5$ $\tilde{G}'_k = \alpha\tilde{G}_k;\ \alpha \geq -0.5$
Stability conditions	None

of the algorithm for any initial values of the estimated parameters and of the generalized error (for class A integral + proportional identification algorithms and for class B integral + proportional identification algorithms with $|\alpha| \ll 1$).

5. The sufficient stability condition for the parallel identifier depends on the process parameters, which are unknown. An ad hoc solution which has been extensively tested in practice is to choose $d_i = -\hat{a}_i(0)$, where $\hat{a}_i(0)$ is the initial value of the estimated parameters obtained from a prior identification using a series-parallel identifier with integral identification (the least-squares algorithm). Another solution (indicated in Sec. 5.3) is to replace the constant coefficients d_i by the adjustable coefficients $\hat{d}_i(k)$, which are adjusted such that

$$\hat{d}_i(k) = -\hat{a}_i^{\,I}(k), \quad i = 1, 2, \ldots, n$$

The problem of selecting the coefficients d_i can be rigorously circumvented using the results of Theorems 5.3-3 and 5.3-4, as shown next.

6. To completely remove the restriction on parallel recursive identifiers with regard to the selection of the coefficients d_i one can use a parallel recursive identifier with an extended adjustable parameter vector of dimension $2n + m + 1$ instead of $n + m + 1$ for the case of identifiers presented in Table 7.3-1. The n additional components of the adjustable parameter vector are formed by the coefficients d_i, which appear in the expression of $\nu_k^{\,0}$ (see Table 7.3-1) and which now are adjustable. The corresponding structure for such an identifier with the various possible algorithms is summarized in Table 7.3-2 (it was derived using Theorems 5.3-3 and 5.3-4). We note that in this case there is no stability condition depending on the parameters to be identified, but this was obtained by augmenting the number of adjustable parameters (i.e., the number of computations to be done at each step will augment). See Ref. 17 for experimental results.

7. In the presence of a measurement noise $\beta(k)$ in Table 7.3-1, $\theta_p(k)$ and $\underline{x}_{k-1}$ will be replaced by

$$\hat{\theta}_p(k) = \theta_p(k) + \beta(k)$$

$$\hat{\underline{x}}_{k-1} = \underline{x}_{k-1} + \underline{b}_{k-1}$$

where

$$\underline{b}_{k-1}^T = [\beta(k-1) \cdots \beta(k-n), 0 \cdots 0]$$

8. The convergence of the parallel identifier is in general speeded up if one takes $\underline{y}_0 = \underline{x}_0$ (this is always true in the absence of measurement noise and for low levels of the measurement noise).
9. The coefficient λ weights the rate of decreasing of the integral adaptation gain matrix G_k. For $\lambda \to \infty$ the class B algorithms tend toward class A algorithms. The most used value for λ is 1.

The block diagrams for parallel and series-parallel recursive identifiers using a class B integral identification algorithm are shown in Figs. 7.3-1(a) and (b).

The extension of the use of the algorithms presented in this section for the identification of multiinput/multioutput systems is discussed in Refs. 17, 18, and 19 using various input/output representations.

7.3-3 *The Effect of Measurement Noise*

Theoretical Basis

In Sec. 7.2-2, it was qualitatively shown that measurement noise has a higher effect upon the series-parallel identifier than on the parallel identifier for low adaptation gain. With low adaptation gain, by using a parallel identifier it is possible to obtain a satisfactory parameter identification in the presence of measurement noise at the expense of slow convergence. These phenomena have also been observed experimentally in the discrete case [20]. The above observations have suggested the possibility of obtaining an unbiased

(a)

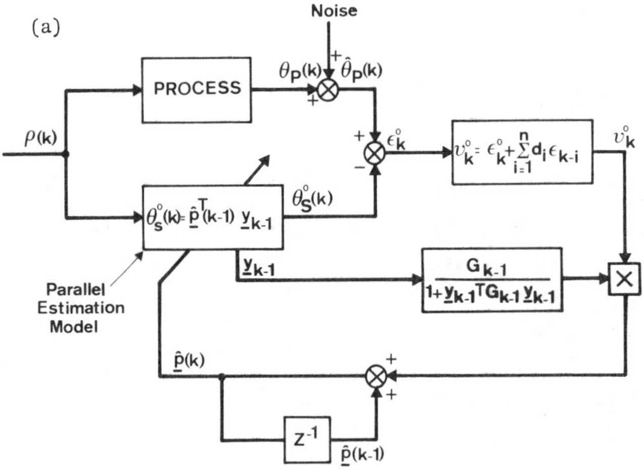

(b)

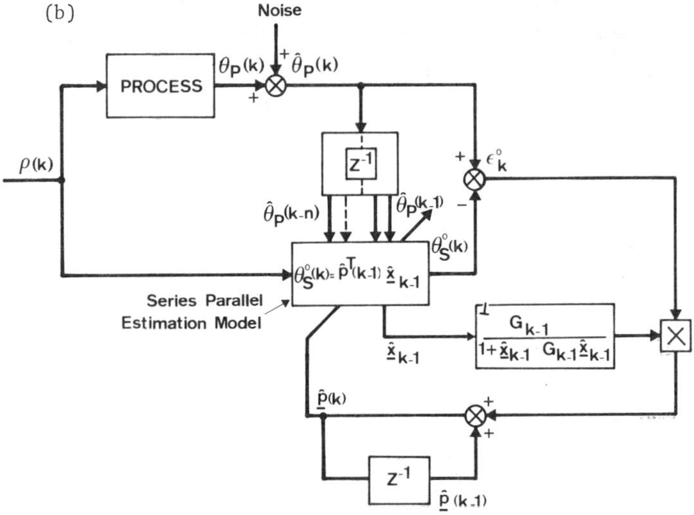

Fig. 7.3-1 Discrete-time recursive identifiers with integral identification and decreasing adaptation gains. (a) Parallel identifier. (b) Series-parallel identifier.

parameter estimation and a good speed of convergence by using a parallel identifier with decreasing adaptation gain (class B identification algorithm) [14, 21, 22].

Next we shall examine analytically the effect of measurement noise upon the identifiers presented in Table 7.3-1. The analysis will be done for the two basic configurations (series-parallel and parallel) with integral identification and decreasing adaptation gain (class B). The results for the case of integral + proportional identification with decreasing adaptation gain or with constant adaptation gain can be obtained straightforwardly, and they will only be mentioned

We are interested in finding what identifiers among the four given in Table 7.3-1 lead to an *asymptotic unbiased parameter estimation* characterized by the following property:

$$\lim_{k \to \infty} E\{\hat{\underline{p}}(k)\} = \underline{p} \tag{7.3-1}$$

where $E\{\hat{\underline{p}}(k)\}$ is the mean or expected value of the estimated parameter vector, defined as

$$E\{\hat{\underline{p}}(k)\} = \lim_{(k-i) \to \infty} \frac{1}{2(k - i)} \sum_{k-i}^{k+i} \hat{\underline{p}}(i) \tag{7.3-2}$$

Equation (7.3-1) can be read as follows: The mean value of the estimated parameter vector tends asymptotically toward the true value of the parameter vector characterizing the mathematical model of the process. Since $\underline{p}$ has a constant value, one can define the *parameter error vector*

$$\Delta\hat{\underline{p}}(k) \triangleq \hat{\underline{p}}(k) - \underline{p} \tag{7.3-3}$$

and the asymptotic unbiasedness of the estimated parameter vector can be expressed by the following condition:

$$\lim_{k \to \infty} E\{\Delta\hat{\underline{p}}(k)\} = 0 \tag{7.3-4}$$

The property of Eq. (7.3-4) can immediately be related to a stochastic stability concept: the *asymptotic stability in the mean*.

Consider the linear time-varying stochastic system

$$\underline{z}_{k+1} = A(k + 1, k)\underline{z}_k + B(k + 1, k)\underline{v}_k \tag{7.3-5}$$

where A(k + 1, k), B(k + 1, k) are time-varying matrices and $\underline{v}_k$ is a zero mean random vector $[E(\underline{v}_k) = 0]$. The system of Eq. (7.3-5) is termed *asymptotically stable in the mean* if and only if

$$\lim_{k\to\infty} E\{\underline{z}_k\} = 0 \tag{7.3-6}$$

If one can obtain for $\Delta\hat{\underline{p}}(k)$ an equation similar to that for $\underline{z}_k$ given in Eq. (7.3-5), the asymptotic stability in the mean of the corresponding stochastic system will imply the asymptotic unbiasedness of the estimated parameter vector.

The following stochastic stability theorem will be useful for the examination of the asymptotic properties of the identifiers given in Table 7.3-1 in the presence of noise.

THEOREM 7.3-1 [2]. The system of Eq. (7.3-5) is *asymptotically stable in the mean* if

1. The free dynamic system

$$\underline{z}_{k+1} = A(k + 1, k)\underline{z}_k \tag{7.3-7}$$

   is asymptotically stable.

2. $$\lim_{k\to\infty} E\{B(k + 1, k)\underline{v}_k\} = 0 \tag{7.3-8}$$

3. $$\lim_{k\to\infty} E\left\{\prod_{i=j}^{k} A(i + 1, i)B(j, j - 1)\underline{v}_{j-1}\right\} = 0 \quad \text{for all}$$

$$j = 1, 2, \ldots, k \tag{7.3-9}$$

In fact, condition 2 of Theorem 7.3-1 says that asymptotically (as $k \to \infty$) B(k + 1, k) and $\underline{v}_k$ are statistically independent, and condition 3 says that (as $k \to \infty$) $\Pi_{i=j}^{k}$ A(i + 1, i)B(j, j - 1) and $\underline{v}_{j-1}$ become statistically independent. The proof of this theorem is given in Ref. 2 and can easily be obtained by writing $\underline{z}_{k+1}$ and imposing the condition $\lim_{k\to\infty} E\{\underline{z}_{k+1}\} = 0$. [Equations (7.3-7) to (7.3-9) represent the three components of $E\{\underline{z}_{k+1}\}$.]

To use Theorem 7.3-1 for analyzing the asymptotic behavior of the identifiers given in Table 7.3-1 in the presence of noise, we shall describe the identifiers under the form of Eq. (7.3-5) in which the vector $\underline{z}_k$ will involve the parameter error vector $\Delta\hat{\underline{p}}(k)$, defined in Eq. (7.3-3), and eventually the error vector

$$\underline{e}_k^T = [\varepsilon_k, \varepsilon_{k-1}, \ldots, \varepsilon_{k-n+1}] \tag{7.3-10}$$

i.e.,

$$\underline{z}_k^T = [\Delta\hat{\underline{p}}(k)^T, \underline{e}_k^T] \tag{7.3-11}$$

The vector given in Eq. (7.3-11) will be called the *identification error vector*, and the corresponding equation will be called the *identification error system equation.*

Checking that an identification algorithm does or does not lead to an asymptotically unbiased estimate represents only the first step in the analysis of the effects of measurement noise. If the algorithm under consideration leads to an asymptotically unbiased estimate which is in fact a property in the mean of the estimate one must also examine the convergence in probability of the identified parameters to the right values. A good identifier should converge with probability 1 (almost sure convergence) to the right values. This condition is expressed as

$$\text{Prob}\{\lim_{k\to\infty} \hat{\underline{p}}(k) = \underline{p}\} = 1 \tag{7.3-12}$$

or equivalently, using the parameter error vector defined in Eq. (7.3-3), as

$$\lim_{k\to\infty} \text{Prob}\{\sup_{k>k_1} ||\Delta\hat{\underline{p}}(k)|| > \varepsilon\} = 0 \quad \text{for any } \varepsilon > 0 \tag{7.3-13}$$

To study the convergence with probability 1 (or, more briefly, w.p.1 convergence) of the recursive identification algorithms presented in Table 7.3-1, one can use a theorem due to Ljung which allows us to describe the asymptotic behavior of a class of recursive identification algorithms by the properties of an associated ordinary differential equation (ODE) [23-25]. To introduce this theorem, consider a process to be identified whose output is disturbed by noise and which can be described by

$$\hat{\theta}_p(k) = \underline{p}^T \hat{\underline{x}}_{k-1} + \alpha(k) \tag{7.3-14}$$

with

$$\underline{p}^T = [a_1, \ldots, a_n, b_0, \ldots, b_m] \tag{7.3-15}$$

$$\hat{\underline{x}}_{k-1} = [\hat{\theta}_p(k-1), \ldots, \hat{\theta}_p(k-n), \rho(k), \ldots, \rho(k-m)] \tag{7.3-16}$$

where $\hat{\theta}_p(k)$ is the measured output and $\alpha(k)$ is the *residual* modeled (in general) by

$$\alpha(k) = \sum_{i=1}^{n} h_i \alpha(k-i) + \sum_{i=1}^{n} c_i \xi_{k-1} + \xi_k \tag{7.3-17}$$

In Eq. (7.3-17), $\{\xi_k\}$ is a stationary sequence of zero mean independent random variables, and in Eq. (7.3-16) $\rho(k)$ is a stationary external input signal independent of ξ_k.

For the identification of the parameter vector $\underline{p}$ of the process described by Eq. (7.3-14), consider a recursive identification algorithm of the form

$$\begin{aligned}\hat{\underline{p}}(k) &= \hat{\underline{p}}(k-1) + \frac{G_{k-1}\underline{y}(k-1)}{1 + \underline{y}^T(k-1)G_{k-1}\underline{y}(k-1)} \varepsilon^0(k) \\ &= \hat{\underline{p}}(k-1) + \frac{(1/k)R_{k-1}^{-1}\underline{y}(k-1)}{1 + (1/k)[\underline{y}^T(k-1)R_{k-1}^{-1}\underline{y}(k-1) - 1]} \varepsilon^0(k)\end{aligned} \tag{7.3-18}$$

where

$$R_k^{-1} = kG_k \tag{7.3-19}$$

$$G_k^{-1} = G_{k-1}^{-1} + \underline{y}(k-1)\underline{y}^T(k-1) \tag{7.3-20}$$

$$R_k = R_{k-1} + \frac{1}{k}[\underline{y}(k-1)\underline{y}^T(k-1) - R_{k-1}] \tag{7.3-21}$$

and where $\underline{y}(k-1)$ and $\varepsilon^0(k)$ are variables to be specified for each particular type of identification algorithm. For the algorithms of the form of Eq. (7.3-18), one has the following result.

THEOREM 7.3-2 (Ljung). Assume that the stationary processes $\{\bar{\varepsilon}^0(k, \hat{p})\}$ and $\{\underline{\bar{y}}(k, \hat{p})\}$ can be defined for any possible values of the estimated parameter vector $\hat{\underline{p}}$. Define

$$\underline{f}(\hat{p}) = E\{\underline{\bar{y}}(k, \hat{p})\bar{\varepsilon}^0(k, \hat{\underline{p}})\} \tag{7.3-22}$$

$$G(\hat{p}) = E\{\underline{\bar{y}}(k, \hat{\underline{p}}), \underline{\bar{y}}^T(k, \hat{\underline{p}})\} \tag{7.3-23}$$

where the expectation is over the corresponding stochastic processes involved. Consider the system of ordinary differential equations

$$\frac{d}{dt}\hat{\underline{p}} = R^{-1}\underline{f}(\hat{\underline{p}}) \tag{7.3-24}$$

$$\frac{d}{dt}R = G(\hat{\underline{p}}) - R \tag{7.3-25}$$

Then, if $\underline{p}^*$ is a globally asymptotically stable equilibrium point of Eqs. (7.3-24) and (7.3-25), $\hat{\underline{p}}(k)$ as given by Eq. (7.3-18) converges with probability 1 to $\underline{p}^*$; i.e.,

$$\text{Prob}\{\lim_{k\to\infty} \hat{\underline{p}}(k) = \underline{p}^*\} = 1 \tag{7.3-26}$$

We shall subsequently use Theorems 7.3-1 and 7.3-2 for analyzing the properties of the identifiers given by Table 7.3-1 in the presence of measurement noise.

Series-Parallel Identifier

In the absence of measurement noise the identification error equation for the case of integral adaptation and decreasing adaptation gain (recursive least-squares algorithm) is

$$\Delta\hat{\underline{p}}(k + 1) = G_{k+1}G_k^{-1}\,\Delta\hat{\underline{p}}(k) = [I + G_k\underline{x}_k\underline{x}_k^T]^{-1}\,\Delta\hat{\underline{p}}(k) \tag{7.3-27}$$

and if the input is sufficiently rich, one has

$$\lim_{k\to\infty} \Delta\hat{\underline{p}}(k + 1) = \lim_{k\to\infty} \prod_{i=0}^{k} G_{i+1}G_i^{-1}\,\Delta\hat{\underline{p}}(0) = 0 \tag{7.3-28}$$

When the measurement of the output is obscured by a zero mean random noise of finite variance $\beta(k)$ [$E\{\beta(k)\} = 0$, $0 < E\{\beta^2(k)\} < \infty$], the identification error equation becomes

$$\Delta\hat{\underline{p}}(k+1) = G_{k+1}G_k^{-1}\,\Delta\hat{\underline{p}}(k) + G_{k+1}(\underline{x}_k + \underline{b}_k)$$
$$[\beta(k+1) - b_k^T\underline{p}] \tag{7.3-29}$$

where b_k is the vector noise given by

$$\underline{b}_k^T = [\beta(k), \beta(k-1), \ldots, \beta(k-n+1), \overbrace{0, 0, \ldots, 0}^{m+1}] \tag{7.3-30}$$

and one has the following result.

THEOREM 7.3-3. The series-parallel identifier with class B integral identification (decreasing adaptation gain) provides an asymptotically biased parameter estimation except for the case when the measurement noise β_k is a colored noise given by

$$\beta_k = \underline{b}_{k-1}^T\underline{p} + \xi_k = \sum_{i=1}^{n} a_i\beta_{k-i} + \xi_k \tag{7.3-31}$$

where ξ_k is a stationary sequence of zero mean independent random variables of finite variance.

This result can be obtained by explicitly expressing $\lim_{k\to\infty} E\{\Delta\hat{\underline{p}}(k+1)\}$ and also using Theorem 7.3-1.

Remark. For series-parallel identifiers a reduction of the bias on the estimated parameters can be obtained if the output of the process is filtered by a state-variable filter as well as the inputs applied to the estimation model (a similar technique is used in the continuous-time case; see Sec. 7.2-3).

Parallel Identifier

For the parallel identifier with integral adaptation and decreasing adaptation gain in the presence of the random measurement noise β_k [$E\{\beta_k\} = 0$, $0 < E\{\beta_k^2\} < \infty$, $E\{\beta_k\rho(k)\} = 0$], one has the following result.

THEOREM 7.3-4. In the presence of measurement noise β_k, the parallel identifier with a class B integral identification algorithm (decreasing adaptation gain) provides an asymptotic unbiased parameter estimation characterized by

$$\lim_{k\to\infty} E\{\hat{\underline{p}}(k)\} = \underline{p} \tag{7.3-32}$$

if the parameters of the estimation model converge to those of the process mathematical model in the absence of noise.

The proof of Theorem 7.3-4 can be found in Ref. 21.

Note. The result is valid for any values of λ verifying that $0.5 < \lambda < \infty$ as well as for the integral + proportional identification algorithm with decreasing adaptation gains. In the case of algorithms with constant adaptation gains, Theorem 7.3-4 does not hold except for very low adaptation gains (with some approximations).

Since the parallel identifier with decreasing integral adaptation gains provides an asymptotic unbiased estimation of the process parameter vector $\underline{p}$ (convergence in the mean), it is interesting to examine in more detail whether the estimated parameter vector $\hat{\underline{p}}(k)$ converges with probability 1 to the correct value. This investigation can be done by applying Theorem 7.3-2, and a detailed analysis can be found in [25]. We shall restrict ourselves here to the presentation and discussion of the main results.

In the presence of measurement noise β_k the measured output of the process to be identified is denoted by $\theta_p(k)$ and is given by

$$\begin{aligned}\hat{\theta}_p(k) &= \theta_p(k) + \beta_k = \underline{p}^T\underline{x}_{k-1} + \beta_k \\ &= \underline{p}^T\hat{\underline{x}}_{k-1} + \beta_k - \sum_{i=1}^{n} a_i\beta_{k-i}\end{aligned} \tag{7.3-33}$$

where

$$\hat{\underline{x}}_{k-1}^T = [\hat{\theta}_p(k-1), \ldots, \hat{\theta}_p(k-n), \rho(k), \ldots, \rho(k-m)] \tag{7.3-34}$$

We note that Eq. (7.3-33) is of the form of Eq. (7.3-14) considered in Theorem 7.3-2, where $\alpha(k)$ is given by

$$\alpha(k) = \beta_k - \sum_{i=1}^{n} a_i\beta_{k-i} \tag{7.3-35}$$

which corresponds to Eq. (7.3-17) for $h_i = 0$, and $c_i = -a_i$. $i = 1, 2, \ldots, n$. (The assumptions about $\{\beta_k\}$ will be specified next.)

Furthermore, we note the following correspondence between the variables involved in Eqs. (7.3-18) to (7.3-21) and those appearing in the parallel identifier with decreasing integral adaptation gain (see Table 7.3-1):

$$\underline{y}(k-1) = \underline{y}_{k-1}, \quad \varepsilon_k^0 = \nu_k^0 \tag{7.3-36}$$

Now, in the parameter space define a compact subset D_S which contains all the values of $\hat{\underline{p}}$ for which the frozen adjustable system

$$\theta_S(k, \hat{\underline{p}}) = \hat{\underline{p}}^T \underline{y}_{k-1}(\hat{p}) \tag{7.3-37}$$

is asymptotically stable; i.e.,

$$D_S = \{\hat{\underline{p}} \mid (1 - \sum_{i=1}^{n} \hat{a}_i z^{-i}) = 0 \Rightarrow |z| < 1\} \tag{7.3-38}$$

The assumption that $\hat{p}$ lies in D_S will imply that the stationary processes $\{\bar{\underline{y}}_k\}$ and $\{\bar{\underline{\nu}}_k^0\}$ can be defined.

The use of Theorem 7.3-2 leads to the following result.

THEOREM 7.3-5 [25]. Consider the parallel identifier with integral identification algorithm and decreasing adaptation gain (class B, Table 7.3-1). Assume that $\hat{\underline{p}}(k)$ belongs to D_S infinitely often with probability 1. Assume that in the absence of measurement noise $\hat{\underline{p}}(k)$ converges to $\underline{p}$. Assume that $\{\beta_k\}$ and $\{\rho_k\}$ are stationary stochastic processes with rational spectral densities such that all moments exist and $\{\beta_k\}$ and $\{\rho_k\}$ are independent. Then, in the presence of measurement noise $\{\beta_k\}$ $(E\{\beta_k\} = 0)$, $\hat{\underline{p}}_k$ converges to $\underline{p}$ with probability 1, i.e.,

$$\text{Prob}\{\lim_{k\to\infty} \hat{\underline{p}}(k) = \underline{p}\} = 1 \tag{7.3-39}$$

if the discrete transfer function

$$h'(z) = \frac{1 + \Sigma_{i=1}^{n} d_i z^{-i}}{1 - \Sigma_{i=1}^{n} a_i z^{-i}} - \frac{1}{2\lambda} \tag{7.3-40}$$

is strictly positive real.

Remark 1. The satisfaction of the sufficient condition for deterministic asymptotic stability of the parallel identifier with decreasing integral adaptation gain [h'(z) given by Eq. (7.3-40) should be strictly positive real] implies that the condition of Theorem 7.3-5 concerning h'(z) is also satisfied.

Remark 2. Theorem 7.3-5 is valid if all possible values of $\hat{\underline{p}}(k)$ are restricted within the domain D_S. This can be achieved, for example, by introducing a stability test upon $\hat{\underline{p}}(k)$ at each instant k. However, this is not necessary in practice for the reasons discussed below.

Suppose that at a certain instant $\hat{\underline{p}}(k)$ is outside D_S. In this case, $\underline{y}_k$ and ν_k^0 will tend to increase. This of course implies that the effect of measurement noise will be negligible compared to the values of ν_k^0 in the absence of noise, and we can return to the case where the deterministic analysis may be applied. But in the absence of measurement noise the condition that h'(z) be strictly positive real implies that one has a globally asymptotically stable adaptive system and as a consequence $\hat{\underline{p}}(k)$ will tend toward a value within D_S. Since in the meantime the steps of adaptation become smaller and smaller, $\hat{\underline{p}}(k)$ will remain within D_S (except perhaps for isolated values of k).

7.4 CASE STUDY: REAL-TIME IDENTIFICATION AND ADAPTIVE CONTROL OF A STATIC DC/AC CONVERTER [26]

The dc/ac converter considered has been presented in Sec. 1.3-5, and a conventional voltage control scheme has been shown in Fig. 1.3-14. From the analysis presented in Sec. 1.3-5 it was concluded that

1. To have a control system with good dynamic performance, the controller must be adjusted in accordance with the load characteristics. This requires a real-time identification of the load characteristics.
2. Since an important load change also provokes a large output voltage disturbance, to obtain a fast voltage recovery an

adaptive feedforward control must be added to the control scheme.

We shall first focus our attention on the real-time identification of the load characteristics.

In the case of a resistive-inductive load, the process to be controlled can be described for a given load by a first-order transfer function:

$$H_P(s) = \frac{\alpha}{As + B} \tag{7.4-1}$$

where α is a constant term depending on the gain of the linearization block included in the SCR control device, A is approximately constant if the power factor of the load is larger than 0.6, and B is given by

$$B = \frac{1}{(R_c^2 + L_c^2\omega^2)^{1/2}} = \frac{1}{R_c} \cos \phi \tag{7.4-2}$$

where R_c and L_c are the resistive and inductive components, respectively, of the load.

We are therefore interested in identifying R_c and L_c. Note that the load is in fact a first-order dynamic system which is always excited since the input to the load is the sinusoidal output voltage of the converter. Also note that, as shown in Theorem 7.2-2, a sinusoidal input with only one frequency is enough for the identification of this first-order dynamic system. For a given resistance and inductance, the load is characterized by the transfer function (admittance)

$$H_L(s) = \frac{1}{L_c s + R_c} \tag{7.4-3}$$

The parameters of this transfer function can be identified using, for example, the parallel identifier shown in Fig. 7.4-1 and derived from the Table 7.2-2. Note that this is not the only configuration possible for the parallel estimation model (another one can be obtained, for example, using a state-variable description of the load). In Fig. 7.4-1, U_S is the converter sinusoidal voltage, which represents the input in the process to be identified, and the load current I_S is the corresponding output.

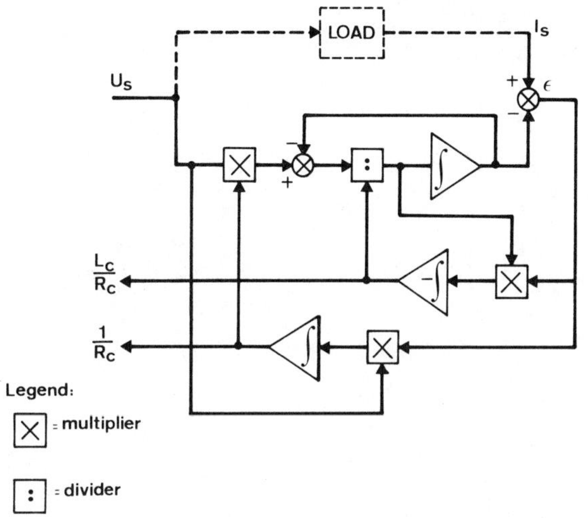

Fig. 7.4-1 Real-time identifier of the converter load parameters.

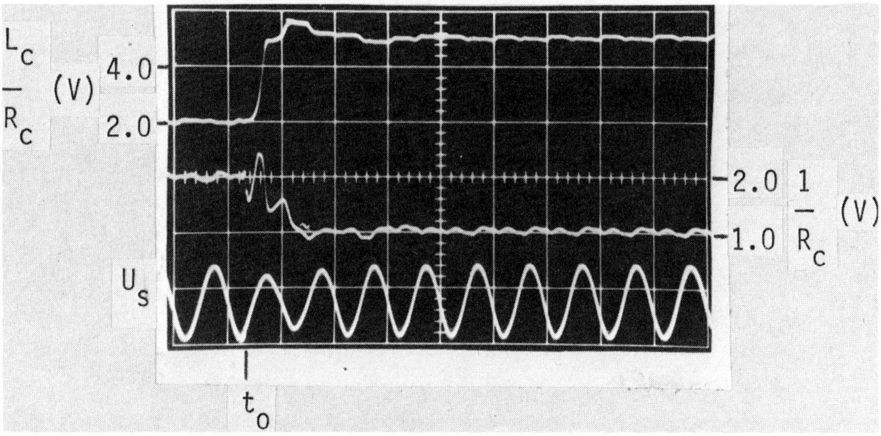

Fig. 7.4-2 Identification results obtained for a load change on a 500-W power converter. Time base: 20 msec/unit.

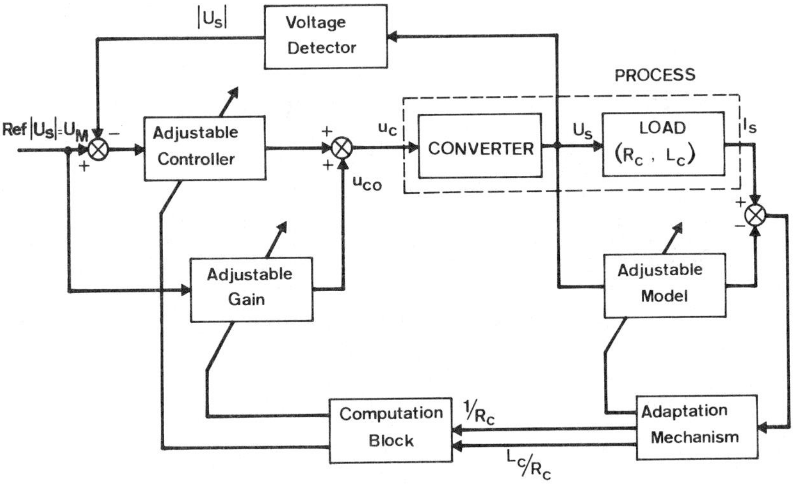

Fig. 7.4-3 Block diagram of the adaptive voltage control of the dc/ac converter.

Figure 7.4-2 illustrates the performance of the identifier when used with a dc/ac power converter of 500-W nominal power [26].[2] Because of the continuous testing feature of the converter sinusoidal voltage, very fast and precise identification is obtained (even in closed loop). One can see that only a period of the sinusoidal wave (20 msec for a 50-Hz dc/ac converter) is necessary to obtain a good load parameter estimation.

From the load estimated parameters and the values of the parameters α and A, a *computing block* (see Fig. 7.4-3) automatically computes the parameters of an adjustable PI controller (it modifies the integral component). This allows us to obtain the desired closed-loop characteristics of the voltage control around any values of the resistance and inductance of the load. However, it was observed that this adaptive control system, which adjusts only the parameters of

[2]The subsequent result has been provided by courtesy of ALSTHOM--Directions des Recherches, Department d'Automatique et d'Electronique, Grenoble.

the controller from the identified values of the load parameters, does not assure a fast voltage recovery after a rapid and large variation of the load characteristics.

Since the driving voltage u_c is a function of the desired voltage and load characteristics, one can approximately determine off line the law

$$u_c = f(U_M, R_c, L_c) = K_U(U_M, R_c, L_c)U_M \tag{7.4-4}$$

where U_M is the desired output voltage. Since R_c and L_c are continuously identified, one can immediately compute the gain K_U, assuring the suitable value of the driving voltage u_c. Using the adjustable feedforward gain K_U, one obtains the complete scheme of the adaptive voltage control shown in Fig. 7.4-3. For more experimental details, see [26].

7.5 CASE STUDY: COMPARISON OF SEVERAL RECURSIVE ALGORITHMS FOR IDENTIFICATION OF DISCRETE LINEAR TIME-INVARIANT PROCESSES

We shall compare several recursive identification techniques on a fourth-order simulated process whose output is highly disturbed by measurement noise and on a real industrial process (a paper machine).

We shall consider the identification algorithms with decreasing adaptation gains derived from MRAS design (for parallel and series-parallel identifiers) as well as other well-known recursive identification techniques derived from statistical considerations.

The identification techniques to be compared are

1. A series-parallel identifier (MRAS) using the integral identification algorithm and decreasing adaptation gain with $\lambda = 1$ (this corresponds to the recursive least-squares algorithm)
2. A parallel identifier (MRAS) with decreasing adaptation gain (the integral and integral + proportional identification algorithm with $\lambda = 1$)
3. An instrumental variable identifier with an auxiliary model (Young [27])

Table 7.5-1 Instrumental Variable Identifier

The process to be identified	$\theta_P(k) = \underline{p}^T \underline{x}_{k-1}$
	$\underline{p}^T = [a_1, \ldots, a_n, b_0, \ldots, b_m]$
	$\underline{x}_{k-1}^T = [\theta_P(k-1), \ldots, \theta_P(k-n), \rho(k), \ldots, \rho(k-m)]$
The estimation model	$\theta_S(k) = \hat{\underline{p}}^T(k)\underline{x}_{k-1}$
	$\theta_S^0(k) = \hat{\underline{p}}^T(k-1)\underline{x}_{k-1}$
	$\hat{\underline{p}}^T(k) = [\hat{a}_1(k), \ldots, \hat{a}_n(k), \hat{b}_0(k), \ldots, \hat{b}_m(k)]$
A priori generalized error (equation error)	$\varepsilon_k^0 = \theta_P(k) - \theta_S^0(k)$
Instrumental variable vector	$\hat{\underline{y}}_k$
Identification algorithm	$\hat{\underline{p}}(k) = \hat{p}(k-1) + \dfrac{G_{k-1}\hat{\underline{y}}_{k-1}}{1 + \hat{\underline{y}}_{k-1}^T G_{k-1}\underline{x}_{k-1}} \varepsilon_k^0$
	$G_k = G_{k-1} - \dfrac{G_{k-1}\underline{x}_{k-1}\hat{\underline{y}}_{k-1}^T G_{k-1}}{1 + \hat{\underline{y}}_{k-1}^T G_{k-1}\underline{x}_{k-1}}$

4. An instrumental variable identifier with a delayed auxiliary model (Béthoux [6, 28, 29])
5. An instrumental variable identifier with delayed observations (Banon and Aguilar-Martin [30])
6. A series-parallel identifier with the extended recursive least-squares algorithm (Béthoux [6, 23, 29]).

For a detailed discussion of the convergence properties in the presence of measurement noise of methods 1 to 4 and 6, see Refs. 23-25, 29, and 31 as well as Sec. 7.3-3. Experimental results concerning the comparison of methods 1 and 3 with other methods not discussed here can be found in [23].

Table 7.5-2 Generation of the Instrumental Variable Vector

Auxiliary adjustable model (3)	$\hat{\theta}_S(k) = \underline{\hat{p}}(k)^T \underline{\hat{y}}_{k-1}$
	$\underline{\hat{y}}_{k-1}^T = [\hat{\theta}_S(k-1), \ldots, \hat{\theta}_S(k-n),$
	$\rho(k), \ldots, \rho(k-n)]$
Auxiliary adjustable model with delayed parameters (4)	$\hat{\theta}_S(k) = \underline{\hat{p}}(k-r)^T \underline{y}_{k-1}, \quad r = 1, 2, \ldots$
	$\underline{\hat{y}}_{k-1}^T = [\hat{\theta}_S(k-1), \ldots, \hat{\theta}_S(k-n),$
	$\rho(k), \ldots, \rho(k-n)]$
Delayed observations (5)	$\underline{\hat{y}}_{k-1} = \underline{\hat{x}}_{k-1-r}, \quad r = 1, 2, \ldots$

The first two identifiers have already been presented in Sec. 7.3-2 and summarized in Table 7.3-1 (class B) [see also Figs. 7.3-1(a) and (b)]. We shall next briefly present identifiers 3 to 6.

The *instrumental variable* identifiers 3, 4, and 5 are all derived from the series-parallel identifier using the recursive least-squares algorithm [see Table 7.3-1 for the class B series-parallel identifier with the integral identification algorithm ($G_k' = 0$)]. They are all based on the idea of replacing a variable disturbed by a measurement noise by an auxiliary one which is correlated with the noise-free variable but not with the measurement noise. This allows us to remove the undesired correlations in the product of two variables disturbed by the same measurement noise. Therefore, if one denotes the instrumental variable vector by $\underline{\hat{y}}_k$ (which must be correlated with the noise-free measurement of the vector $\underline{x}_k$), one obtains the structure of instrumental variable identifiers given in Table 7.5-1.

Table 7.5-2 indicates how the instrumental variable vector is generated.

The series-parallel identifier with the *extended recursive least-squares* algorithm is also an improvement of the series-parallel identifier with the recursive least squares algorithm allowing one to obtain better parameter estimates in the presence of measurement noise. This type of identifier is derived under the assumption that the effect of the measurement noise $\beta(k)$ together with the process to be identified can be modeled by [3, 29]

$$\hat{\theta}_p(k) = \theta_p + \beta_k = \sum_{i=1}^{n} a_i\hat{\theta}_p(k - i) + \sum_{i=0}^{m} b_i\rho(k - i)$$

$$+ \sum_{i=1}^{n} d_i\alpha_{k-i} + \sum_{i=1}^{n} c_i\xi_{k-i} + \xi_k \tag{7.5-1}$$

where the last three terms form the *residual* α_k and ξ_k is a white noise (or in general a stationary sequence of independent random variables). The particular form of the model considered in the algorithm to be discussed here uses for α_k the expression [6, 29]

$$\alpha_k = \sum_{i=1}^{n} d_i\alpha_{k-i} + \xi_k \tag{7.5-2}$$

The identifier is obtained by applying the least-squares algorithm for the identification of the extended model given in Eq. (7.5-1).

Experimental Results

To illustrate the performance of the above-considered identifiers in the presence of noise-obscured measurements, two cases will be considered: (1) identification of a fourth-order simulated process and (2) identification of a real plant (fiber flow → moisture transfer in a paper machine).

Identification of Fourth-Order Simulated Process. The simulated system is described by

$$\theta_p(k) = \sum_{i=1}^{4} a_i\theta_p(k - i) + b_0\rho(k) \tag{7.5-3}$$

where

$$a_1 = -1.3, \quad a_2 = 0.22, \quad a_3 = 0.832, \quad a_4 = 0.269$$

$$b_0 = 1$$

The system is characterized by a low damped oscillatory response. The input is a pseudo-random binary sequence (PRBS) (+1, -1), and the output is obscured by a zero mean Gaussian noise with variance $\sigma^2 = 0.81$.

The parametric distance defined as

$$D(k) \triangleq [\Delta\hat{\underline{p}}(k)^T \Delta\hat{\underline{p}}(k)]^{1/2} = \left\{ \sum_{i=1}^{4} [a_i - \hat{a}_i(k)]^2 + [b_0 - \hat{b}_0(k)]^2 \right\}^{1/2} \tag{7.5-4}$$

is used in order to evaluate the performance of the various identifiers.

One should mention that for this example the *instrumental variable* identifier with delayed observations diverges because, the system having a very low damped response, by delaying the measurements one cannot assure a correlation between $\hat{\underline{x}}_{k-r}$ and the noise-free measurement of $\underline{x}_k$.

In Fig. 7.5-1 the evolution of the parametric distance is given for identifiers 1 to 4 and 6. The estimation models have been similarly initialized as well as the integral adaptation gain (G_0 has been made a diagonal matrix with all elements equal to 100).

For parallel identifier 2 an integral + proportional identification algorithm has been used, and G_k' is given by

$$G_k' = \alpha G_k \quad \text{with } \alpha = -0.4 \tag{7.5-5}$$

(This value of α leads to the fastest convergence in the initial stage of the identification process, and the proportional term has an influence up to $k = 500$ [21].)

Concerning the bias of the estimated parameters, one could conclude that the performance of parallel identifier 2 is comparable to that of the instrumental variable identifiers with auxiliary model 3 and with delayed auxiliary model 4. The performances of these three identifiers are far better than those of the series-parallel identifiers with the recursive least-squares algorithm (1) and with the extended recursive least-squares algorithm (6). One can also see that the parallel identifier with integral + proportional identification assures a slightly better convergence than the other identifiers at the beginning of the identification process.

Identification of Fiber Flow → Moisture Transfer in a Paper Machine. The input is a zero mean signal obtained by exciting the control flow valve by a PRBS. The magnitude of the input signal

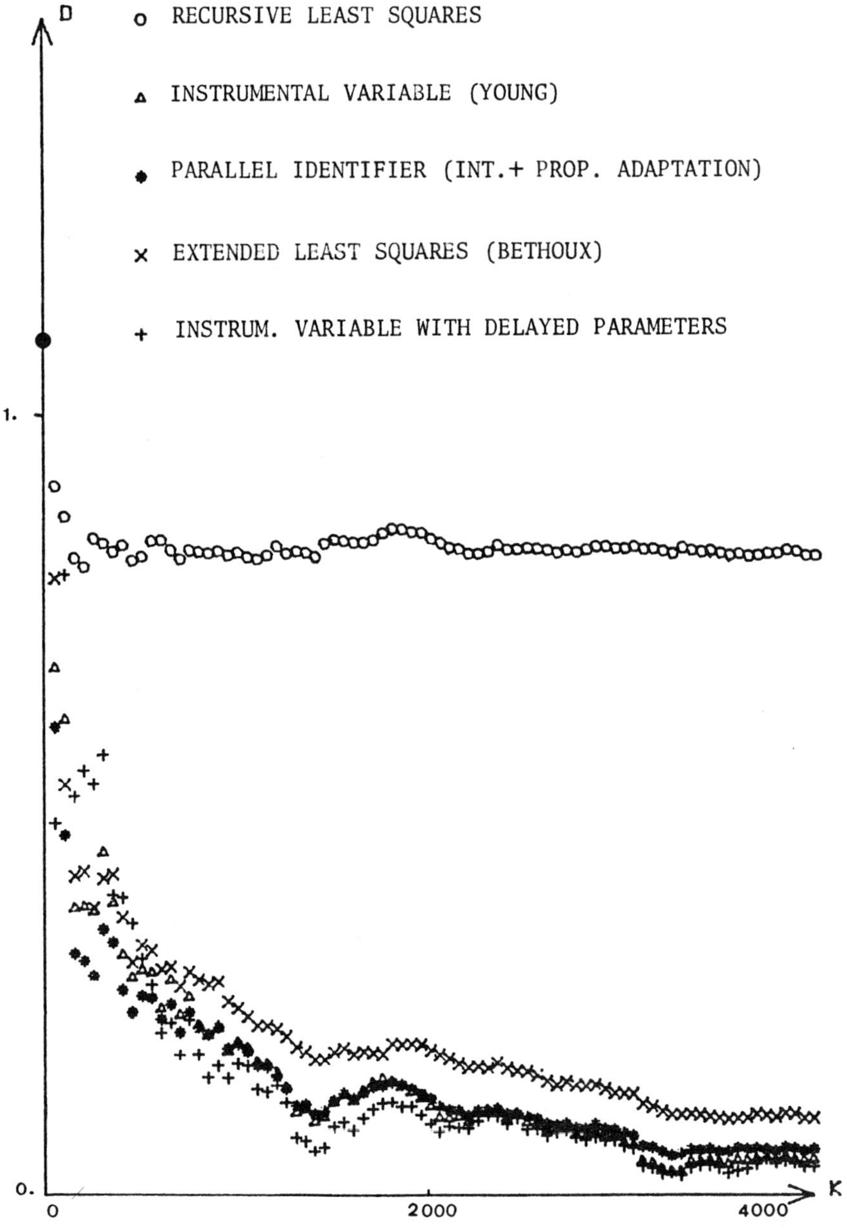

Fig. 7.5-1 Evolution of the parametric distance for various identifiers.

varies between ±0.01 m^3/hr, and the input signal is superimposed on a mean value fiber flow corresponding to a mean value of the process output (moisture) of 5.44% H_2O. The same identifiers that were compared above were used.

The best results were obtained by considering that the mathematical model of the process to be identified was characterized by the discrete transfer function

$$H_p(z) = z^{-r} \frac{b_0 + b_1 z^{-1}}{1 - a_1 z^{-1} - a_2 z^{-2}} \tag{7.5-6}$$

where r = 13 samples (the sampling period was 5 sec).

The quality of the identification was evaluated by using the following criterion:

$$J = \frac{1}{N} \sum_{k=1}^{N} [\theta_P(k) - \tilde{\theta}_S(k)]^2 \tag{7.5-7}$$

where

$$\tilde{\theta}_S(k) = \sum_{i=1}^{2} \hat{a}_i \tilde{\theta}_S(k - i) + \sum_{i=0}^{1} \hat{b}_i \rho(k - r - i) \tag{7.5-8}$$

In Eq. (7.5-8), $\hat{a}_i$, $\hat{b}_i$ are the final values of the estimated parameters. $\rho(k)$ and $\theta_P(k)$ in Eqs. (7.5-7) and (7.5-8) represent the sequence of process input and output measurements used for identification. The values of the estimated parameters $\hat{a}_i$ and $\hat{b}_i$ obtained with various identifiers on the basis of 245 measurements are given in Ref. 21.

The series-parallel identifier using recursive least squares was initialized with $\hat{a}_1(0) = \hat{a}_2(0) = \hat{b}_0(0) = \hat{b}_1(0) = 0$. All the other identifiers were initialized with the parameter values obtained from the series-parallel identifier using recursive least squares.

For the considered quality criterion, the parallel identifier with integral adaptation gives J = 0.081, which is slightly better than the results obtained with the instrumental variable identifiers (J = 0.084) and with the series-parallel identifier using extended the least-squares algorithm (J = 0.089), while the series-parallel identifier gave J = 0.105.

These results obtained on a real process confirm the results obtained for the identification of the simulated fourth-order process.

7.6 CONCLUDING REMARKS

Concerning the use of MRAS's for identification, we wish to emphasize the following basic ideas:

1. Model reference adaptive systems are essentially used for recursive parameter identification on on-line and real-time (parameter tracking) identification.
2. When MRAS's are used for parametric identification, the three basic structures parallel, series-parallel, and series correspond to the identification structures known as output error, equation error, and input error, respectively.
3. The uniqueness of the identification depends on the characteristics of the input signal. To obtain a unique set of identified parameters, the input signal should contain, for example, in the case of a completely controllable and observable SISO system, a number of components with distinct frequencies, which is higher than half of the number of the parameters which must be identified.
4. The series-parallel structure allows us to build an asymptotically stable identifier for any initial values of the parameters of the estimation model, and no prior knowledge about the range of the values of the parameters to be identified is necessary. However, in the presence of measurement noise, the estimated parameters are always biased.
5. The parallel structure allows us to build an asymptotically stable identifier for a certain region in the parameter space around the initial values of the parameters of the estimation model. Therefore information about the range of the values of the parameters to be identified is in general necessary. This limitation can be overcome by using algorithms which also adapt the parameter of the compensator acting on the generalized error, but the total

number of parameters to be adapted will be augmented by n ($2n + m + 1$ parameters instead of $n + m + 1$).

6. In the presence of measurement noise the parallel identifier with decreasing integral adaptation gain (class B) leads to asymptotically unbiased parameter estimates which converge with probability 1 to the right values.
7. Series-parallel identifiers can be used in order to provide initializations for parallel identifiers (the last being more accurate in the presence of measurement noise).
8. The identifiers derived from model reference adaptive techniques can be used for the identification of certain classes of nonlinear processes as well as for the identification of processes containing delays [14, 32].
9. MRAS's can be used for the identification of multi input/multi output systems. The performance of the identifier, however, as for other types of identifiers will depend on prior structural identification [18, 19].
10. MRAS identifiers can be used for implementing adaptive control schemes [8, 26, 33-35].

PROBLEMS

7.1 The discrete-time system formed by a sampler, a zero-order hold element, and a continuous system having the transfer function $h_p(s) = K/(1 + s\tau)$ can be described by the first-order difference equation $\theta_p[kT] = a_1\theta_p[(k - 1)T] + b_1\rho[(k - 1)T]$, where $\theta_p(kT)$ and $\rho(kT)$ are the values of the process output and input, respectively, at the sampling instant kT.

(a) Find the expressions for K and τ in terms of a_1, b_0, and the sampling period T.

(b) For $K = 2$, $\tau = 0.5$ sec, and $T = 0.1$, determine the required precision for the identification of a_1 and b_1 so that K and τ will be identified with 10% precision. Also discuss the influence of the sampling period upon the precision required for the identification of a_1 and b_1.

7.2 In certain dc electrical drives, the parameters of the PI speed controller are adjusted proportionally with J and f, where J is the moment of inertia of the drive and f is the friction. The transfer function relating the motor armature current and the speed is $h_M(s) = K/(Js + f)$. Assuming K constant and known and J and f variable during operation, construct an identifier allowing one to directly obtain J and f without requiring dividers for its implementation. (Hint: Use one of the structures given in Table 7.2-2.)

7.3 [2] Show that in the identification error system equation corresponding to the series-parallel identifier with decreasing gain given in Eq. (7.3-27) the eigenvalues of the matrix $G_{k+1}G_k^{-1}$ are always between 0 and 1 if $\underline{x}_k \neq 0$ and if for each k and q, $\underline{r}_q^T \underline{x}_k \underline{x}_k^T \underline{r}_q \neq 0$, where $\underline{r}_q$ is any of the eigenvectors of the matrix $G_{k+1}G_k^{-1}$. Show also that this implies $\lim_{k\to\infty} \Delta\hat{\underline{p}}(k+1) = 0$.

7.4 When using a digital computer for implementing a recursive identification algorithm the sampling frequency can sometimes be limited by the number of computations to be done at each step. For the identification algorithms given in Table 7.3-1, compute the number of additions, multiplications, and divisions as a function of the number of parameters to be identified.

7.5 To reduce the effect of measurement noise, one can use filtered measurements of the process output, denoted by $\theta_{Pf}(k)$. The filtered output is governed by the difference equation $(1 + \Sigma_{i=1}^{n-1} c_i z^{-1})\theta_{Pf}(k) = \theta_P(k)$, where $\theta_P(k)$ is the process output at instant k and the c_i are the coefficients of the transfer function of the filter, which is assumed to be asymptotically stable. Modify the algorithms given in Table 7.3-1 for the use of filtered values of the process output.

7.6 Extend the identification algorithms given in Table 7.3-1 for the case of a series identifier when the process to be identified is described by $\theta_P(k) = \Sigma_{i=1}^{n} a_i\theta_P(k - i) + \rho(k) + \Sigma_{i=1}^{m} b_i\rho(k - i)$.

7.7 Consider a two-input/two-output process described in terms of input/output relations using polynomial matrices as

$$\begin{bmatrix} z^2 - a_1^{11}z - a_2^{11} & -a_1^{12} \\ -a_1^{21}z - a_2^{21} & z - a_1^{22} \end{bmatrix} \begin{bmatrix} \theta_p^1(k) \\ \theta_p^2(k) \end{bmatrix} = \begin{bmatrix} b_1^{11}z + b_2^{11} & b_1^{12}z + b_2^{12} \\ b_1^{21} & b_1^{22} \end{bmatrix} \begin{bmatrix} u_1(k) \\ u_2(k) \end{bmatrix}$$

where $u_1(k)$ and $u_2(k)$ are the two inputs and $\theta_p^1(k)$ and $\theta_p^2(k)$ are the two outputs. Write in detail the identification algorithm corresponding to a parallel identifier with decreasing adaptation gain.

REFERENCES

1. J. Richalet, A. Rault, and R. Pouliquen, *Identification des Processus par la Méthode du Modèle*, Gordon & Breach, London, 1971.
2. J. M. Mendel, *Discrete Techniques of Parameter Estimation--The Equation Error Formulation*, Marcel Dekker, New York, 1973.
3. K. J. Aström and P. Eykoff, System identification, a survey, Automatica, *7*, 123-167 (1971).
4. I. D. Landau, A survey of model reference adaptive techniques--theory and applications, Automatica, *10*, 353-379 (1974).
5. I. D. Landau, Algorithme d'identification récursive utilisant le concept de positivité, Rev. RAIRO. Jaune, *J-2*, 69-87 (May 1975).
6. G. Béthoux, Identification en temps réel des procédés multidimensionnels par la méthode des moindres carrés généralisés dans l'optique des systèmes adaptatifs discrets hyperstables, Internal Report ALSTHOM--Direction des Recherches, Grenoble, Aug. 1973.

7. I. D. Landau, Hyperstability and identification, in *Proceedings of the 1970 IEEE Symposium on Adaptive Processes Decision and Control*, University of Texas, Austin, Texas, X.4.1-X.4.8, Dec. 1970, IEEE, New York, 1970.
8. K. S. Narendra and P. Kudva, Stable adaptive schemes for system identification and control, Parts I and II, IEEE Trans. Syst. Man Cybern., *SMC-4*, 542-560 (1974).
9. H. J. Kushner, On the convergence of Lion's identification method with random inputs, IEEE Trans. Autom. Control, *AC-15*, 652-654 (1970).
10. C. C. Hang, Design studies of model reference adaptive control and identification systems, Ph.D. thesis, University of Warwick, Warwick, Nov. 1973.
11. I. D. Landau, Design of discrete model reference adaptive systems using the positivity concept, in *Proceedings of the Third IFAC Symposium on Sensitivity Adaptivity and Optimality*, Ischia, June 1973, pp. 307-314, Instruments Society of America, Pittsburgh, 1973.
12. I. D. Landau, E. Sinner, and B. Courtiol, Model reference adaptive systems--some examples, in *Proceedings of the 93rd ASME Winter Meeting*, paper 72 WA/AUT-13, ASME, New York, Dec. 1972.
13. J. S. Pazdera and H. J. Pottinger, Linear system identification via Lyapunov design techniques, in *Proceedings of the JACC*, 1969, pp. 795-801.
14. I. D. Landau, Sur une méthode de synthèse des systemes adaptatifs avec modèle utilisés pour la commande et l'identification d'une classe de procédés physiques, Thèse d'Etat-ès-Sciences, No. C.N.R.S.: A.0.8495, University of Grenoble, Grenoble, June 1973.
15. C. C. Hang, The design of model reference parameter estimation systems using hyperstability theories, in *Proceedings of the Third IFAC Symposium on Parameter Estimation and System Identification*, North Holland (American-Elsevier), The Hague, 1973, pp. 741-744.
16. P. M. Lion, Rapid identification of linear and nonlinear systems, AIAA J., *5*, 1835-1842 (1967).

17. I. D. Landau, Elimination of the real positivity condition in the design of parallel MRAS, IEEE Trans. on Autom. Control, *AC-23*, no. 6, 1015-1020 (1978).

18. A. Gauthier, Identification réccurente des systèmes multi-entrees, multi-sorties, Thèse 3me Cycle, Institut National Polytechnique de Grenoble, Grenoble, March 1977.

19. A. Gauthier and I. D. Landau, On the recursive identification of multi-input, multi-output systems, Automatica, *14*, no. 6 (1978).

20. J. J. Hirsh and P. Peltié, Real time identification using adaptive discrete models, in *Proceedings of the Third IFAC Symposium on Sensitivity, Adaptivity and Optimality,* Ischia, 1973, pp. 290-297, Instruments Society of American, Pittsburgh, 1973.
Sensitivity, Adaptivity and Optimality, Ischia, 1973, pp. 290-297.
21. I. D. Landau, Unbiased recursive identification using model reference adaptive techniques, IEEE Trans. Autom. Control, *AC-21*, No. 2, 194-202 (1976), and *AC-23*, No. 1, 97-99 (1978).

22. I. D. Landau, Algorithme pour l'identification à l'aide d'un modèle ajustable parallèle, C. R. Acad. Sci., Ser. A, *t. 277*, 197-200 (July 16, 1973).

23. T. Söderstrom, L. Ljung, and I. Gustavsson, A comparative study of recursive identification methods, Report 7427, Lund Institute of Technology, Department of Automatic Control, Dec. 1974.

24. L. Ljung, Analysis of recursive stochastic algorithms, IEEE Trans. Autom. Control, *AC-22*, no. 4, 551-575 (1977).

25. L. Ljung, On positive real transfer functions and the convergence of some recursive schemes, IEEE Trans. Autom. Control, *AC-22*, No. 4, 539-551 (1977).

26. B. Courtiol, Applying model reference adaptive techniques for the control of electromechanical systems, in *Proceedings of the Sixth IFAC World Congress*, Part ID, International Federation of Automatic Control, Boston, Aug. 1976, pp. 58.2-1 to 58.2-9.

27. P. C. Young, An instrumental variable method for real time identification of a noisy process, Automatica, *6*, 271-288 (1970).

28. G. Béthoux, Etude comparative de plusieures méthodes d'identification, Internal Report, ALSTHOM--Directions des Recherches, Grenoble, July 1973.

29. G. Béthoux, Approche unitaire des méthodes d'identification et de commande adaptative des procedes dynamiques, Thèse 3me Cycle, Institut National Polytechnique de Grenoble, Grenoble, July 1976.

30. G. Banon and J. Aguilar-Martin, Estimation lineaire récurrente de paramètres de processus dynamiques soumis à des perturbations aléatoires, Rev. CETHEDEC, *9*, 39-86 (1972).

31. L. Ljung, T. Söderstrom, and I. Gustavsson, Counterexamples to general convergence of a commonly used recursive identification method, IEEE Trans. Autom. Control, *AC-20*, No. 5, 643-652 (1975).

32. I. D. Landau, L. Muller, G. Dolle, and G. Bianchi, A new method for carbon control in basic oxygen furnace, *Second IFAC Symposium on Automation in Mining, Mineral and Metal Processing*, Johannesburg, Sept. 1976. International Federation of Automatic Control, 1976.

33. E. Sinner, Régulateur adaptif à variables d'Etat pour processes industriels, Rev. RAIRO, Jaune, *1*, 100-123 (1975).

34. L. Ljung and I. D. Landau, Model reference adaptive systems and self tuning regulators--some connections, *Proceedings Seventh IFAC Congress*, Helsinki, June 1978. International Federation of Automatic Control and Pergamon Press, Oxford, 1978.

35. I. D. Landau, Dualité asymptotique entre les systèmes de commande adaptative avec modèle et les régulateurs à variance minimale auto-ajustables, *Proc. Int. Symposium on Systems Optimization and Analysis (IRIA, IFAC, AFCET)*, IRIA, Roquencourt, France, Dec. 1978. Springer-Verlag, Heidelberg, 1979.

Chapter Eight

SIMULTANEOUS ADAPTIVE STATE OBSERVATION AND PARAMETER IDENTIFICATION

8.1 INTRODUCTION

As mentioned in Chap. 1, when the process to be controlled with inaccessible states has unknown parameters or they vary during operation, adaptive state observers must be built. In this case from the linear asymptotic observer one can derive a model reference adaptive system (MRAS) structure (see Sec. 1.3) which permits simultaneous observation of the states of the process (or their estimation, if the available measurements are obscured by noise) and identification of the process parameters.

The possibility of using MRAS's for building adaptive state observers and identifiers has been recognized since 1970, and approximate design solutions for them have been derived and tested [1-4]. However, credit for the first exact analytic design of an adaptive state observer and identifier goes to Caroll and Lindorff (1973) [5]. Subsequently other designs have been proposed by Lüders and Narendra [6] and Kudva and Narendra [7]. The last three adaptive observer designs concern the adaptive observation of single-input/single-output continuous, linear, time-invariant systems. The designs are based on the use of Lyapunov functions, and they all lead to a series-parallel MRAS configuration. Since these developments extension to some multiinput/multioutput situations and to the discrete-time case (with integral adaptation and constant adaptation gains) has been discussed in [8-10].

Because of their structure, these adaptive observers have the advantages and the drawbacks of series-parallel MRAS's (as examined in Chap. 7). They do not require any prior knowledge about the dynamics of the process to be observed, but their performance in the presence of measurement noise is poor. On the other hand, starting from the configuration of a linear asymptotic observer, it is possible to derive adaptive observers which have a parallel MRAS structure. If these adaptive observers use adaptation algorithms with decreasing adaptation gains, very good performance results are obtained even when the output of the process is obscured by noise.

Our first objective in this chapter is to develop a unified design for the two basic structures of discrete-time adaptive state observers and identifiers as an extension of the linear asymptotic observer. We shall briefly review the linear asymptotic observer in Sec. 8.2 while indicating the two structures which will be used next for deriving adaptive observers.

The design of adaptive state observers will be presented in Sec. 8.3. The tools to be used will be the design techniques discussed in Chap. 5 for discrete-time MRAS's. Conclusions about the performance of each type of adaptive state observer are presented based on a theoretical analysis as well as on extensive comparative experimental evaluation of various designs.

Section 8.4 summarizes the main characteristics of the adaptive observers presented in this chapter. Concerning the use of adaptive state observers and identifiers for building adaptive state-variable controllers, see Refs. 3, 10, 11, and 12.

8.2 THE LINEAR ASYMPTOTIC STATE OBSERVER

Consider a single-input/single-output, discrete, time-invariant linear system (P) described by

$$\underline{x}'_{k+1} = A\underline{x}'_k + \underline{b}\rho(k) \tag{8.2-1}$$

$$\theta_p(k) = \underline{c}^T\underline{x}'_k \tag{8.2-2}$$

where $\underline{x}'_k$ is the state vector (n-dimensional), $\rho(k)$ is the input sequence (one-dimensional), and $\theta_p(k)$ is the scalar output. A is assumed to be an $(n \times n)$-dimensional matrix, and $\underline{b}$ and $\underline{c}$ are n-dimensional vectors.

One assumes that the pair $(\underline{c}^T, A)$ is completely observable, and to simplify the presentation of the various designs we shall also asume that the pair $(A, \underline{b})$ is completely controllable. This implies that Eqs. (8.2-1) and (8.2-2) are a minimal realization of the system (P) under consideration. Since the system of Eqs. (8.2-1) and (8.2-2) is completely observable, it can be put in an observable canonical form using an equivalence transformation $\underline{x}_k = T\underline{x}'_k$ [13].

For the remainder of this chapter we shall focus our attention on the observation of the states of an observable, canonical realization of the system (P) under consideration. This canonical observable form is [5, 13]

$$\underline{x}_{k+1} = \begin{bmatrix} a_1 & & I & \\ \vdots & & & \\ a_n & 0 & \cdots & 0 \end{bmatrix} \underline{x}_k + \begin{bmatrix} b_1 \\ \vdots \\ b_n \end{bmatrix} \rho(k) \qquad (8.2\text{-}3)$$

$$\theta_p(k) = [1, 0, \ldots, 0]\underline{x}_k = x_1(k) \qquad (8.2\text{-}4)$$

where I is the identity matrix of dimension $(n - 1) \times (n - 1)$ and $x_1(k)$ is the first component of the state vector $\underline{x}_k$. Recall that the parameters a_i and b_i allow us to write the discrete transfer function of the system (P) under consideration directly. The principle of the asymptotic linear state observer [13-15] is to construct an observation model having the same structure as the observable canonical state-space realization of the system whose states must be observed and, in addition to the input $\rho(k)$, to feed in the difference (denoted ε_k below) between the output of the process and the output of the observation model. Therefore in this case the structure of the asymptotic linear state observer will be given by the following equations:

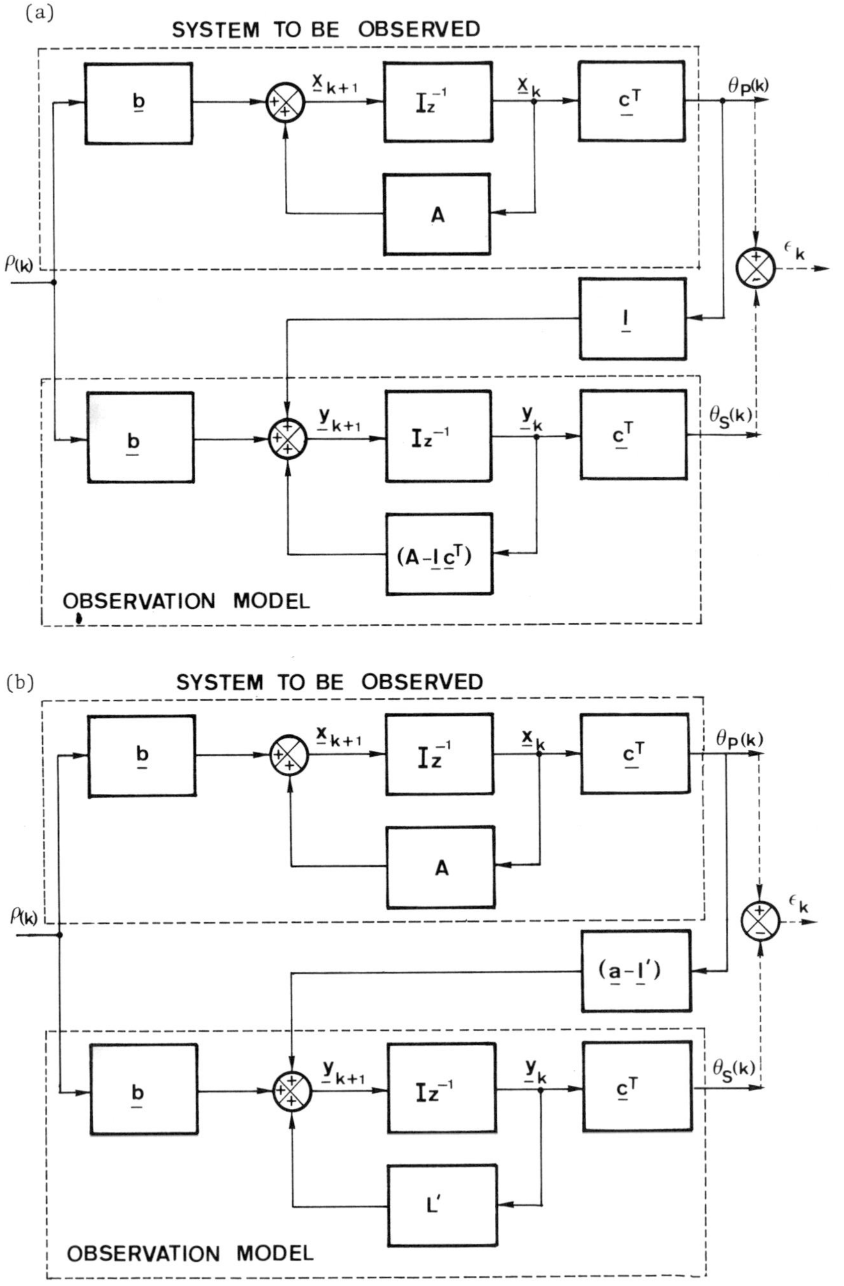

Fig. 8.2-1 Two alternative realizations of an asymptotic linear state observer. (a) Structure I. (b) Structure II.

$$\underline{y}_{k+1} = \begin{bmatrix} a_1 & & \\ \vdots & & I \\ a_n & 0 & \cdots & 0 \end{bmatrix} \underline{y}_k + \begin{bmatrix} b_1 \\ \vdots \\ b_n \end{bmatrix} \rho(k) + \begin{bmatrix} \ell_1 \\ \vdots \\ \ell_n \end{bmatrix} \varepsilon_k \qquad (8.2\text{-}5)$$

$$\theta_S(k) = [1, \ldots, 0]\underline{y}_k = y_1(k) \qquad (8.2\text{-}6)$$

and

$$\varepsilon_k = \theta_P - \theta_S(k) \qquad (8.2\text{-}7)$$

In Eqs. (8.2-5) and (8.2-6), $\underline{y}_k$ is the observed state (the state vector of the observation model), $y_1(k)$ is its first component, $\theta_S(k)$ is the output of the observation model, and $\ell_1, \ldots, \ell_n$ is a set of gains which will determine the characteristics of the observer (for a block diagram, see Fig. 1.3-5).

By taking into account Eqs. (8.2-6) and (8.2-7), Eq. (8.2-5) can be rewritten as

$$\underline{y}_{k+1} = \begin{bmatrix} a_1 - \ell_1 & & \\ \vdots & & I \\ a_n - \ell_n & 0 & \cdots & 0 \end{bmatrix} \underline{y}_k + \begin{bmatrix} b_1 \\ \vdots \\ b_n \end{bmatrix} \rho(k) + \begin{bmatrix} \ell_1 \\ \vdots \\ \ell_n \end{bmatrix} \theta_P(k) \qquad (8.2\text{-}8)$$

which, combined with the output equation

$$\theta_S(k) = [1, 0, \ldots, 0]\underline{y}_k = \underline{c}^T \underline{y}_k \qquad (8.2\text{-}9)$$

yields the block diagram shown in Fig. 8.2-1(a). This arrangement will be called *structure I*. Subtracting Eq. (8.2-8) from Eq. (8.2-3) and defining the generalized state error vector as

$$\underline{e}_k \triangleq \underline{x}_k - \underline{y}_k \qquad (8.2\text{-}10)$$

one obtains

$$\underline{e}_{k+1} = \begin{bmatrix} a_1 - \ell_1 & & \\ \vdots & & I \\ a_n - \ell_n & 0 & \cdots & 0 \end{bmatrix} \underline{e}_k = L\underline{e}_k \qquad (8.2\text{-}11)$$

One notes that the evolution of the generalized error vector, which characterizes the error between the real state vector and the observed one, is governed by a free vector difference equation, and we shall obtain an asymptotic observer characterized by

$$\lim_{k\to\infty} \underline{e}_k = \underline{0} \tag{8.2-12}$$

if and only if the eigenvalues of the matrix L are inside the unit circle. Since the eigenvalues of the matrix L will depend on the gains $\ell_1, \ldots, \ell_n$ which are at designer disposal, the eigenvalues of the matrix L can be arbitrarily chosen.

Consider now another set of gains $(a_1 - \ell_1', \ldots, a_n - \ell_n')$ in Eq. (8.2-5) instead of $\ell_1, \ldots, \ell_n$. Equation (8.2-5) can then be rewritten in the form

$$\underline{y}_{k+1} = \begin{bmatrix} \ell_1' & & & \\ \vdots & & I & \\ & & & \\ \ell_n' & 0 & \cdots & 0 \end{bmatrix} \underline{y}_k + \begin{bmatrix} b_1 \\ \vdots \\ b_n \end{bmatrix} \rho(k) + \begin{bmatrix} a_1 - \ell_1 \\ \vdots \\ a_n - \ell_n \end{bmatrix} \theta_p(k) \tag{8.2-13}$$

which, combined with the output equation

$$\theta_S(k) = [1, 0, \ldots, 0]\underline{y}_k = \underline{c}^T \underline{y}_k \tag{8.2-14}$$

yields the observer shown in Fig. 8.2-1(b), which will be called *structure II*.

Subtracting Eq. (8.2-13) from Eq. (8.2-3) and also using Eq. (8.2-10), one obtains

$$\underline{e}_{k+1} = \begin{bmatrix} \ell_1' & & & \\ \vdots & & I & \\ & & & \\ \ell_n' & 0 & \cdots & 0 \end{bmatrix} \underline{e}_k = L'\underline{e}_k \tag{8.2-15}$$

One sees that the generalized state error vector is governed by an equation similar to that of Eq. (8.2-11) and that an asymptotic observer is obtained if and only if the eigenvalues of L' are inside the unit circle. If $\ell_i' = a_i - \ell_i$ for all $i = 1 \ldots n$, then the two observers shown in Figs. 8.2-1(a) and (b) behave identically. However, as will be shown in the forthcoming sections, *because of the different locations of the parameters* a_i *in the equations describing the observation model, two different types of adaptive observers will be obtained when these parameters are adjustable.*

We shall conclude this section by recalling that if the parameters of the system to be observed are not exactly known, the evolution of the generalized state error vector is no longer governed by a free difference equation but by a forced difference equation. The forcing term will depend on the error between the parameters of the observation model and those of the system to be observed. In such a case it is no longer possible, in general, to satisfy Eq. (8.2-12). This is the reason the adaptive state observers must be employed when the parameters of the system to be observed are either unknown or vary during operation.

8.3 DESIGN OF ADAPTIVE STATE-VARIABLE OBSERVERS AND IDENTIFIERS

8.3.1 *The Parallel Adaptive State Observer*

We assume that the system whose states we wish to observe is in the observable canonical form described by

$$\underline{x}_{k+1} = \begin{bmatrix} a_1 & & I & \\ \vdots & & & \\ a_n & 0 & \cdots & 0 \end{bmatrix} \underline{x}_k + \begin{bmatrix} b_1 \\ \vdots \\ b_n \end{bmatrix} \rho(k) \tag{8.3-1}$$

$$\theta_p(k) = [1, 0, \ldots, 0]\underline{x}_k = x_1(k) \tag{8.3-2}$$

where $\underline{x}_k$ is the state vector, $\rho(k)$ is the scalar input sequence, and $\theta_p(k)$ is the scalar output of the system (process). In contrast to the case discussed in Sec. 8.2, the parameters $a_1, \ldots, a_n, b_1, \ldots, b_n$ are assumed to be unknown, and furthermore their values can change during operation.

We would like to build an adaptive state observer derived from a structure I linear asymptotic state observer [Eqs. (8.2-8) and (8.2-9) and Fig. 8.2-1(a)] which will assure that

$$\lim_{k\to\infty} \underline{e}_k = \lim_{k\to\infty} (\underline{x}_k - \underline{y}_k) = \underline{0} \tag{8.3-3}$$

despite the fact that the exact values of the parameters a_i and b_i are not known [in Eq. (8.3-3), $\underline{y}_k$ is the observed state).

To do this, we shall consider that the *system to be observed* represents the *reference model*, and we shall replace the linear time-invariant observation model of Eqs. (8.2-8) and (8.2-9) by an *adjustable observation model (adjustable system)*. The parameters of this adjustable observation model and the adaptation signals will be generated by the adaptation mechanism in order to satisfy Eq. (8.3-3). Since we shall have to build a discrete-time MRAS using only input and output measurements for adaptation (the vector $\underline{e}_k$ is not measurable), we must use the designs discussed in Sec. 5.5. These designs suggest (as in the continuous case) that we choose the following structure for the adjustable observation model:

$$\underline{y}_{k+1} = \begin{bmatrix} \hat{a}_1(k+1) - \ell_1 & & \\ \vdots & & I \\ \hat{a}_n(k+1) - 1_n & 0 & \cdots & 0 \end{bmatrix} \underline{y}_k + \begin{bmatrix} \ell_1 \\ \vdots \\ \ell_n \end{bmatrix} \theta_p(k)$$

$$+ \begin{bmatrix} \hat{b}_1(k+1) \\ \vdots \\ \hat{b}_n(k+1) \end{bmatrix} \rho(k) + \underline{u}_a(k+1) + \underline{u}_b(k+1) \qquad (8.3\text{-}4)$$

and

$$\theta_S(k) = [1, 0, \ldots, 0]\underline{y}_k = y_1(k) \qquad (8.3\text{-}5)$$

where $\underline{y}^T = [y_1(k), \ldots, y_n(k)]$ is the state of the adjustable observation model, $\theta_S(k)$ is the output of the observation model, $\hat{a}_i(k+1)$, $\hat{b}_i(k+1)$ are the adjustable parameters, $\ell_1, \ldots, \ell_n$ is a set of gains which will influence the characteristics of the observer, and $\underline{u}_a(k+1)$ and $\underline{u}_b(k+1)$ are additional transient adaptation signals having the property that

$$\underline{u}_a(k)\Big|_{\varepsilon_k \equiv 0} = 0, \quad \underline{u}_b(k)\Big|_{\varepsilon_k \equiv 0} = 0 \qquad (8.3\text{-}6)$$

and where

$$\varepsilon_k = \theta_p(k) - \theta_S(k) \tag{8.3-7}$$

These transient adaptation signal vectors have the form

$$\underline{u}_a(k+1) = \begin{bmatrix} u_a^{\,1}(k+1) \\ \vdots \\ u_a^{\,n-1}(k+1) \\ 0 \end{bmatrix}, \quad \underline{u}_b(k+1) = \begin{bmatrix} u_b^{\,1}(k+1) \\ \vdots \\ u_b^{\,n-1}(k+1) \\ 0 \end{bmatrix} \tag{8.3-8}$$

and will be determined as part of the design.

Our objective now is to design an adaptation mechanism such that the condition of Eq. (8.3-3) is satisfied and, in addition (under some supplementary conditions), that

$$\lim_{k\to\infty} a_i(k) = a_i, \quad \lim_{k\to\infty} b_i(k) = b_i, \quad i = 1, 2, \ldots, n \tag{8.3-9}$$

If the conditions of Eqs. (8.3-3) and (8.3-9) are both satisfied, one has an *asymptotic adaptive state observer and identifier.*

From this point, two designs can be developed based on cases 1 and 2, respectively, from Sec. 5.5. The design derived from case 1, which uses delay operators, is of interest for the parallel adaptive observer. However, applied to the series-parallel observer (derived from a structure II linear asymptotic observer), the design leads to an observer with very poor results in the presence of measurement noise. For this reason the derivation of an adaptation mechanism using such a design is left as an exercise for the reader, and we shall focus our attention on the use of the design derived from case 2, which uses state-variable filters and allows us to directly use the output generalized error ε_k for adaptation. The resulting algorithms leading to an asymptotic adaptive state observer are given in Table 8.3-1 (given later in this section), and their derivation is briefly shown next.

The Equivalent Feedback System of the Parallel Adaptive Observer

Subtracting Eq. (8.3-4) from Eq. (8.3-1) and also using Eqs. (8.3-2), (8.3-3), and (8.3-5), one obtains

$$\underline{e}_{k+1} = \begin{bmatrix} a_1 - \ell_1 & \vert & I \\ \vdots & \vert & \\ a_n - \ell_n & 0 & \cdots \; 0 \end{bmatrix} \underline{e}_k - \begin{bmatrix} a_{11}(k+1) \\ \vdots \\ a_{n1}(k+1) \end{bmatrix} \theta_S(k) - \begin{bmatrix} b_{11}(k+1) \\ \vdots \\ b_{n1}(k+1) \end{bmatrix} \rho(k) - \underline{u}_a(k+1) - \underline{u}_b(k+1) \qquad (8.3\text{-}10)$$

$$\varepsilon_k = [1, 0, \ldots, 0]\underline{e}_k = \underline{c}^T \underline{e}_k \qquad (8.3\text{-}11)$$

where

$$a_{i1}(k+1) = \hat{a}_i(k+1) - a_i \qquad (8.3\text{-}12)$$

$$b_{i1}(k+1) = \hat{b}_i(k+1) - b_i \qquad (8.3\text{-}13)$$

and

$$\varepsilon_k = \theta_P(k) - \theta_S(k) \qquad (8.3\text{-}14)$$

Equations (8.3-10) and (8.3-11) lead to the following difference equation characterizing ε_k (where z^{-1} has the signification of the unit delay operator):

$$\left[1 - \sum_{i=1}^{n} (a_i - l_i) z^{-i}\right] \varepsilon_k = -\left[\sum_{i=1}^{n} a_{i1}(k - i + 1)\theta_S(k - i) + \sum_{i=1}^{n} b_{i1}(k - i + 1)\rho(k - i) + \sum_{i=0}^{n-2} u_a^{\,i+1}(k - i) + \sum_{i=0}^{n-2} u_b^{\,i+1}(k - 1)\right] \qquad (8.3\text{-}15)$$

On the other hand, consider the difference equation

$$\left[1 - \sum_{i=1}^{n} (a_i - l_i)z^{-i}\right]\varepsilon_k = -\left(1 + \sum_{i=1}^{n-1} d_i z^{-i}\right)\left[\sum_{i=1}^{n} a_{i1}(k)w_i(k) + \sum_{i=1}^{n} b_{i1}(k)r_i(k)\right] \tag{8.3-16}$$

We assume that the above equation corresponds to an asymptotically hyperstable feedback system, and therefore we shall have

$$\lim_{k\to\infty} \varepsilon_k = 0 \tag{8.3-17}$$

Now if ε_k given by Eq. (8.3-15) is zero-state equivalent to ε_k given by Eq. (8.3-16), then ε_k given by Eq. (8.3-15) will also satisfy Eq. (8.3-17). If the zero-state equivalence between Eqs. (8.3-15) and (8.3-16) can be assured by a convenient choice of $\underline{u}_a(k + 1)$, $\underline{u}_b(k + 1)$, $w_i(k)$, and $r_i(k)$, then Eq. (8.3-16) can be used to derive the adaptation algorithm for $a_{i1}(k)$ and $b_{i1}(k)$ and to specify the stability conditions.

However, before going further we must show that zero-state equivalence between Eqs. (8.3-15) and (8.3-16) will imply not only Eq. (8.3-17) but also Eq. (8.3-3). Zero-state equivalence between Eqs. (8.3-15) and (8.3-16) implies that ε_k given by Eqs. (8.3-10) and (8.3-11) satisfies Eq. (8.3-17). Now since the system of Eqs. (8.3-10) and (8.3-11) is completely observable, Eq. (8.3-17) will also imply that Eq. (8.3-3) is satisfied if asymptotic identification of the parameters is achieved, i.e.,

$$\begin{aligned} \lim_{k\to\infty} a_{i1}(k + 1) &= \lim_{k\to\infty}[a_i(k + 1) - a_i] = 0, \quad i = 1, \ldots, n \\ \lim_{k\to\infty} b_{i1}(k + 1) &= \lim_{k\to\infty}[b_i(k + 1) - b_i] = 0, \quad i = 1, \ldots, n \end{aligned} \tag{8.3-18}$$

because $\lim_{k\to\infty} \underline{u}_a(k + 1) = 0$ and $\lim_{k\to\infty} \underline{u}_b(k + 1) = 0$ if Eq. (8.3-17) holds.

Using arguments similar to those used in Chap. 7 when designing identifiers, one concludes (under the hypothesis that the process to be observed and identified is completely observable and completely controllable) that if the input sequence $\rho(k)$ is a sum of at least n + 1 periodic signals of distinct frequencies (since there are a maximum number of 2n parameters to be identified), then Eq. (8.3-17) will imply the verification of Eq. (8.3-18). Therefore under the above conditions Eq. (8.3-17) will also imply that $\lim_{k\to\infty} \underline{e}_k = 0$.

Under the hypothesis that Eqs. (8.3-15) and (8.3-16) are zero-state equivalent, the design of an asymptotically adaptive state observer is transformed into the design of an asymptotically hyperstable feedback system, which leads to Eq. (8.3-16).

The remaining design contains several steps:

Determination of the adaptation signals $\underline{u}_a(k + 1)$ and $\underline{u}_b(k + 1)$ and of the auxiliary variables $w_i(k)$ and $r_i(k)$ such that Eqs. (8.3-15) and (8.3-16) are zero-state equivalent

Determination of the adaptation algorithm for $\hat{a}_i(k)$ and $\hat{b}_i(k)$ and of the conditions assuring that Eq. (8.3-16) corresponds to an asymptotically hyperstable feedback system

The use of the a priori value of ε_k computed with the parameters $\hat{a}_i(k - 1)$, $\hat{b}_i(k - 1)$ for expressing the adaptation algorithm

We shall only develop the determination of the transient adaptation signals and of the auxiliary variables; the other steps are direct extensions of the designs presented in Chap. 5.

Determination of $\underline{u}_a(k + 1)$, $\underline{u}_b(k + 1)$, $w_i(k)$, and $r_i(k)$

Equating the two right-hand sides of Eqs. (8.3-15) and (8.3-16) and separating the terms which contain $a_{i1}(k)$ from those which contain $b_{i1}(k)$, one obtains (after changing the summation indices)

$$\sum_{\ell=0}^{n-2} \underline{u}_a^{\ell+1}(k - \ell) = \left(\sum_{i=0}^{n-1} d_i z^{-i}\right)\left[\sum_{i=1}^{n} a_{i1}(k) w_i(k)\right] - \sum_{i=1}^{n} a_{i1}(k - i + 1)\theta_S(k - i) \qquad (d_0 = 1) \tag{8.3-19}$$

and

$$\sum_{\ell=0}^{n-2} \underline{u}_b^{\ell+1}(k - \ell) = \left(\sum_{i=0}^{n-1} d_i z^{-i}\right)\left[\sum_{i=1}^{n} b_{i1}(k) r_i(k)\right] - \sum_{i=1}^{n} b_{i1}(k - i + 1)\rho(k - i) \qquad (d_0 = 1) \tag{8.3-20}$$

We shall now define the auxiliary variables $w_i(k)$ and $r_i(k)$ as

$$\sum_{j=0}^{n-1} d_j z^{-j} w_i(k) = \theta_S(k - i) \qquad (d_0 = 1) \tag{8.3-21}$$

and

$$\sum_{j=0}^{n-1} d_j z^{-j} r_i(k) = \rho(k - i) \qquad (d_0 = 1) \tag{8.3-22}$$

From Eqs. (8.3-21) and (8.3-22) one observes also that

$$w_i(k - 1) = w_{i+1}(k) \tag{8.3-23}$$

and

$$r_i(k - 1) = r_{i+1}(k) \tag{8.3-24}$$

The auxiliary variables $w_i(k)$ and $r_i(k)$ can be obtained from two discrete state-variable filters having the discrete transfer function

$$h_F(z) = \frac{z^{-1}}{1 + \Sigma_{i=1}^{n-1} d_i z^{-i}} \tag{8.3-25}$$

with inputs $\theta_S(k)$ and $\rho(k)$, respectively.

Using Eq. (8.3-21), one can rewrite Eq. (8.3-19) as

$$\begin{aligned}\sum_{\ell=0}^{n-2} u_a^{\ell+1}(k - \ell) &= \sum_{i=1}^{n} \sum_{j=0}^{n-1} \{d_j z^{-j} [a_{i1}(k) w_i(k)] \\ &\quad - a_{i1}(k - i + 1) d_j z^{-j} w_i(k)\} \\ &= \sum_{i=1}^{n} \sum_{j=0}^{n-1} [a_{i1}(k - j) \\ &\quad - a_{i1}(k - i + 1)] d_j w_i(k - j)\end{aligned} \tag{8.3-26}$$

By decomposing the summation over the index i, Eq. (8.3-26) can also be written as

$$\begin{aligned}\sum_{\ell=0}^{n-2} u_a^{\ell+1}(k - \ell) &= - \sum_{j=0}^{n-1} [a_{11}(k - 1 + 1) - a_{11}(k - j)] d_j w_1(k - j) \\ &\quad + \sum_{i=2}^{n-1} \left\{ \sum_{j=0}^{i-2} [a_{i1}(k - j) - a_{i1}(k - i + 1)] d_j w_i(k - j) \right. \\ &\quad \left. - \sum_{j=1}^{n-1} [a_{i1}(k - i + 1) - a_{i1}(k - j)] d_j w_i(k - j) \right\} \\ &\quad + \sum_{j=0}^{n-1} [a_{n1}(k - j) - a_{n1}(k - n + 1)] d_j w_n(k - j)\end{aligned} \tag{8.3-27}$$

since for $j = i - 1$ the term of the sum on the right-hand side of Eq. (8.3-26) is null.

Introducing the notation

$$\Delta a_{i1}(k + 1) \triangleq a_{i1}(k + 1) - a_{i1}(k) \tag{8.3-28}$$

one has

$$a_{i1}(k + p) - a_{i1}(k + q) = \sum_{\ell=0}^{p-q-1} \Delta a_{i1}(k + q + \ell + 1) \tag{8.3-29}$$

Then Eq. (8.3-27) can be rewritten as

$$\begin{aligned} \sum_{\ell=0}^{n-2} u_a^{\ell+1}(k - \ell) = & \sum_{i=2}^{n} \sum_{j=0}^{i-2} \sum_{\ell=0}^{-j+i-2} \Delta a_{i1}(k - i + \ell + 2) d_j w_i(k - j) \\ & - \sum_{i=1}^{n-1} \sum_{j=i}^{n-1} \sum_{\ell=0}^{j-i} \Delta a_{i1}(k - j + \ell + 1) d_j w_i(k - j) \end{aligned} \tag{8.3-30}$$

By changing the order of summation on the right-hand side of Eq. (8.3-30) several times in order to make it similar to the form of summation appearing on the left-hand side, one finally gets

$$\begin{aligned} \sum_{l=0}^{n-2} u_a^{l+1}(k - l) = & \sum_{i=2}^{n} \sum_{j=0}^{i-2} \sum_{\ell=j}^{i-2} \Delta a_{i1}(k - \ell) d_j w_i(k - j) \\ & - \sum_{i=1}^{n-1} \sum_{j=1}^{n-1} \sum_{\ell=i-1}^{j-1} \Delta a_{i1}(k - \ell) d_j w_i(k - j) \\ = & \sum_{i=2}^{n} \sum_{\ell=0}^{i-2} \sum_{j=0}^{\ell} \Delta a_{i1}(k - \ell) d_j w_i(k - j) \\ & - \sum_{i=1}^{n-1} \sum_{\ell=i-1}^{n-2} \sum_{j=\ell+1}^{n-1} \Delta a_{i1}(k - \ell) d_j w_i(k - j) \\ = & \sum_{\ell=0}^{n-2} \sum_{i=\ell+2}^{n} \sum_{j=0}^{\ell} \Delta a_{i1}(k - \ell) d_j w_i(k - j) \\ & - \sum_{\ell=0}^{n-2} \sum_{i=1}^{\ell+1} \sum_{j=\ell+1}^{n-1} \Delta a_{i1}(k - \ell) d_j w_i(k - j) \end{aligned} \tag{8.3-31}$$

Also, taking into account that

$$w_i(k - j + \ell) = w_{i+j-\ell}(k) \tag{8.3-32}$$

one obtains

$$u_a^{\ell+1}(k) = \sum_{i=\ell+2}^{n} \sum_{j=0}^{\ell} \Delta a_{i1}(k) d_j w_{i+j-\ell}(k) - \sum_{i=1}^{\ell+1} \sum_{j=\ell+1}^{n-1} \Delta a_{i1}(k) d_j w_{i+j-\ell}(k), \quad \ell = 0, 1, \ldots, n-2 \tag{8.3-33}$$

Similarly, for $u_b^{\ell+1}(k)$ one obtains the expression

$$u_b^{\ell+1}(k) = \sum_{i=\ell+2}^{n} \sum_{j=0}^{\ell} \Delta b_{i1}(k) d_j r_{i+j-\ell}(k), \quad \ell = 0, 1, \ldots, - \sum_{i=1}^{\ell+1} \sum_{j=\ell+1}^{n-1} \Delta b_{i1}(k) d_j r_{i+j-\ell}(k), \quad \ell = 0, 1, \ldots, n-2 \tag{8.3-34}$$

Once $w_i(k)$, $r_i(k)$, $\underline{u}_a(k)$, and $\underline{u}_b(k)$ have been determined, one proceeds to the next step, the determination of the parametric adaptation laws.

A block diagram of the parallel adaptive state observer using state variable filters is shown in Fig. 8.3-1. The structure of the observer and the adaptation algorithms are summarized in Table 8.3-1.

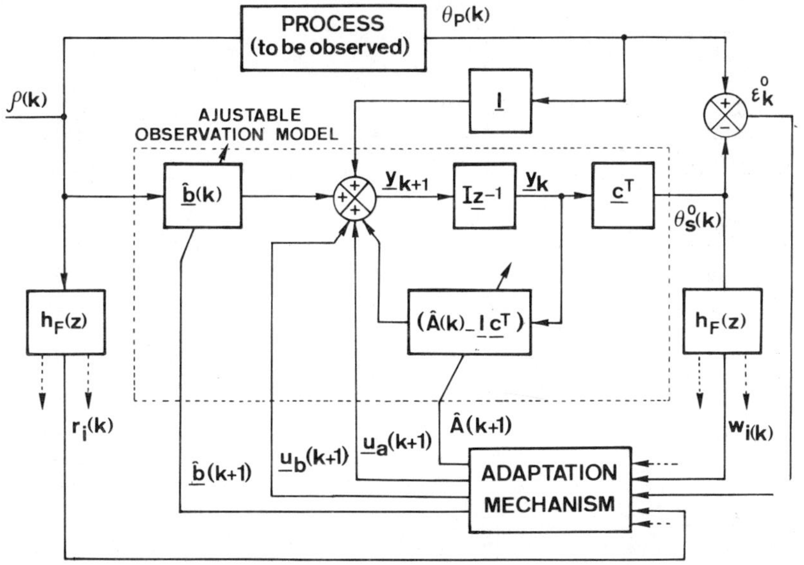

Fig. 8.3-1 Block diagram of a parallel adaptive state observer.

Table 8.3-1 Parallel Adaptive State Observer (with State-Variable Filters)

Process to be observed	$\underline{x}_{k+1} = \begin{bmatrix} a_1 & & I \\ \vdots & & \\ a_n & 0 & \cdots & 0 \end{bmatrix} \underline{x}_k + \begin{bmatrix} b_1 \\ \vdots \\ b_n \end{bmatrix} \rho(k), \quad \theta_P(k) = [1, 0, \ldots, 0]\underline{x}_k = \underline{c}^T \underline{x}_k = x_1(k)$
	$\underline{x}_k^T = [x_1(k), \ldots, x_n(k)], \quad \theta_P(k) = \text{process output}, \; \rho(k) = \text{process input}$
Adjustable observation model	$\underline{y}_{k+1} = \begin{bmatrix} \hat{a}_1(k+1) - \ell_1 & & I \\ \vdots & & \\ \hat{a}_n(k+1) - \ell_n & 0 & \cdots & 0 \end{bmatrix} \underline{y}_k + \begin{bmatrix} \ell_1 \\ \vdots \\ \ell_n \end{bmatrix} \theta_P(k) + \begin{bmatrix} \hat{b}_1(k+1) \\ \vdots \\ \hat{b}_n(k+1) \end{bmatrix} \rho(k) + \begin{bmatrix} u_a^1(k+1) \\ \vdots \\ u_a^{n-1}(k+1) \\ 0 \end{bmatrix} + \begin{bmatrix} u_b^1(k+1) \\ \vdots \\ u_b^{n-1}(k+1) \\ 0 \end{bmatrix}$
	$\theta_S(k) = \underline{c}^T \underline{y}_k = [1, 0, \ldots, 0]\underline{y}_k = y_1(k)$
	$\underline{y}_k^T = [y_1(k), \ldots, y_n(k)]$
	$\theta_S^0(k) = [\hat{a}_1(k-1) - \ell_1]\theta_S(k-1) + y_2(k-1) + \ell_1\theta_P(k-1) + \hat{b}_1(k-1)\rho(k-1)$
Output error	A posteriori: $\varepsilon_k = \theta_P(k) - \theta_S(k)$
	A priori: $\varepsilon_k^0 = \theta_P(k) - \theta_S^0(k)$
State-variable filter	$h_F(z) = z^{-1}/(1 + \Sigma_{i=1}^{n-1} d_i z^{-i})$

Auxiliary variables	$$\left(1 + \sum_{i=1}^{n-1} d_i z^{-i}\right) w_i(k) = \theta_S(k - i), \quad i = 1, \ldots, n$$
	$$\left(1 + \sum_{i=1}^{n-1} d_i z^{-i}\right) r_i(k) = \rho(k - i), \quad i = 1, \ldots, n$$
Adaptation algorithm (integral + proportional)	$$\underline{\hat{p}}^T(k) = [\hat{a}_1(k), \ldots, \hat{a}_n(k), \hat{b}_1(k), \ldots, \hat{b}_n(k)]$$
	$$\underline{\tilde{y}}_{k-1}^T = [w_1(k), \ldots, w_n(k), r_1(k), \ldots, r_n(k)]$$

$$\varepsilon_k = \frac{\varepsilon_k^0 + \underline{\tilde{y}}_{k-1}^T G'_{k-2} \underline{\tilde{y}}_{k-2} \varepsilon_{k-1}}{1 + \underline{\tilde{y}}_{k-1}^T (G_{k-1} + G'_{k-1}) \underline{\tilde{y}}_{k-1}}$$

$$\underline{\hat{p}}(k) = \underline{\hat{p}}^I(k) + \underline{\hat{p}}^P(k)$$

$$\underline{\hat{p}}^I(k) = \underline{\hat{p}}^I(k - 1) + G'_{k-1} \underline{\tilde{y}}_{k-1} \varepsilon_k$$

$$\underline{\hat{p}}^P(k) = G'_{k-1} \underline{\tilde{y}}_{k-1} \varepsilon_k$$

$$u_a^{\ell+1}(k) = \sum_{i=\ell+2}^{n} \sum_{j=0}^{\ell} [\hat{a}_i(k) - \hat{a}_i(k - 1)] d_j w_{i+j-\ell}(k) - \sum_{i=1}^{\ell+1} \cdot \sum_{j=\ell+1}^{n-1} [\hat{a}_i(k) - \hat{a}_i(k - 1)] d_j w_{i+j-\ell}(k)$$

$$u_b^{\ell+1}(k) = \sum_{i=\ell+2}^{n} \sum_{j=0}^{\ell} [\hat{b}_i(k) - \hat{b}_i(k - 1)] d_j r_{i+j-\ell}(k) - \sum_{i=1}^{\ell+1} \sum_{j=\ell+1}^{n-1} [\hat{b}_i(k) - \hat{b}_i(k - 1)] d_j w_{i+j-\ell}(k),$$

$$\ell = 0, 1, \ldots, n - 2$$

Table 8.3-1 (Continued)

Constant adaptation gain	$G_k = G > 0,\ G'_k = \alpha G,\quad -0.5 \le \alpha < \infty$
Decreasing adaptation gain	$G_k = G_{k-1} - \dfrac{1}{\lambda} \dfrac{G_{k-1}\underline{\tilde{y}}_{k-1}\underline{\tilde{y}}_{k-1}^T G_{k-1}}{1 + (1/\lambda)\underline{\tilde{y}}_{k-1}^T G_{k-1}\underline{\tilde{y}}_{k-1}},\quad \lambda > 0.5,\ G_0 > 0,\ G'_k = \alpha G_k,\ -0.5 \le \alpha < \infty$
Stability condition	(a) Constant adaptation gain: $h(z) = \dfrac{1 + \Sigma_{i=1}^{n-1} d_i z^{-i}}{1 - \Sigma_{i=1}^{n}(a_i - \ell_i) z^{-i}}$ = strictly positive real discrete transfer function.
	(b) Decreasing adaptation gain: $h'(z) = h(z) - \dfrac{1}{2\lambda}$ = strictly positive real discrete transfer function.

For adaptive observers using delay operators the parametric adaptation algorithm (for integral adaptation) is that given in Table 7.3-1 and the adaptation signals are given by

$$u_a^{\ell+1} = \sum_{i=\ell+2}^{n} [\hat{a}_i(k) - \hat{a}_i(k-1)]\theta_S(k - i + \ell)$$

and

$$u_b^{\ell+1} = \sum_{i=\ell+2}^{n} [\hat{b}_i(k) - \hat{b}_i(k-1)]\rho(k - i + \ell)$$

Discussion of the Results

1. To obtain an asymptotic adaptive state observer, a sufficient condition which must be satisfied is that

$$h(z) = \frac{1 + \Sigma_{i=1}^{n-1} d_i z^{-i}}{1 - \Sigma_{i=1}^{n} (a_i - \ell_i) z^{-i}} \quad \text{or} \quad h'(z) = h(z) - \frac{1}{2\lambda}$$

   be a strictly positive real transfer function (depending on the use of either constant adaptation gains or decreasing adaptation gains, respectively). In fact this condition implies first (see Appendix B) that the coefficients ℓ_i are chosen such that the polynomial $1 - \Sigma_{i=1}^{n} (a_i - \ell_i)z^{-i}$ has all its roots inside the unit circle. This is exactly the condition appearing in the case of the linear asymptotic observer. Therefore, some a priori information about the values of the parameters a_i is necessary to enable us to choose ℓ_i satisfying the above-mentioned condition. Then one can proceed to compute the coefficients d_i using the methods indicated in Sec. 5.3-3, in particular Eqs. (5.3-125), (5.3-126), and (5.3-128), where the a_i, i = 1, 2, ..., n, are replaced by $\hat{a}_i(0) - \ell_i$, i = 1, 2, ..., n, $\hat{a}_i(0)$ being the a priori estimation of a_i.
2. Note that the parallel adaptive state observer can satisfy the condition of Eq. (8.3-3) even if $\ell_i = 0$, i = 1, 2, ..., n, because the error between the output of the system to be observed and the output of the observation model is fed through the adaptation mechanism such that $\lim_{k\to\infty} \underline{e}_k = \underline{0}$ for any initial parameter or state error.
3. Two state-variable filters are necessary for the design given in Table 8.3-1 in order to generate the auxiliary variables $w_i(k)$ and $r_i(k)$ required for adaptation.

Table 8.3-2 Series-Parallel Adaptive State Observer (with State-Variable Filters)

Process to be observed	$\underline{x}_{k+1} = \begin{bmatrix} a_1 & & \\ \vdots & & I \\ a_n & 0 & \cdots & 0 \end{bmatrix} \underline{x}_k + \begin{bmatrix} b_1 \\ \vdots \\ b_n \end{bmatrix} \rho(k), \quad \theta_P(k) = [1, 0, \ldots, 0]\underline{x}_k = \underline{c}^T \underline{x}_k = x_1(k)$ $\underline{x}_k^T = [x_1(k), \ldots, x_n(k)], \quad \theta_P(k) = \text{process output}, \ \rho(k) = \text{process input}$
Adjustable observation model	$\underline{y}_{k+1} = \begin{bmatrix} \ell_1' & & \\ \vdots & & I \\ \ell_n' & 0 & \cdots & 0 \end{bmatrix} \underline{y}_k + \begin{bmatrix} \hat{a}_1(k+1) - \ell_1' \\ \vdots \\ \hat{a}_n(k+1) - \ell_n' \end{bmatrix} \theta_P(k) + \begin{bmatrix} \hat{b}_1(k+1) \\ \vdots \\ \hat{b}_n(k+1) \end{bmatrix} \rho(k) + \begin{bmatrix} u_a^1(k+1) \\ \vdots \\ u_a^{n-1}(k+1) \\ 0 \end{bmatrix} + \begin{bmatrix} u_b^1(k+1) \\ \vdots \\ u_b^{n-1}(k+1) \\ 0 \end{bmatrix}$ $\theta_S(k) = \underline{c}^T \underline{y}_k = [1, 0, \ldots, 0]\underline{y}_k = y_1(k)$ $\underline{y}_k^T = [y_1(k), \ldots, y_n(k)]$ $\theta_S^0(k) = [\hat{a}_1(k-1) - \ell_1']\theta_P(k-1) + y_2(k-1) + \ell_1'\theta_S(k-1) + \hat{b}_1(k-1)\rho(k-1)$
Generalized output error	A posteriori: $\varepsilon_k = \theta_P(k) - \theta_S(k)$ A priori: $\varepsilon_k^0 = \theta_P(k) - \theta_S^0(k)$
State-variable filter	$h_F(z) = z^{-1}/(1 + \Sigma_{i=1}^{n-1} d_i z^{-i})$

Auxiliary variables	$\left(1 + \sum_{i=1}^{n-1} d_i z^{-i}\right) w_i(k) = \theta_p(k - i), \quad i = 1, \ldots, n$ $\left(1 + \sum_{i=1}^{n-1} d_i z^{-i}\right) r_i(k) = \rho(k - i), \quad i = 1, \ldots, n$
Adaptation algorithm (integral + proportional)	Same as in Table 8.3-1
Constant adaptation gain	Same as in Table 8.3-1
Decreasing adaptation gain	Same as in Table 8.3-1
Stability condition	(a) Constant adaptation gains: $h(z) = \dfrac{1 + \sum_{i=1}^{n-1} d_i z^{-i}}{1 - \sum_{i=1}^{n} \ell_i' z^{-i}}$ = strictly positive real discrete transfer function. (b) Decreasing adaptation gains: $h'(z) = h(z) - \dfrac{1}{2\lambda}$ = strictly positive real discrete transfer function.

4. If the adaptation algorithm with decreasing adaptation gain is used, one can obtain unbiased parameter estimates in the presence of measurement noise. [The parallel adaptive state observer is similar, to a certain extent, to the parallel identifier analyzed in Sec. 7.3 except that for the design given in Table 8.3-1 one uses instead of $\rho(k - i)$ and $\theta_S(k - i)$ their filtered values $w_i(k)$ and $r_i(k)$.] The adaptive state observer will converge to the linear asymptotic state observer.
5. The state-variable filters used in the design given in Table 8.3-1 will have a beneficial effect in the presence of noise (especially if the frequency spectrum of the noise is located mostly outside the band pass of the state-variable filters), but in the absence of noise, this can slightly reduce the speed of convergence of the generalized state error at the beginning of the adaptation process in the case of an important initial generalized state error.
6. The use of the results of Theorems 5.3-3 and 5.3-4 allows one to design parallel asymptotic adaptive state observers which do not require any prior information about the parameters a_i [12].

8.3.2 The Series-Parallel Adaptive State Observer

We assume that the system whose states we wish to observe is again described by Eqs. (8.3-1) and (8.3-2) and that the parameters $a_1, \ldots, a_n, b_1, \ldots, b_n$ are unknown. As in Sec. 8.3-1 we shall set up an adjustable observation model; however, instead of using structure I as a basic for a linear asymptotic observer, we shall use structure II, given by Eqs. (8.2-13) and (8.2-14).

The corresponding adaptation algorithms are summarized in Table 8.3-2, and their derivation follows closely the methodology used in Sec. 8.3-1. A block diagram of the series-parallel adaptive observer using state-variable filters is shown in Fig. 8.3-2. Note that, *in this case, the auxiliary variables* $w_i(k)$ *are obtained from* $\theta_P(k)$ *and not from* $\theta_S(k)$ *as for the parallel adaptive state observer.*

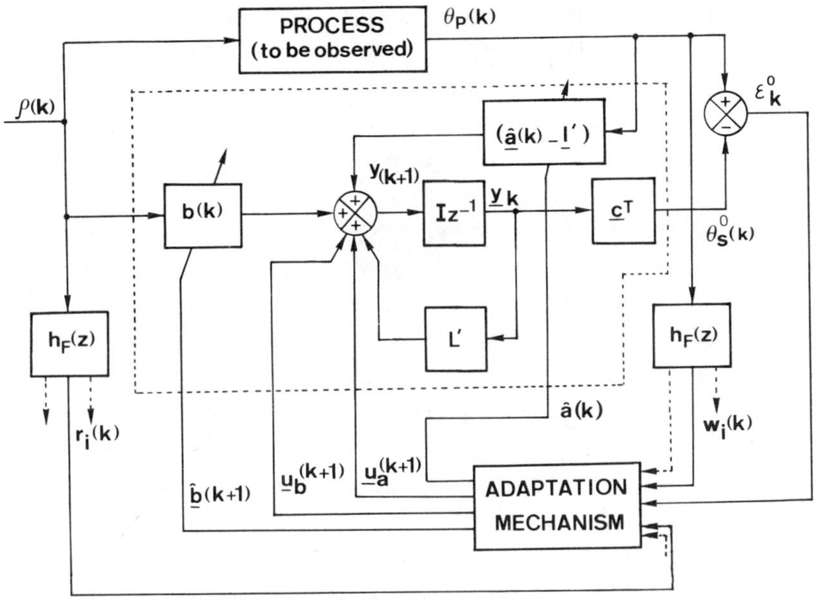

Fig. 8.3-2 Block diagram of a series-parallel adaptive state observer.

Discussion of the Results

1. To obtain an asymptotic adaptive state observer a sufficient condition which must be satisfied is that

$$h'(z) = \frac{1 + \Sigma_{i=1}^{n-1} d_i z^{-i}}{1 - \Sigma_{i=1}^{n} \ell_i' z^{-i}} - \frac{1}{2\lambda}$$

be a strictly positive real discrete transfer function (with $\lambda = \infty$ in the case of constant adaptation gains and with $0.5 < \lambda < \infty$ in the case of decreasing adaptation gains). To satisfy this condition no prior information about the parameters a_i of the system to be observed is necessary. The strictly real positivity condition can be satisfied either by first choosing the coefficients ℓ_i' which define the dynamics of the linear observer or by first choosing the coefficients d_i which define the dynamics of the state-variable filters and then computing the coefficients ℓ_i'.

2. The auxiliary variable $w_i(k)$ is obtained from $\theta_P(k)$ through a low-band-pass filter (the state-variable filter) which in some cases can have a filtering effect upon the measurement noise affecting $\theta_P(k)$.
3. In the presence of noise, $w_i(k)$ and $\varepsilon(k)$ will in general be highly correlated, leading in all the cases to a *biased* identification. The bias can be reduced if the noise spectrum is essentially outside the band pass of the state-variable filter used to generate $w_i(k)$.
4. The adaptive state observer can work with $\ell_i' = 0$, $i = 1, \ldots, n$. In this case the coefficients d_i are no longer necessary, but performance in the presence of measurement noise will be very poor.
5. In the absence of measurement noise, with a decreasing adaptation gain, the series-parallel adaptive state observer tends toward the linear asymptotic observer. However, in the presence of measurement noise, the identified parameters will be biased, and this property does not hold.
6. To improve the performance of the series-parallel observer in the presence of noise, one can derive approximate algorithms, but $\lim_{k\to\infty} \underline{e}_k = \underline{0}$ can no longer be guaranteed for all initial parameter and state errors.
7. A first approximate algorithm allowing us to improve performance in the presence of measurement noise consists in just replacing $\theta_P(k)$ by $\theta_S(k)$ for generating the auxiliary variable $w_i(k)$.
8. A second approximate algorithm allowing us to improve the performance in the presence of measurement noise is the use of an instrumental variable technique (like that used in recursive identification; see Sec. 7.7).

8.4 CONCLUDING REMARKS

We wish to emphasize the following basic ideas from this chapter:

1. There exist two equivalent linear asymptotic observer structures, each of them leading to a different adaptive state observer. One has the structure of a parallel MRAS, and the other one has the structure of a series-parallel MRAS.
2. The adaptive state observers also provide simultaneously an identification of the parameters of the process to be observed.
3. For each structure of the adaptive state observer one can distinguish two types of design (for the discrete-time case): one using state-variable filters and one using delay operators. (In the continuous-time case one also has two types of design, both using state-variable filters, but one will directly use the output generalized error for adaptation, while the other will use a linear combination of the filtered output generalized error and its derivatives.)
4. Adaptive state observers can be used either with constant adaptation gains or with decreasing adaptation gains. The first type can be used for observing the state for processes where frequent parameter changes occur or for processes with unknown but constant parameters. The second type can be used when the parameters of the process are constant but unknown. To use it when the parameters change often, it is necessary to reinitialize the adaptation gains frequently (or to use other algorithms with time varying adaptation gains [12, 16]).
5. *In the presence of measurement noise the parallel adaptive state observer with state-variable filters and decreasing adaptation gain provides the best results*. Even for a high level of measurement noise it converges to the linear asymptotic observer.

6. For a reduced level of measurement noise, a series-parallel adaptive state observer with state-variable filters and decreasing adaptation gain can be used satisfactorily as can the parallel adaptive state observer with state-variable filters and constant adaptation gain inferior or equal to ≤ 0.1 (i.e., $G \leq 0.1I$).
7. The design of parallel adaptive state observers requires either some information about the dynamics of the process to be observed or the use of a more complex adaptation algorithm. (This problem is similar to that appearing in the case of identifiers; see Sec. 7.3.)
8. The performance of a series-parallel adaptive state observer in the presence of measurement noise can eventually be improved by using approximate algorithms derived from identification techniques.
9. The convergence of the parallel adaptive state observer can in general be improved by taking $\theta_S(k_0 - i) = \theta_P(k_0 - i)$, where k_0 corresponds to the beginning of the adaptation (this is always true in the absence of measurement noise).
10. Adaptive state observers can also work without the linear gains ℓ_i acting on the output generalized error since the output generalized error is fed through the adaptation mechanism to the observation model.
11. The designs presented can be extended for adaptive state observation of multiinput/multioutput systems.
12. Adaptive state observers can be designed to observe either the states of an observable canonical form or the states of a controllable canonical form [12].
13. Adaptive state observers can be used in adaptive control schemes for poles and zeros placement or for poles placement only. The corresponding controller which acts simultaneously upon the plant and the adaptive observer is such that it forms with the adaptive observer an *implicit* reference model whose state is the state of the observer and whose output is the output of the observer [12, 16, 17].

PROBLEMS

8.1 Consider a continuous-time linear system described by

$$\begin{bmatrix} \dot{x}_1 \\ \dot{x}_2 \end{bmatrix} = \begin{bmatrix} -a_1 & 1 \\ -a_0 & 0 \end{bmatrix} \begin{bmatrix} x_1 \\ x_2 \end{bmatrix} + \begin{bmatrix} b_1 \\ b_0 \end{bmatrix} \rho$$

$$\theta_p = x_1$$

Construct adaptive state observers with parallel and series-parallel structures using

(a) The design methodology indicated in Sec. 4.2-3.

(b) The design methodology indicated in Sec. 4.2-4.

8.2 Extend the results of Problem 8.1 for an n-dimensional single-input/single-output continuous-time linear process.

8.3 Consider a plant to be controlled that has the transfer function

$$h_p(s) = \frac{\hat{b}_0}{s(s + \hat{a}_1)}$$

where $\hat{b}_0$ and $\hat{a}_1$ are unknown or their values can change during operation. We want to construct an adaptive state-variable controller which will generate the plant input as

$$u_p = -[k_1, k_2] \begin{bmatrix} x_1 \\ \hat{y}_2 \end{bmatrix} + k_3 u_{ext}$$

where x_1 is the measured state (the output of the process) and $\hat{y}_2$ is the observed component of the plant state vector obtained through an adaptive state observer, so that the transfer function of the closed-loop system will be

$$h(s) = \frac{1}{s^2 + 2\xi\omega_n s + \omega_n^2}$$

with $\xi = 0.7$ and $\omega_n = \hat{a}_1$.

(a) Design an adaptive state observer allowing us to observe the inaccessible component of the plant state vector and to identify the parameters $\hat{a}_1$, and $\hat{b}_0$.

(b) Express the coefficients k_1, k_2, and k_3 as a function of identified parameters (hint: use the implicit model reference concept).

(c) Give a complete scheme of the adaptive control system.

8.4 Consider a reference model described by

$$\begin{bmatrix} \dot{x}_1 \\ \dot{x}_2 \end{bmatrix} = \begin{bmatrix} 0 & 1 \\ -a_0 & -a_1 \end{bmatrix} \begin{bmatrix} x_1 \\ x_2 \end{bmatrix} + \begin{bmatrix} 0 \\ b_0 \end{bmatrix} u_M$$

and a plant to be controlled which is described by

$$\begin{bmatrix} \dot{y}_1 \\ \dot{y}_2 \end{bmatrix} = \begin{bmatrix} -\hat{a}_1 & 1 \\ -\hat{a}_0 & 0 \end{bmatrix} \begin{bmatrix} y_1 \\ y_2 \end{bmatrix} + \begin{bmatrix} 0 \\ \hat{b}_0 \end{bmatrix} u_P$$

$$\theta_P = y_1$$

where the parameters $\hat{a}_0$, $\hat{a}_1$, and $\hat{b}_0$ are unknown or their values can change during operation.

(a) Construct an adaptive state observer for the controlled plant allowing us to obtain the observed state $\hat{y}_2$ and to identify the values of the parameters $\hat{a}_0$, $\hat{a}_1$, and $\hat{b}_0$.

(b) Compute the linear transformation (which will depend on the identified parameters) allowing us to obtain $\underline{\bar{y}} = Qy$, where the state vector $\underline{\bar{y}}$ corresponds to the following state-space realization:

$$\begin{bmatrix} \dot{\bar{y}}_1 \\ \dot{\bar{y}}_2 \end{bmatrix} = \begin{bmatrix} 0 & 1 \\ -\hat{a}_0 & -\hat{a}_1 \end{bmatrix} \begin{bmatrix} \bar{y}_1 \\ \bar{y}_2 \end{bmatrix} + \begin{bmatrix} 0 \\ \hat{b}_0 \end{bmatrix} u_P$$

$$\bar{\theta}_S = \bar{y}_1$$

(c) Design an AMFC system (see Sec. 6.3) with $u_P = u_{P1} + u_{P2}$, where

$$u_{P1} = K_U u_M - K_P \begin{bmatrix} y_1 \\ \bar{y}_2 \end{bmatrix}$$

$$u_{P2} = \Delta K_P(\underline{e}, t) \begin{bmatrix} y_1 \\ \bar{y}_2 \end{bmatrix} + \Delta K_u(\underline{e}, t) u_M$$

and

$$\underline{e} = \begin{bmatrix} x_1 \\ x_2 \end{bmatrix} - \begin{bmatrix} y_1 \\ \overline{y}_2 \end{bmatrix}$$

(one assumes that $\hat{b}_0$ is always positive).

(d) Give a complete scheme of the adaptive control system.

REFERENCES

1. I. D. Landau, Sur la théorie et les applications des systèmes adaptatifs avec modèle, in *Proceedings of "Journée d'Etude sur les Systèmes Adaptatifs,"* Institut Polytechnique de Grenoble, 1-33, Feb. 1970. (ALSTHOM--Dir. des Recherches, Grenoble, 1970)
2. I. D. Landau, Sur une méthode de synthèse des systèmes adaptatifs avec modéle utilisés pour la commande et l'identification d'une classe de procédés physiques, Thèse d'Etat-ès-Sciences, No. C.N.R.S.: A.0.8495, Université Scientifique et Médicale de Grenoble, Grenoble, June 1973.
3. E. Sinner, Etude d'un régulateur adaptatif pour la commande d'une classe de procédés industriels, Thèse de Docteur-Ingénieur, No. C.N.R.S.: A.0.8631, Université Scientifique et Médicale de Grenoble, Grenoble, July 1973.
4. C. C. Hang, A new form of stable adaptive observer, IEEE Trans. Autom. Control, *AC-21*, no. 4, 544-547 (Aug. 1976).
5. R. L. Caroll and D. P. Lindorff, An adaptive observer for single-input single-output linear systems, IEEE Trans. Autom. Control, *AC-18*, no. 5, 428-435 (Oct. 1973).
6. G. Lüders and K. S. Narendra, An adaptive observer and identifier for a linear system, IEEE Trans. Autom. Control, *AC-18*, no. 5, 496-499 (Oct. 1973).
7. P. Kudva and K. S. Narendra, Synthesis of an adaptive observer using Lyapunov's direct method, Int. J. Control, *18*, 1201-1210 (Dec. 1973).
8. G. Lüders and K. S. Narendra, A new canonical form for an adaptive observer, IEEE Trans. Autom. Control, *AC-19*, no. 2, 117-119 (April 1974).

9. G. Luders and K. S. Narendra, Stable adaptive schemes for state estimation and identification of linear systems, IEEE Trans. Autom. Control, *AC-19*, no. 6, 841-847 (Dec. 1974).

10. K. S. Narendra and P. Kudva, Stable adaptive schemes for system identification and control, Parts I and II, IEEE Trans. Syst. Man Cybern., *SMC-4*, no. 6, 542-560 (Nov. 1974).

11. E. Sinner, Régulateur adaptatif à variables d'etat pour processus industriel, Rev. RAIRO, *J-1*, 103-123 (1975).

12. H. M. Silveira, Contributions à la synthèse des systèmes adaptatifs avec modèle sans accès aux variables d'état, Thèse d'Etat ès Sciences, Institut National Polytechnique de Grenoble, Grenoble, March 1978.

13. C. T. Chen, *Introduction to Linear System Theory*, Holt, Rinehart and Winston, New York, 1970.

14. G. Luenberger, Observers for multivariable systems, IEEE Trans. Autom. Control, *AC-11*, no. 2, 190-197 (April 1966).

15. R. E. Kalman, P. L. Falb, and M. A. Arbib, *Topics in Mathematical System Theory*, McGraw-Hill, New York, 1969.

16. I. D. Landau and H. M. Silveira, A stability theorem with application to adaptive control, in *Proceedings of the 17th IEEE Control and Decision Conference*, San Diego, January 1979 (IEEE, New York, 1979).

17. I. D. Landau, Dualité asymptotique entre les systèmes de commande adaptative avec modèle et les regulateurs à variance minimale auto-ajustables, *Proc. Int. Symp. on Systems Optimization and Analysis (IRIA, IFAC, AFCET)*, Rocqencourt, France, Dec. 1978 (Springer Verlag, Heidelberg, 1979).

Appendix A

STABILITY[1]

This appendix reviews several definitions of the stability of dynamic systems. These definitions (in particular, Definition A-3) are necessary for understanding of the work based on them throughout the book. Also given are some of the main theorems concerning the study of the stability of dynamic systems via the second method of Lyapunov. These results are either used directly in the design of MRAS's (Theorem A-1) or indirectly through the properties of positive dynamic systems (see Appendix B). The basic reference for the results presented in this appendix is the seminal paper by Kalman and Bertram [1]. (For definitions concerning the stability of discrete-time dynamic systems, see Refs. 1 and 3.)

Consider a free (unforced) dynamic system described by the vector differential equation

$$\dot{\underline{x}} = \underline{f}(\underline{x}, t), \quad -\infty < t < +\infty \tag{A-1}$$

where $\underline{x}$ is the n-dimensional state vector of the system. Consider the vector function $\underline{\phi}(t, \underline{x}_0, t_0)$, a solution of the system of Eq. (A-1) which is unique and differentiable with respect to t and such that for any fixed $\underline{x}_0$, t_0 it satisfies the equations

[1]References 1-5.

$$\underline{\phi}(t_0, \underline{x}_0, t_0) = \underline{x}_0 \tag{A-2}$$

$$\dot{\underline{\phi}}(t, \underline{x}_0, t_0) = \underline{f}[\underline{\phi}(t, \underline{x}_0, t_0), t] \tag{A-3}$$

A state $\underline{x}_e$ of the free dynamic system given by Eq. (A-1) is called an *equilibrium state* if

$$\underline{f}(\underline{x}_e, t) = \underline{0} \quad \text{for all } t \tag{A-4}$$

DEFINITION A-1 (stability). An equilibrium state $\underline{x}_e$ of the system in Eq. (A-1) is *stable* if for every real number $\varepsilon > 0$ there exists a real number $\delta(\varepsilon, t_0) > 0$ such that

$$||\underline{x}_0 - \underline{x}_e|| \le \delta(\varepsilon, t_0) \Longrightarrow ||\underline{\phi}(t, \underline{x}_0, t_0) - \underline{x}_e|| \le \varepsilon \quad \text{for all } t \ge t_0 \tag{A-5}$$

DEFINITION A-2 (asymptotic stability). An equilibrium state $\underline{x}_e$ of the system in Eq. (A-1) is *asymptotically stable* if

1. $\underline{x}_e$ is stable.
2. There exists a real number $\delta(t_0) > 0$ such that

$$||\underline{x}_0 - \underline{x}_e|| \le \delta(t_0) \Longrightarrow \lim_{t\to\infty} ||\underline{\phi}(t, \underline{x}_0, t_0) - \underline{x}_e|| = 0 \tag{A-6}$$

[i.e., every solution $\underline{\phi}(t, \underline{x}_0, t_0)$ starting sufficiently near $\underline{x}_e$ converges to $\underline{x}_e$ as $t \to \infty$].

DEFINITION A-3 (global asymptotic stability). An equilibrium state $\underline{x}_e$ of the system in Eq. (A-1) is *globally asymptotically stable* if for all $\underline{x}_0 \in R^n$

1. $\underline{x}_e$ is stable.
2. $\lim_{t\to\infty} ||\underline{\phi}(t, \underline{x}_0, t_0) - \underline{x}_e|| = 0.$

Note: If $\underline{x}_0$ is restricted to a certain region $r \in R^n$, then one has asymptotic stability *in the large* [2].

THEOREM A-1 (Lyapunov). Consider the free dynamic system given by Eq. (A-1) where

$\underline{f}(0, t) = \underline{0}, \quad -\infty < t < +\infty$

If there exists a real scalar function $V(\underline{x}, t)$ with continuous first partial derivatives with respect to $\underline{x}$ and t such that

1. $$V(0, t) = 0, \quad -\infty < t < +\infty \tag{A-8}$$
2. $$V(\underline{x}, t) \geq \alpha(||\underline{x}||) > 0 \quad \text{for all } \underline{x} \neq 0, \ \underline{x} \in R^n \text{ and for all } t \tag{A-9}$$

   where $\alpha(.)$ is a continuous nondecreasing scalar function such that $\alpha(0) = 0$,
3. $$V(\underline{x}, t) \to \infty \text{ with } ||\underline{x}|| \to \infty \quad \text{for all } t \tag{A-10}$$
4. $$\dot{V} = \frac{dV}{dt} = \frac{\partial V}{\partial t} + (\text{grad } V)^T \underline{f}(\underline{x}, t) \leq -\gamma(||\underline{x}||) < 0$$
   $$\text{for all } \underline{x} \neq 0, \ \underline{x} \in R^n \text{ and for all } t \tag{A-11}$$

   where $\gamma(.)$ is a continuous scalar function such that $\gamma(0) = 0$,

then the equilibrium state $\underline{x}_e = \underline{0}$ is *globally asymptotically stable*, and $V(\underline{x}, t)$ is a Lyapunov function for the system given by Eq. (A-1). Note: If in addition to the conditions of Theorem A-1 $V(\underline{x}, t) \leq \beta(||\underline{x}||)$ where $\beta(.)$ is a continuous, nondecreasing scalar function such that $\beta(0) = 0$, $\underline{x}_c = \underline{0}$, is *globally uniformly asymptotically stable*. (The term *uniform* means that the stability properties of the system are independent of the initial time t_0.)

COROLLARY A-1. If $V(\underline{x}, t)$ in Theorem A-1 is replaced by $V(x)$, which is a real scalar function with continuous first partial derivatives with respect to $\underline{x}$, the equilibrium state $\underline{x}_e = 0$ of the autonomous dynamic system $\underline{x} = \underline{f}(\underline{x})$ is *globally* (uniformly) *asymptotically stable* if

1. $$V(0) = 0 \tag{A-12}$$
2. $$V(\underline{x}) > 0 \quad \text{for all } \underline{x} \neq \underline{0}, \ \underline{x} \in R^n \tag{A-13}$$
3. $$V(\underline{x}) \to \infty \quad \text{with } ||\underline{x}|| \to \infty \tag{A-14}$$
4. $$\dot{V}(\underline{x}) = \frac{dV(\underline{x})}{dt} < 0 \quad \text{for all } \underline{x} \neq \underline{0}, \ \underline{x} \in R^n \tag{A-15}$$

COROLLARY A-2. In Corollary A-1, condition 4 may be replaced by

4'. $\dot{V}(\underline{x}) \leq 0$ for all $\underline{x} \neq \underline{0}$, $\underline{x} \in R^n$ (A-16)

4". $\dot{V}(\underline{x})$ is not identically zero along any trajectory of the system in Eq. (A-1) for any t_0 or any $\underline{x}_0 \neq \underline{0}$.

THEOREM A-2 (Lyapunov) [1]. The equilibrium state $\underline{x}_e = \underline{0}$ of a linear time-invariant free system

$$\dot{\underline{x}} = A\underline{x} \tag{A-17}$$

is (globally uniformly) *asymptotically stable* if and only if given any positive definite matrix Q there exists a symmetric positive definite matrix P which is the unique solution of the matrix equation

$$A^T P + PA = -Q \tag{A-18}$$

and $V = \underline{x}^T P \underline{x}$ is a Lyapunov function for the system in Eq. (A-17).

Notes

1. There are several methods for solving Eq. (A-18)(known also as the *Lyapunov equation*). For a detailed presentation of various methods, see Ref. 5.
2. Many of the methods for solving the Lyapunov equation fail when matrix A is "ill-conditioned" (i.e., the largest and the smallest elements of the matrix differ by several orders of magnitude). For these cases Jameson's method [5] can be successfully used.

COROLLARY A-3 (Kalman) [1]. The real parts of the eigenvalues of a constant matrix A are < μ if and only if given any symmetric positive definite matrix Q there exists a symmetric positive definite matrix P which is the solution of the matrix equation

$$-2\mu P + A^T P + PA = -Q \tag{A-19}$$

THEOREM A-3 [1]. The equilibrium state $\underline{x}_e = \underline{0}$ of the linear discrete-time free dynamic system

$$\underline{x}_{k+1} = A\underline{x}_k \tag{A-20}$$

is (globally uniformly) *asymptotically stable* if and only if given any positive definite matrix Q there exists a symmetric positive definite matrix P which is the unique solution of the matrix equation

$$A^T PA - P = -Q \tag{A-21}$$

and $V(\underline{x}) = \underline{x}_k{}^T P\underline{x}_k$ is a Lyapunov function for the system in Eq. (A-20).

COROLLARY A-4 [1]: The modulus of the eigenvalues of a constant matrix A are $< \rho$, if and only if given any symmetric, positive definite matrix Q, there exists a symmetric positive definite matrix P which is the unique solution of the matrix equation

$$\rho^{-2} A^T PA - P = -Q \tag{A-22}$$

REFERENCES

1. R. E. Kalman and J. E. Bertram, Control system analysis and design via the second method of Lyapunov, Parts I and II, Trans ASME, J. Basic Eng., *82*, 371-400 (1960).
2. J. C. Hsu and A. U. Meyer, *Modern Control Principles and Applications*, McGraw-Hill, New York, 1968.
3. J. M. Mendel, *Discrete Techniques of Parameter Estimation—The Equation Error Formulation*, Marcel Dekker, New York, 1973.
4. K. S. Narendra and J. H. Taylor, *Frequency Domain Criteria for Absolute Stability*, Academic Press, New York, 1973.
5. S. Barnett and C. Storey, *Matrix Methods in Stability Theory*, Nelson, London, 1970.

Appendix B

POSITIVE DYNAMIC SYSTEMS

B.0 INTRODUCTION

The term *positive dynamic system* is an extension to dynamic systems of the mathematical concept of *positivity*. This is a necessary and sufficient condition for a mathematical object to be *factorizable* as a product (for example, a *positive definite* matrix Q can be factored as $Q = LL^T$). The concepts of positivity and factorization can be extended to more complicated mathematical objects (e.g., functions, operators) and to more complicated types of products (e.g., convolution).

In control theory, a factorization problem—and associated with it, a positivity problem for a dynamic system—appeared first in connection with the Wiener filter. Since then, the concept of a positive dynamic system has appeared explicitly or implicitly in many control problems (as, for example, in the stability of nonlinear time-varying feedback systems and optimal control). The works of Popov, Kalman, and Yakubovitch are the basis of the theory of positive dynamic systems, which is now an important part of systems theory [1]. It is important to note that the work done in this area has played an important role in establishing the links between the results of control theory obtained in the time and frequency domains. We note also that the theory of positive systems also plays a key role in network theory (passive networks) and stochastic processes (Markov realizations) [2-4].

In this appendix, we shall cover some aspects of the various types of positive dynamic systems (continuous linear time-invariant and time-varying systems as well as discrete linear time-invariant and time-varying systems) to the extent necessary to understand and utilize the results presented in the main body of the book. For those interested in a deeper study of the positive dynamic systems, we recommend the monograph by Popov [1] and the book by Anderson and Vongpanitlerd [5] (the last one covers only continuous linear time-invariant positive systems).

We shall start by introducing some basic mathematical definitions which will be used subsequently for describing positive dynamic systems.

B.1 POSITIVE REAL (MATRIX) FUNCTIONS OF A COMPLEX VARIABLE

Formally a *positive real function of a complex variable* is defined as follows.

DEFINITION B.1-1 [6]. A rational function h(s) of the complex variable $s = \sigma + j\omega$ is *positive real* if

1. h(s) is real for real s.
2. $\mathrm{Re}[h(s)] \geq 0$ for all $\mathrm{Re}[s] > 0$.

Because the examination of Re[h(s)] over the half plane $\mathrm{Re}[s] > 0$ is a tedious operation, one uses the following equivalent definition.

DEFINITION B.1-2. A rational function h(s) of the complex variable $s = \sigma + j\omega$ is *positive real* if

1. h(s) is real for real s.
2. h(s) has no poles in the open right half plane, $\mathrm{Re}[s] > 0$.

3. The eventual poles of $h(s)$ on the axis $Re[s] = 0$ (i.e., $s = j\omega$) are distinct, and the associated residues are real and positive (or null).
4. For all real ω for which $s = j\omega$ is not a pole of $h(s)$ one has

$$Re[h(j\omega)] \geq 0$$

DEFINITION B.1-3. A rational function $h(s)$ of the complex variable $s = \sigma + j\omega$ is *strictly positive real* if

1. $h(s)$ is real for real s.
2. $h(s)$ has no poles in the closed right half plane $Re[s] \geq 0$.
3. $Re[h(j\omega)] > 0, \ -\infty < \omega < +\infty$.

Therefore the difference between *positive real* transfer functions and *strictly positive real* transfer functions is that in the second case the poles on $Re[s] = 0$ are no longer tolerated and that $Re[h(j\omega)]$ is greater than zero for all real ω (instead of greater or equal to zero).

We shall now focus our attention on some of the properties of the positive real functions $h(s)$ of the form

$$h(s) = \frac{M(s)}{N(s)}$$

where $M(s)$ and $N(s)$ are relatively prime polynomials of the complex variable s.

LEMMA B.1-1. If $h(s) = M(s)/N(s)$ is a positive real function,

1. $M(s)$ and $N(s)$ have real coefficients.
2. $1/h(s)$ is also a positive real function.
3. $M(s)$ and $N(s)$ are Hurwitz polynomials (i.e., they verify the Hurwitz criterion, and their zeros have negative real parts).
4. The order of $N(s)$ does not differ from the order of $M(s)$ by more than ± 1.

For testing the *real positivity* of a function of the complex variable $h(s) = M(s)/N(s)$, see Refs. 7 and 8.

Before discussing the real positivity of a matrix H(s) of real rational functions of the complex variable s, we shall introduce the concept of a *Hermitian matrix* and mention some of its properties.

DEFINITION B.1-4 [9]. A matrix function H(s) of the complex variable $s = \sigma + j\omega$ is a *Hermitian matrix* (or simply *Hermitian*) if

$$H(s) = H^T(s^*) \qquad \text{(B.1-1)}$$

(where the asterisk means *conjugate*).

Hermitian matrices feature several properties, including

1. A Hermitian matrix is a square matrix, and the diagonal terms are real.
2. The eigenvalues of a Hermitian matrix are always real.
3. If H(s) is a Hermitian matrix and $\underline{x}$ is a vector with *complex components*, the quadratic form $\underline{x}^T H \underline{x}^*$ is always real.

One can now introduce the following definitions.

DEFINITION B.1-5 [5]. An m × m matrix H(s) of real rational functions is *positive real* if

1. All elements of H(s) are analytic in the open right half plane Re[s] > 0 (i.e., they do not have poles in Re[s] > 0).
2. The eventual poles of any element of H(s) on the axis Re[s] = 0 are distinct, and the associated residue matrix of H(s) is a positive semidefinite Hermitian.
3. The matrix $H(j\omega) + H^T(-j\omega)$ is a *positive semidefinite Hermitian* for all real values of ω which are not a pole of any element of H(s)

Note: Conditions 2 and 3 in Definition B.1-5 can be replaced by (3 + 2): The matrix $H(s) + H^T(s^*)$ is a positive semidefinite Hermitian for all Re[s] > 0.

DEFINITION B.1-6. An $m \times m$ matrix $H(s)$ of real rational functions is *strictly positive real* if

1. All elements of $H(s)$ are analytic in the closed right half plane $\mathrm{Re}[s] \geq 0$ (i.e., they do not have poles in $\mathrm{Re}[s] \geq 0$).
2. The matrix $H(j\omega) + H^T(-j\omega)$ is a *positive definite Hermitian* for all real ω.

Positive Definite Integral Kernels

DEFINITION B.1-7 [10]. The square matrix kernel $K(t, \tau)$ is termed *positive definite* if for each interval $[t_0, t_1]$ and all the vector functions $\underline{f}(t)$ that are piecewise continuous in $[t_0, t_1]$ the following inequality holds:

$$\eta(t_0, t_1) = \int_{t_0}^{t_1} \underline{f}^T(t) \left| \int_{t_0}^{t} K(t, \tau)\underline{f}(\tau)\, d\tau \right| dt \geq 0 \qquad \text{(B.1-2)}$$

The term $\int_{t_0}^{t} K(t, \tau)\underline{f}(\tau)\, d\tau$ can be interpreted as the output of a block whose input is $\underline{f}(t)$, and therefore the expression (B.1-2) can be interpreted as the integral of the input/output inner product, which is positive or null if $K(t, \tau)$ is a *positive definite kernel*.

If $K(t, \tau)$ depends only on the argument $t - \tau$ and the elements of $K(t, \tau) = K(t - \tau)$ are bounded, then $K(t - \tau)$ has a Laplace transform

$$H(s) = \int_0^{\infty} K(t)e^{-st}\, dt \qquad \text{(B.1-3)}$$

For this situation one has the following result, which allows us to characterize a positive definite kernel.

LEMMA B.1-2 [1, 10]. For the class of kernels $K(t - \tau)$ for which the Laplace transform exists, the necessary and sufficient condition for $K(t - \tau)$ to be a *positive definite kernel* is that its Laplace transform be a *positive real matrix of rational functions of the complex variable* $s = \sigma + j\omega$ (i.e., a positive real transfer matrix).

B.2 CONTINUOUS LINEAR TIME-INVARIANT POSITIVE SYSTEMS

Consider the linear time-invariant multivariable system

$$\dot{\underline{x}} = A\underline{x} + B\underline{u} \tag{B.2-1}$$

$$\underline{v} = C\underline{x} + J\underline{u} \tag{B.2-2}$$

where $\underline{x}$ is an n-dimensional state vector; $\underline{u}$ and $\underline{v}$ are m-dimensional vectors representing the input and the output, respectively; and A, B, C, and J are matrices of appropriate dimension. One assumes also that the pair (A, B) is *completely controllable* and that the pair (C, A) is *completely observable*.

The system of Eqs. (B.2-1) and (B.2-2) is also characterized by the *square* transfer matrix H(s):

$$H(s) = J + C(sI - A)^{-1}B \tag{B.2-3}$$

We can now introduce the following definition [1].

DEFINITION B.2-1. The system of Eqs. (B.2-1) and (B.2-2) is *positive* if the integral of the input/output scalar product can be expressed for all $t_1 \geq 0$ by

$$\int_0^{t_1} \underline{v}^T\underline{u}\, dt = [\xi(\underline{x})]_0^{t_1} + \int_0^{t_1} \lambda(\underline{x}, \underline{u})\, dt \tag{B.2-4}$$

with

$$\lambda(\underline{x}, \underline{u}) \geq 0 \quad \text{for all } \underline{x} \in R^n,\ \underline{u} \in C^m \tag{B.2-5}$$

and with the functions $\xi(\underline{x})$ and $\lambda(\underline{x}, \underline{u})$ being defined for all $\underline{x}$ and $\underline{u}$.

We shall give below some of the equivalent formulations of the positivity properties for a system of the form of Eqs. (B.2-1) and (B.2-2) which are of direct interest in this book. These properties are necessary and sufficient conditions for a linear time-invariant system to be positive. For other equivalent formulations, see Refs. 1, 4, and 5.

The importance of Theorem B.2-1, which is given below, resides in the fact that it provides tools either for constructing positive systems characterized by positive real transfer matrices or for checking if a given system is positive (or if a given transfer matrix is positive real). The various equivalent formulations of the positivity properties will allow us to switch from one to another, trying to use the most appropriate when dealing with applications.

THEOREM B.2-1. The following propositions concerning the system of Eqs. (B.2-1) and (B.2-2) are equivalent to each other:

1. The system of Eqs. (B.2-1) and (B.2-2) is *positive* (in the sense of Definition B.2-1).
2. H(s) given by Eq. (B.2-3) is a *positive real transfer matrix*.
3. There exist a symmetric positive definite matrix P, a symmetric positive semidefinite matrix Q, and matrices S and R such that

$$PA + A^TP = -Q \tag{B.2-6}$$

$$B^TP + S^T = C \tag{B.2-7}$$

$$J + J^T = R \tag{B.2-8}$$

$$\begin{bmatrix} Q & S \\ S^T & R \end{bmatrix} \geq 0 \text{ [1]} \tag{B.2-9}$$

4. (positive real lemma)[2] There exist a symmetric positive matrix P and matrices K and L such that

$$PA + A^TP = -LL^T \tag{B.2-10}$$

$$B^TP + K^TL^T = C \tag{B.2-11}$$

$$K^TK = J + J^T \tag{B.2-12}$$

5. The Hermitian matrix $Z(-s, s) = H^T(-s) + H(s)$ is factorizable as

---

[1] The notation $M \geq 0$ means that M is a positive semidefinite matrix, and $M > 0$ means that M is a positive definite matrix.

[2] The equivalency between propositions 2 and 4 is called the *positive real lemma* or the Kalman-Yakubovitch-Popov lemma [1, 3, 5].

$$Z(-s, s) = H^T(-s) + H(s) = W^T(-s)W(s)$$
$$= [K^T + B^T(-sI - A^T)^{-1}L][K + L^T(sI - A)^{-1}B] \tag{B.2-13}$$

6. The Hermitian matrix $Z(-s, s) = H^T(-s) + H(s)$ is *positive semidefinite* for all $s = j\omega$ for which $\det(j\omega I - A) \neq 0$.
7. Every solution $\underline{x}[\underline{x}(0), \underline{u}, t]$ of the system given by Eqs. (B.2-1) and (B.2-2) verifies the following equality:

$$\int_0^{t_1} \underline{v}^T\underline{u}\, dt = \frac{1}{2}\underline{x}(t_1)^T P\underline{x}(t_1) - \frac{1}{2}\underline{x}(0)^T P\underline{x}(0) + \frac{1}{2}\int_0^{t_1} (\underline{x}^T Q\underline{x} + 2\underline{u}^T S^T\underline{x} + \underline{u}^T R\underline{u})dt \tag{B.2-14}$$

where P is a positive definite matrix and matrices P, Q, S, and R satisfy Eqs. (B.2-6) to (B.2-9).
8. For $\underline{x}(0) = \underline{0}$ and for any input vector function $\underline{u}(t)$ and its corresponding solution $\underline{x}(0, \underline{u}, t)$ of the system given by Eqs. (B.2-1) and (B.2-2), the following inequality is verified:

$$\int_0^{t_1} \underline{v}^T\underline{u}\, dt \geq 0 \tag{B.2-15}$$

9. The matrix kernel (impulse response matrix) $K(t - \tau)$,

$$K(t - \tau) = J\delta(t - \tau) + Ce^{A(t-\tau)}B \cdot 1(t - \tau) \tag{B.2-16}$$

is a *positive definite kernel*. In Eq. (B.2-16), $\delta(t - \tau)$ is the Dirac δ-function, which has the property that $\int_{-\infty}^{+\infty} \underline{f}(\tau)\delta(t - \tau)\, d\tau = \underline{f}(t)$, and $1(t - \tau)$ is the unit step function $[1(t - \tau) = 0$ for $(t - \tau) < 0$, $1(t - \tau) = 1$ for $(t - \tau) \geq 0]$.

In Theorem B.2-1, one can note that Eq. (B.2-14) corresponds to Definition B.2-1 of a linear time-invariant positive system with

$$\xi(\underline{x}) = \frac{1}{2}\underline{x}^T P\underline{x} \tag{B.2-17}$$

$$\lambda(\underline{x}, \underline{u}) = \frac{1}{2}(\underline{x}^T Qx + 2\underline{u}^T S^T\underline{x} + \underline{u}^T R\underline{u})\ dt \tag{B.2-18}$$

and from the equivalency of propositions 7 and 3 one concludes from Eq. (B.2-9) that $\lambda(\underline{x}, \underline{u})$ given by Eq. (B.2-18) verifies the conditions of Eq. (B.2-5).

If J = 0 in Eq. (B.2-2), propositions 3 and 4 of Theorem B.2-1 become the following.

LEMMA B.2-1. The linear time-invariant system

$$\dot{\underline{x}} = A\underline{x} + B\underline{u} \tag{B.2-19}$$

$$v = C\underline{x} \tag{B.2-20}$$

is positive, and the transfer matrix

$$H(s) = C^T(sI - A)^{-1}B \tag{B.2-21}$$

is a *positive real transfer matrix* if and only if there exist a symmetric positive definite matrix P and a symmetric positive semidefinite matrix Q (or a matrix L) such that

$$PA + A^TP = -Q = -LL^T \tag{B.2-22}$$

$$B^TP = C \tag{B.2-23}$$

For the case where $J + J^T = R$ is a positive definite matrix [i.e., K in Eq. (B.2-12) is a nonsingular matrix], Eqs. (B.2-10), (B.2-11), and (B.2-12) can be reduced to a single equation.

We shall not go into the proofs of the equivalency of the various propositions of Theorem B.2-1; they can be found in Ref. 1 or 5. We shall only show that proposition 3 implies proposition 4.

Because the matrix appearing in Eq. (B.2-9) must be at least positive semidefinite, it can be factored as

$$\begin{bmatrix} Q & S \\ S^T & R \end{bmatrix} = MM^T = \begin{bmatrix} L \\ K^T \end{bmatrix} [L^T \;\; K] = \begin{bmatrix} LL^T & LK \\ K^TL^T & K^TK \end{bmatrix} \tag{B.2-24}$$

where L is an $(n \times q)$-dimensional matrix and K^T is an $(m \times q)$-dimensional matrix, q being arbitrary. Replacing Q by LL^T, S^T by K^TL^T, and R by K^TK in Eqs. (B.2-6), (B.2-7), and (B.2-8), one obtains Eqs. (B.2-10), (B.2-11), and (B.2-12).

Note: From Eq. (B.2-14) one obtains the following: For $\underline{x}(0) \neq \underline{0}$, the integral of the input/output inner product verifies the inequality

$$\int_0^{t_1} \underline{v}^T\underline{u}\, dt \geq -\frac{1}{2}\underline{x}(0)^T P\underline{x}(0) = -\gamma_0^2 \quad \text{for all } t_1 \geq 0 \tag{B.2-25}$$

where γ_0^2 is a finite positive constant.

Theorem B.2-1 provides conditions allowing us to test or to construct positive systems characterized by real positive transfer matrices. However, in many stability studies concerning time-varying nonlinear feedback systems we are interested in defining a state-space realization of a system characterized by a strictly positive real transfer matrix. One then has the following results.

LEMMA B.2-2 [6, 11]. The transfer matrix H(s) given in Eq. (B.2-3) is *strictly positive real*, and the elements of H(s) are analytic in $\text{Re}[s] > -\mu$ if there exists a symmetric positive definite matrix P, a matrix L, and a scalar $\mu > 0$ (or a symmetric positive definite matrix Q') and a matrix K such that

$$PA + A^TP = -LL^T - 2\mu P = -Q' \tag{B.2-26}$$

$$B^TP + K^TL^T = C \tag{B.2-27}$$

$$K^TK = J + J^T \tag{B.2-28}$$

Note: Because Q' and P are positive definite matrices, one can always find a scalar $\mu > 0$ such that

$$Q' - 2\mu P \geq 0 \tag{B.2-29}$$

which, of course, implies the existence of a matrix L such that $LL^T = Q' - 2\mu P$.

LEMMA B.2-3. The transfer matrix H(s) given in Eq. (B.2-21) is *strictly positive real* if there exist a symmetric positive definite matrix P and a symmetric positive definite matrix Q (or a nonsingular matrix L) such that the system of Eqs. (B.2-22) and (B.2-23) can be verified.

Note: $Q > 0$ implies from Corollary A-3 that the eigenvalues of A have negative real parts, and therefore the elements of H(s) will be analytic in $\text{Re}[s] \geq 0$.

B.3 CONTINUOUS LINEAR TIME-VARYING POSITIVE SYSTEMS

Consider the linear time-varying multivariable system

$$\dot{\underline{x}} = A(t)\underline{x} + B(t)\underline{u} \tag{B.3-1}$$

$$\underline{v} = C(t)\underline{x} + J(t)\underline{u} \tag{B.3-2}$$

where $\underline{x}$ is an n-dimensional state vector; $\underline{u}$ and $\underline{v}$ are m-dimensional vectors representing the input and the output, respectively; and A(t), B(t), C(t), and J(t) are time-varying matrices with piecewise continuous elements defined for all $t \geq t_0$

DEFINITION B.3-1. The system of Eqs. (B.3-1) and (B.3-2) is *positive* if the integral of the input/output inner product can be expressed by

$$\int_{t_0}^{t_1} \underline{v}^T\underline{u}\, dt = [\xi(\underline{x}, t)]_{t_0}^{t_1} + \int_{t_0}^{t_1} \lambda(\underline{x}, \underline{u}, t)\, dt \tag{B.3-3}$$

where

$$\lambda(\underline{x}, \underline{u}, t) \geq 0 \quad \text{for all } t \geq t_0,\ \underline{u} \in C^m,\ \underline{x} \in R^n \tag{B.3-4}$$

In the case of continuous time-varying linear systems *one has only sufficient conditions* for the positivity of the systems in the form of Eqs. (B.3-1) and (B.3-2). Two of these sufficient conditions are given below, and they represent a direct extension of part of the results given in Theorem B.2-1.

LEMMA B.3-1. The system of Eqs. (B.3-1) and (B.3-2) is *positive* if there exist a symmetric time-varying positive definite matrix P(t), differentiable with respect to t, a symmetric time-varying semidefinite matrix Q(t), and matrices S(t) and R(t) such that

$$\dot{P}(t) + A^T(t)P(t) + P(t)A(t) = -Q(t) \tag{B.3-5}$$

$$B^T(t)P(t) + S^T(t) = C(t) \tag{B.3-6}$$

$$J(t) + J^T(t) = R(t) \tag{B.3-7}$$

$$\begin{bmatrix} Q(t) & S(t) \\ S^T(t) & R(t) \end{bmatrix} \geq 0 \quad \text{for all } t \geq t_0 \tag{B.3-8}$$

LEMMA B.3-2. The system of Eqs. (B.3-1) and (B.3-2) is *positive* if every solution $\underline{x}[\underline{x}(t_0), \underline{u}, t]$ satisfies the following equality:

$$\int_{t_0}^{t_1} \underline{v}^T\underline{u}\, dt = \frac{1}{2}\underline{x}(t_1)^T P(t_1)\underline{x}(t_1) - \frac{1}{2}\underline{x}(t_0)^T P(t_0)\underline{x}(t_0) + \frac{1}{2}\int_{t_0}^{t} [\underline{x}^T Q(t)\underline{x} + 2\underline{u}^T S^T(t)\underline{x} + \underline{u}^T R(t)\underline{u}]\, dt \qquad \text{(B.3-9)}$$

with

$$P(t) > 0, \quad \begin{bmatrix} Q(t) & S(t) \\ S^T(t) & R(t) \end{bmatrix} \geq 0 \quad \text{for all } t \geq t_0$$

B.4 DISCRETE LINEAR TIME-INVARIANT POSITIVE SYSTEMS

The concept of the positive dynamic system introduced for continuous-time linear systems can be extended to discrete-time linear systems. Consider the linear time-invariant discrete system

$$\underline{x}_{k+1} = A\underline{x}_k + B\underline{u}_k \qquad \text{(B.4-1)}$$

$$\underline{v}_k = C\underline{x}_k + J\underline{u}_k \qquad \text{(B.4-2)}$$

where $\underline{x}$, $\underline{u}$, $\underline{v}$, A, B, C, and J have the same meaning as in Sec. B.2. One assumes again that the pair (A, B) is *completely controllable* and that the pair (C, A) is *completely observable*.

The system of Eqs. (B.4-1) and (B.4-2) is also characterized by the discrete *square* transfer matrix

$$H(z) = J + C(zI - A)^{-1}B \qquad \text{(B.4-3)}$$

One introduces the following definitions, which are counterparts of those introduced in the case of continuous linear time-invariant systems.

DEFINITION B.4-1 [12]. An m × m discrete matrix H(z) of real rational functions is *positive real* if

1. All elements of H(z) are analytic outside the unit circle (i.e., they do not have poles in $|z| > 1$).
2. The eventual poles of any element of H(z) on the unit circle $|z| = 1$ are simple, and the associated residue matrix is a positive semidefinite Hermitian.

3. The matrix

$$H(z) + H^T(z^*) = H(e^{j\omega}) + H^T(e^{-j\omega})$$

is a positive semidefinite Hermitian for all real values of ω which are not a pole of any element of $H(e^{j\omega})$ [i.e., for all z on the unit circle $|z| = 1$ which are not a pole of H(z)].

DEFINITION B.4-2. An $m \times m$ discrete matrix H(z) of real rational functions is *strictly positive real* if

1. All the elements of H(z) are analytic in $|z| \geq 1$.
2. The matrix

$$H(z) + H^T(z^*) = H(e^{j\omega}) + H^T(e^{-j\omega})$$

is a positive definite Hermitian for all ω (i.e., for all z on the unit circle $|z| = 1$).

Note: The definition of (strictly) positive real discrete transfer matrices H(z) can be related to the definition of (strictly) positive real transfer matrices H(s) by the transformation $s = (z - 1)/(z + 1)$.

The definitions of positive real and strictly positive real discrete transfer functions are straightforward consequences of the above definitions and are omitted.

DEFINITION B.4-3 [13]. The discrete matrix kernel C(k, l) is termed *positive definite* if for each interval $[k_0, k_1]$ and for all the discrete vectors $\underline{f}(k)$ bounded in $[k_0, k_1]$ the following inequality holds:

$$\sum_{k=k_0}^{k_1} \underline{f}^T(k) \left[\sum_{\ell=k_0}^{k} C(k, \ell)\underline{f}(\ell) \right] \geq 0 \quad \text{for all } k_1 \geq k_0 \qquad \text{(B.4-4)}$$

LEMMA B.4-1 [13]. For the class of discrete kernels $C(k - \ell)$ for which the z-transform exists the necessary and sufficient condition for $C(k - \ell)$ to be a *positive definite discrete matrix kernel* is that its z-transform be a *positive real discrete transfer matrix*.

We can now introduce the definition of discrete linear time-invariant positive systems [1].

DEFINITION B.4-4. The system of Eqs. (B.4-1) and (B.4-2) is *positive* if the sum of the input/output scalar products over the interval $[0, k_1]$ can be expressed by

$$\sum_{k=0}^{k_1} \underline{v}_k^T \underline{u}_k = \xi(\underline{x}_{k_1+1}) - \xi(\underline{x}_0) + \sum_{k=0}^{k_1} \lambda(\underline{x}_k, \underline{u}_k) \quad \text{for all } k_1 \geq 0 \tag{B.4-5}$$

with

$$\lambda(\underline{x}_k, \underline{u}_k) \geq 0 \quad \text{for all } \underline{x}_k \in R^n, \ \underline{u}_k \in R^m \tag{B.4-6}$$

We shall give below some of the equivalent formulations of the properties of the discrete linear time-invariant positive systems. These properties are necessary and sufficient conditions for a discrete linear time-invariant system to be positive.

THEOREM B.4-1. The following propositions concerning the system of Eqs. (B.4-1) and (B.4-2) are equivalent to each other:

1. The system of Eqs. (B.4-1) and (B.4-2) is *positive* (in the sense of Definition B.4-4).
2. $H(z)$ given by Eq. (B.4-3) is a *positive real discrete transfer matrix*.
3. There exist a symmetric positive definite matrix P, a symmetric positive semidefinite matrix Q, and matrices S and R such that

$$A^T PA - P = -Q \tag{B.4-7}$$

$$B^T PA + S^T = C \tag{B.4-8}$$

$$J + J^T - B^T P B = R \tag{B.4-9}$$

$$\begin{bmatrix} Q & S \\ S^T & R \end{bmatrix} \geq 0 \tag{B.4-10}$$

4. (discrete, positive real lemma)[3] There exist a symmetric positive definite matrix P and matrices K and L such that

$$A^T P A - P = -LL^T \tag{B.4-11}$$

$$B^T P A + K^T L^T = C \tag{B.4-12}$$

$$K^T K = J + J^T - B^T P B \tag{B.4-13}$$

5. Every solution $\underline{x}_k(\underline{x}_0, \underline{u}_k)$ of Eqs. (B.4-1) and (B.4-2) satisfies the following equality:

$$\sum_{k=0}^{k_1} \underline{v}_k{}^T \underline{u}_k = \frac{1}{2}\underline{x}_{k_1+1}{}^T P \underline{x}_{k_1+1} - \frac{1}{2}\underline{x}_0{}^T P \underline{x}_0 + \frac{1}{2}\sum_{k=0}^{k_1} (\underline{x}_k{}^T Q \underline{x}_k + 2\underline{u}_k{}^T S^T \underline{x}_k + \underline{u}_k{}^T R \underline{u}_k) \tag{B.4-14}$$

where P is a positive definite matrix and the matrices P, Q, S, and R satisfy Eqs. (B.4-7) to (B.4-10).

6. The discrete matrix kernel (impulse response matrix)

$$C(k - \ell) = J\delta(k - \ell) + CA^{-(\ell-1)}B1(k - \ell) \tag{B.4-15}$$

is a *positive definite discrete kernel*.

Proof of Theorem B.4-1 is given in [1], where other equivalent formulations of the positivity properties are also given. The proof of equivalency between propositions 2 and 4 can also be found in [12] and is based on the use of the transformation s = (z - 1)/(z + 1) and of the *positive real lemma* for continuous time-invariant linear systems. The implication 3 ⟹ 5 will be shown in Sec. B.5 in the more general context of discrete time-varying linear systems.

---

[3]The equivalency between propositions 2 and 4 is called the *discrete positive real lemma* or the Kalman-Szegö-Popov lemma [1, 12].

LEMMA B.4-2. The discrete transfer matrix H(z) given by Eq. (B.4-3) is *strictly positive real* if there exist a symmetric positive definite matrix P, a symmetric positive definite matrix Q, and matrices K and L such that

$$A^T PA - P = -LL^T - Q = -Q' \tag{B.4-16}$$

$$B^T PA + K^T L^T = C \tag{B.4-17}$$

$$K^T K = J + J^T - B^T PB \tag{B.4-18}$$

Note: Q' in Eq. (B.4-16) is a positive definite matrix which implies that the eigenvalues of A have an absolute value less than 1 and that therefore the elements of H(z) will be analytic in $|z| \geq 1$.

B.5 DISCRETE LINEAR TIME-VARYING POSITIVE SYSTEMS

Consider the discrete linear time-varying system

$$\underline{x}_{k+1} = A_k \underline{x}_k + B_k \underline{u}_k \tag{B.5-1}$$

$$\underline{v}_k = C_k \underline{x}_k + J_k \underline{u}_k \tag{B.5-2}$$

where $\underline{x}$, $\underline{u}$, and $\underline{v}$ have the same meaning as in Secs. B.2 to B.4 and A_k, B_k, C_k, and J_k are time-varying matrices of appropriate dimension defined for all $k \geq k_0$.

DEFINITION B.5-1. The system of Eqs. (B.5-1) and (B.5-2) is *positive* if the sum of the input/output scalar products over $[k_0, k_1]$ can be expressed by

$$\sum_{k=k_0}^{k_1} \underline{v}_k^T \underline{u}_k = \xi(\underline{x}_{k_1+1}, k_1+1) - \xi(\underline{x}_{k_0}, k_0) + \sum_{k=k_0}^{k_1} \lambda(\underline{x}_k, \underline{u}_k, k)$$

$$\text{for all } k \geq k_0 \tag{B.5-3}$$

with

$$\lambda(\underline{x}_k, \underline{u}_k, k) \geq 0 \quad \text{for all } \underline{x}_k \in R^n,\ \underline{u}_k \in R^m,\ k \geq k_0 \tag{B.5-4}$$

As in the case of continuous linear time-varying systems, one has only sufficient conditions for positivity [14]. Some of these sufficient conditions are given below.

LEMMA B.5-1. The system of Eqs. (B.5-1) and (B.5-2) is *positive* if there exist a symmetric time-varying positive definite matrix P_k, a symmetric time-varying positive semidefinite matrix Q_k, and time-varying matrices S_k and R_k such that

$$A_k^T P_{k+1} A_k - P_k = -Q_k \tag{B.5-5}$$

$$B_k^T P_{k+1} A_k + S_k^T = C_k \tag{B.5-6}$$

$$J_k + J_k^T - B_k^T P_{k+1} B_k = R_k \tag{B.5-7}$$

$$\begin{bmatrix} Q_k & S_k \\ S_k^T & R_k \end{bmatrix} \geq 0 \quad \text{for all } k \geq k_0 \tag{B.5-8}$$

LEMMA B.5-2. The system of Eqs. (B.5-1) and (B.5-2) is *positive* if there exist a symmetric time-varying positive definite matrix P_k and time-varying matrices K_k and L_k such that

$$A_k^T P_{k+1} A_k - P_k = -L_k L_k^T \tag{B.5-9}$$

$$B_k^T P_{k+1} A_k + K_k^T L_k^T = C_k \tag{B.5-10}$$

$$K_k^T K_k = J_k + J_k^T - B_k^T P_{k+1} B_k \tag{B.5-11}$$

LEMMA B.5-3. If the matrix K_k in Eq. (B.5-11) is restricted to the class of nonsingular matrices (which implies that $K^T K > 0$), then the system of Eqs. (B.5-1) and (B.5-2) is *positive* if there exists a symmetric time-varying positive definite matrix P_k such that

$$A_k^T P_{k+1} A_k - P_k = -(C_k^T - A_k^T P_{k+1} B_k)(J_k^T + J_k - B_k^T P_{k+1} B_k)^{-1} \cdot (C_k - B_k^T P_{k+1} A_k) \tag{B.5-12}$$

LEMMA B.5-4. The system of Eqs. (B.5-1) and (B.5-2) is *positive* if every solution $\underline{x}_k(\underline{x}_{k_0}, \underline{u}_k, k)$ satisfies the following equality:

$$\sum_{k=k_0}^{k_1} \underline{v}_k^T \underline{u}_k = \frac{1}{2}\underline{x}_{k_1+1}^T P_{k_1+1}\underline{x}_{k_1+1} - \frac{1}{2}\underline{x}_{k_0}^T P_{k_0}\underline{x}_{k_0} + \sum_{k=k_0}^{k_1} (\underline{x}_k^T Q_k \underline{x}_k + 2\underline{u}_k^T S_k^T \underline{x}_k + \underline{u}_k^T R_k \underline{u}_k) \qquad (B.5\text{-}13)$$

where P_k is a sequence of positive definite matrices and the matrices P_k, Q_k, S_k, and R_k satisfy Eqs. (B.5-5) to (B.5-8).

We shall now show how Lemma B.5-3 can be obtained from Lemma B.5-2 if in Lemma B.5-2 we restrict K_k to the class of time-varying nonsingular matrices (K_k^{-1} exists for all k). One obtains from Eq. (B.5-10)

$$L_k^T = (K_k^T)^{-1}[C_k - B_k^T P_{k+1} A_k] \qquad (B.5\text{-}14)$$

and, respectively,

$$L_k L_k^T = (C_k^T - A_k^T P_{k+1} B_k)(K_k^T K_k)^{-1}[C_k - B_k^T P_{k+1} A_k]$$

Using Eq. (B.5-11), one then obtains

$$L_k L_k^T = (C_k^T - A_k^T P_{k+1} B_k)(J_k + J_k^T - B_k^T P_{k+1} B_k)^{-1} \cdot (C_k - B_k^T P_{k+1} A_k) \qquad (B.5\text{-}15)$$

Introducing Eq. (B.5-15) into Eq. (B.5-9), one obtains Eq. (B.5-12) of Lemma B.5-3.

Now we shall show that Lemma B.5-1 implies Lemma B.5-4. We shall replace the term $\underline{x}_k^T Q_k \underline{x}_k$ in Eq. (B.5-13) by an equivalent term obtained by making use of Eqs. (B.5-5) and (B.5-1). One obtains

$$\begin{aligned}
\underline{x}_k^T Q_k \underline{x}_k &\equiv -\underline{x}_{k+1}^T P_{k+1}\underline{x}_{k+1} + \underline{x}_k^T P_k \underline{x}_k + \underline{u}_k^T B_k^T P_{k+1}\underline{x}_{k+1} \\
&\quad + \underline{x}_{k+1}^T P_{k+1} B_k \underline{u}_k - \underline{u}_k^T B_k^T P_{k+1} B_k \underline{u}_k \\
&= -\underline{x}_{k+1}^T P_{k+1}\underline{x}_{k+1} + \underline{x}_k^T P_k \underline{x}_k + \underline{u}_k^T B_k^T P_{k+1} (A_k \underline{x}_k + B_k \underline{u}_k) \\
&\quad + (\underline{x}_k^T A_k^T + \underline{u}_k^T B_k^T) P_{k+1} B_k \underline{u}_k - \underline{u}_k^T B_k^T P_{k+1} B_k \underline{u}_k \qquad (B.5\text{-}16)
\end{aligned}$$

Adding the term $2\underline{u}_k^T S_k^T \underline{x}_k + \underline{u}_k^T R_k \underline{u}_k$ to both sides of Eq. (B.5-16) and also using Eqs. (B.5-6), (B.5-7), and (B.5-2), one obtains

$$\underline{x}_k^T Q_k \underline{x}_k + 2\underline{u}_k^T S_k \underline{x}_k + \underline{u}_k^T R_k \underline{u}_k = -\underline{x}_{k+1}^T P_k \underline{x}_{k+1} + \underline{x}_k^T P_k \underline{x}_k + 2\underline{u}_k^T C_k \underline{x}_k + \underline{u}_k^T (J_k + J_k^T)\underline{u}_k$$
$$= -\underline{x}_{k+1}^T P_k \underline{x}_{k+1} + \underline{x}_k^T P_k \underline{x}_k + 2\underline{v}_k^T \underline{u}_k \tag{B.5-17}$$

and

$$\sum_{k=k_0}^{k_1} \underline{v}_k^T \underline{u}_k = \frac{1}{2}\sum_{k=k_0}^{k_1} \underline{x}_{k+1}^T P_{k+1} \underline{x}_{k+1} - \frac{1}{2}\sum_{k=k_0}^{k_1} \underline{x}_k P_k \underline{x}_k + \frac{1}{2}\sum_{k=k_0}^{k_1} (\underline{x}_k^T Q_k \underline{x}_k + 2\underline{u}_k^T S_k^T \underline{x}_k + \underline{u}_k^T R_k \underline{u}_k)$$
$$= \frac{1}{2}\underline{x}_{k+1}^T P_{k+1} \underline{x}_{k+1} - \frac{1}{2}\underline{x}_{k_0}^T P_{k_0} \underline{x}_{k_0} + \frac{1}{2}\sum_{k=k_0}^{k_1} (\underline{x}_k^T Q_k \underline{x}_k + 2\underline{u}_k^T S_k^T \underline{x}_k + \underline{u}_k^T R_k \underline{u}_k) \tag{B.5-18}$$

which is exactly Eq. (B.5-13) of Lemma B.5-4.

Note: A discrete linear time-varying positive system verifying Lemma B.5-4 (or any of the Lemmas B.5-1 to B.5-3) satisfies the following inequality for all x_{k_0}:

$$\sum_{k=k_0}^{k_1} \underline{v}_k^T \underline{u}_k \geq -\frac{1}{2}\underline{x}_{k_0}^T P_{k_0} \underline{x}_{k_0} \geq -\gamma_0^2 \quad \text{for all } k_1 \geq k_0 \tag{B.5-19}$$

where γ_0^2 is a finite positive constant ($\underline{x}_{k_0}$ and P_{k_0} being bounded).

REFERENCES

1. V. M. Popov, *Hyperstability of Automatic Control Systems*, Springer, New York, 1973.
2. R. W. Newcomb, *Linear Multiport Synthesis*, McGraw-Hill, New York, 1966.
3. B. D. O. Anderson, A system theory criterion for positive real matrices, SIAM J. Control, 5, no. 2, 171-182 (1967).

4. P. Faurre, Réalisations markoviennes de processus stationnaires, IRIA Research Report No. 13, Roquencourt, France, March, 1973.
5. B. D. O. Anderson and S. Vongpanitlerd, *Network Analysis and Synthesis. A Modern Systems Approach*, Prentice-Hall, Englewood Cliffs, N.J., 1972.
6. K. S. Narendra and J. H. Taylor, *Frequency Domain Criteria for Absolute Stability*, Academic Press, New York, 1973.
7. D. D. Siljak, Algebraic criterion for absolute stability, optimality and passivity of dynamic systems, in *Proceedings of the IEEE*, Vol. 117, 1970, pp. 2033-2036.
8. J. S. Karmarkar, On nonnegativity of even polynomials, in *Proceedings of the IEEE*, Vol. 58, May 1970, pp. 835-836.
9. F. R. Gantmacher, *The Theory of Matrices*, Vols. 1 and 2, Chelsea Publishing Co., New York, 1959 (trans.).
10. A. Halanay, Noyaux positivement définis et stabilité des systèmes automatiques, C. R. Acad. Sci., *T. 258*, 786-788 (1964).
11. B. D. O. Anderson, A simplified viewpoint on hyperstability, IEEE Trans. Autom. Control, *AC-13*, 292-294 (1968).
12. L. Hitz and B. D. O. Anderson, Discrete positive real functions and their application to system stability, in *Proceedings of the IEEE*, Vol. 116, 1969, pp. 153-155.
13. A. Halanay, Un théorème de stabilité dans la théorie des systèmes non linéaires discrets, C. R. Acad. Sci., *T. 256*, 4818-4821 (1963).
14. T. Ionescu, Hyperstability of linear time varying discrete systems, IEEE Trans. Autom. Control, *AC-15*, 645-647 (1970).

Appendix C

HYPERSTABILITY

C-1 THE HYPERSTABILITY PROBLEM

The *hyperstability problem* was introduced by Popov as a generalization of the *absolute stability problem* [1, 2].

Consider the standard multivariable feedback system depicted in Fig. C-1. The system is formed by a linear time-invariant feedforward block and feedback block which can be linear or nonlinear and time-invariant or time-varying.

In the absolute stability problem one is interested in finding the conditions which must be satisfied by the feedforward block so that the feedback system represented in Fig. C-1 will be globally asymptotically stable for any feedback block of the class satisfying an inequality of the form

$$v_i w_i \geq 0 \quad \text{for all } i = 1, \ldots, m \tag{C-1}$$

(or, more generally, $h_i^m v_i^2 \leq v_i w_i \leq h_i^M v_i^2$, $h_i^M \geq h_i^m \geq 0$),[1] where v_i and w_i are the components of the input and output vectors $\underline{v}$ and $\underline{w}$ of the feedback blocks, both m-dimensional. Popov considered the global asymptotic stability of a system of the form given in Fig. C-1 but for the class of feedback blocks satisfying the inequality

---

[1] $w_i = h_i^M v_i$ and $w_i = h_i^m v_i$ define a sector in the plane v_i-w_i which contains the possible characteristics of the feedback block.

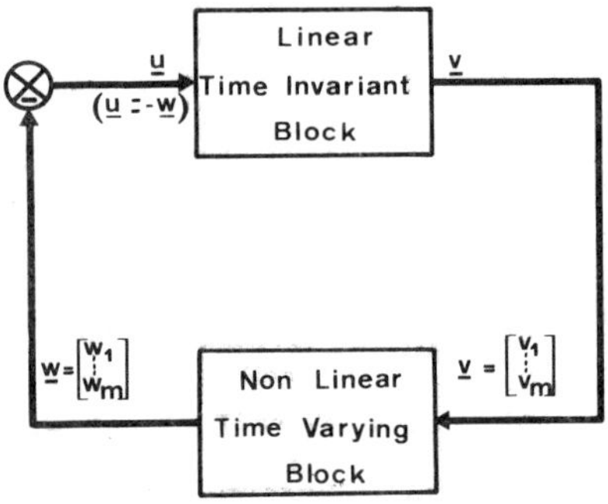

Fig. C-1 Standard multivariable nonlinear time-varying feedback system.

$$\eta(t_0, t_1) \triangleq \int_{t_0}^{t_1} \underline{w}^T \underline{v}\, dt \geq -\gamma_0^2 \quad \text{for all } t_1 \geq t_0 \tag{C-2}$$

called the *Popov integral inequality*. All the blocks for which the integral of the input/output inner product satisfies the inequality of Eq. (C-2) will also be denoted as belonging to the class {P} [or the block $\underline{w} = \underline{f}(\underline{v}) \in \{P\}$]. This class contains as a subclass the feedback blocks considered in the absolute stability problem. Note that Eq. (C-2) is a generalization of Eq. (C-1) in the sense that now the input/output product must be, in the mean, superior to a negative constant instead of being, at each instant, superior to zero.

A feedback system of the form represented in Fig. C-1 when it is globally (asymptotically) stable for all the feedback blocks satisfying the inequality of Eq. (C-2) will be said to be (asymptotically) *hyperstable*. Because the feedback block is defined by Eq. (C-2), the hyperstability properties of the feedback system will depend only on the characteristics of the feedforward block. A feedforward block which assures that the closed-loop system considered is hyperstable will be said to be a *hyperstable block*.

Note: If the feedforward block of the system given in Fig. C-1 admits a state-space representation (state vector: $\underline{x}$), the results of Popov are in fact valid for a slightly more general class of feedback blocks satisfying the inequality

$$\int_{t_0}^{t_1} \underline{w}^T \underline{v}\, dt \geq -\gamma_0^2 - \gamma_0 \sup_{t_0 \leq t \leq t_1} ||\underline{x}(t)|| \quad \text{for all } t_1 \geq t_0 \qquad \text{(C-3)}$$

where γ_0 is a finite positive constant (i.e., the lower bound of the integral of the input/output product may depend also on the maximum value of the norm of the state vector $\underline{x}$ of the feedforward block over the interval $[t_0, t_1]$).

Next we shall present some definitions and basic results of hyperstability theory to the extent necessary for understanding the designs based on them throughout the book.

C-2 HYPERSTABILITY. SOME DEFINITIONS

Consider the closed-loop system having a feedforward block

$$\dot{\underline{x}} = A\underline{x} + B\underline{u} = A\underline{x} - B\underline{w} \qquad \text{(C-4)}$$

$$\underline{v} = C\underline{x} + J\underline{u} = C\underline{x} - J\underline{w} \qquad \text{(C-5)}$$

and a feedback block

$$\underline{w} = \underline{f}(\underline{v}, t, \tau), \quad \tau \leq t \qquad \text{(C-6)}$$

where $\underline{x}$ is the state vector of the feedforward block (n-dimensional); $\underline{u}$ and $\underline{v}$ are the input and the output, respectively, of the feedforward block (both m-dimensional); A, B, C, and J are matrices of appropriate dimension; the pair (A, B) is *completely controllable*; the pair (C, A) is *completely observable*; and $\underline{f}(.)$ denotes a vector functional.

DEFINITION C-1 [3, 4]. The closed-loop system of Eqs. (C-4), (C-5), and (C-6) is *hyperstable* [or the feedforward block defined by Eqs. (C-4) and (C-5) is hyperstable] if there exist a positive constant $\delta > 0$ and a positive constant $\gamma_0 > 0$ such that all the solutions $\underline{x}[\underline{x}(0), t]$ of Eqs. (C-4) and (C-5) satisfy the inequality

$$||\underline{x}(t)|| < \delta[||\underline{x}(0)|| + \gamma_0] \quad \text{for all } t \geq 0 \qquad \text{(C-7)}$$

for any feedback block $\underline{w} = \underline{f}(\underline{v}, t, \tau)$ satisfying the inequality of Eq. (C-2) [or Eq. (C-3)].

DEFINITION C-2 [3]. The closed-loop system of Eqs. (C-4), (C-5), and (C-6) is *asymptotically hyperstable if*

1. It is hyperstable.
2. $\lim_{t\to\infty} \underline{x}(t) = \underline{0}$ for all vector functionals (or feedback blocks) $\underline{w} = \underline{f}(\underline{v}, t, \tau)$ satisfying the inequality of Eq. (C-3) [or Eq. (C-2)].

Definition C-2 can also be restated in the following form.

DEFINITION C-3. The closed-loop system described by Eqs. (C-4), (C-5), and (C-6) is *asymptotically hyperstable* if it is globally asymptotically stable for all the feedback blocks given by Eq. (C-6) which satisfy the inequality of Eq. (C-2) or (C-3).

Definition C-1 of hyperstability can be straightforwardly extended to continuous linear time-varying feedforward blocks. Definitions C-1, C-2, and C-3 can also be straightforwardly extended to discrete-time systems. In this case, the integral inequality of Eq. (C-2) is replaced by [4]

$$\eta(k_0, k_1) \triangleq \sum_{k=k_0}^{k_1} \underline{w}_k^T \underline{v}_k \geq -\gamma_0^2 \quad \text{for all } k_1 \geq k_0 \tag{C-8}$$

where γ_0^2 is a finite positive constant.

For the blocks which satisfy a relation of the form given by Eq. (C-2), the following definition is applicable [5].

DEFINITION C-4. A block described by an input/output relation

$$\underline{w} = H\underline{v} \tag{C-9}$$

where $\underline{w}$ and $\underline{v}$ are piecewise vector functions defined for $t \geq t_0$ and H is an operator acting on the input $\underline{v}$, is termed *hyperstable in the large* if it satisfies the Popov integral inequality of Eq. (C-2).

C-3 HYPERSTABILITY. MAIN RESULTS

THEOREM C-1 [3]. The necessary and sufficient condition for the feedback system described by Eqs. (C-4), (C-5), (C-6), and C-2) or (C-3) to be *hyperstable* [or the block defined by Eqs. (C-4) and (C-5) to be hyperstable] is as follows: The transfer matrix

$$H(s) = J + C(sI - A)^{-1}B \tag{C-10}$$

must be a *positive real transfer matrix.*

COROLLARY C-1. The feedforward block defined by Eqs. (C-4) and (C-5) is *hyperstable* if and only if it is a *positive* block.

Corollary C-1 is a direct consequence of the equivalent formulations of the positivity conditions for continuous linear time-invariant systems (Theorem B.2-1).

The statement "H(s) is a positive real transfer matrix" in Theorem C-1 can be replaced by any of the equivalent statements given in Theorem B.2-1.

THEOREM C-2 [3]. The necessary and sufficient condition for the feedback system described by Eqs. (C-4), (C-5), (C-6), and (C-2) or (C-3) to be *asymptotically hyperstable* is as follows: The transfer matrix of Eq. (C-10) must be a *strictly positive real transfer matrix.*

Consider now a linear time-varying feedforward block described by

$$\dot{\underline{x}} = A(t)\underline{x} + B(t)\underline{u} \tag{C-11}$$

$$\underline{v} = C(t)\underline{x} + J(t)\underline{u} \tag{C-12}$$

where A(t), B(t), C(t), and J(t) are time-varying matrices with piecewise continuous elements defined for all $t \geq t_0$.

THEOREM C-3 [6]. For the feedback system described by Eqs. (C-11), (C-12), (C-6), and (C-2) or (C-3) to be *hyperstable* [or the block defined by Eqs. (C-11) and (C-12) to be hyperstable] it is sufficient that the block of Eqs. (C-11) and (C-12) verifies one of the *positivity* lemmas B.3-1 and B.3-2.

Consider now the (feedforward) discrete linear time-invariant block described by

$$\underline{x}_{k+1} = A\underline{x}_k + B\underline{u}_k \tag{C-13}$$

$$\underline{v}_k = C\underline{x}_k + J\underline{u}_k \tag{C-14}$$

where the pair (A, B) is *completely controllable* and the pair (C, A) is *completely observable*. Consider also the feedback block

$$\underline{w}_k = \underline{f}(\underline{v}, k, \ell), \quad \ell \leq k \tag{C-15}$$

satisfying the inequality of Eq. (C-8).

THEOREM C-4 [7]. The necessary and sufficient condition for the feedback system described by Eqs. (C-13), (C-14), (C-15), and (C-8) to be *(asymptotically) hyperstable* [or the discrete block defined by Eqs. (C-13) and (C-14) to be (asymptotically) hyperstable] is as follows: The discrete transfer matrix

$$H(z) = J + C(zI - A)^{-1}B \tag{C-16}$$

must be a *(strictly) positive real discrete transfer matrix*.

Consider now the discrete linear time varying block described by

$$\underline{x}_{k+1} = A_k\underline{x}_k + B_k\underline{u}_k \tag{C-17}$$

$$\underline{v}_k = C_k\underline{x}_k + J_k\underline{u}_k \tag{C-18}$$

where the discrete time varying matrices A_k, B_k, C_k, and J_k are defined for all $k \geq k_0$.

THEOREM C-5 [10]. In order that the discrete linear time varying feedforward block defined by Eqs. (C-17) and (C-18) be *hyperstable* it is *sufficient* that it verifies one of the *positivity* Lemmas B.5-1 through B.5-4.

Note: The direct application of any of the above theorems either for solving the stability problem related to a feedback system having a particular feedback block $\underline{w} = \underline{f}(\underline{v}, t, \tau) \in \{P\}$ or for solving the stability problem of a feedback system for a subclass of feedback blocks $\underline{w} = \underline{f}(\underline{v}, t, \tau) \in \{P_1\} \in \{P\}$ will provide only sufficient conditions for the stability of the considered system (for example, when the feedback block is a linear positive gain which satisfies the Popov inequality, stability of the closed-loop system can be assured even if the transfer function of the linear feedforward block is not positive real).

When the linear time-invariant feedforward block of Eqs. (C-4) and (C-5) is not completely controllable, there exists a suitable equivalence transformation $\underline{\hat{x}} = F\underline{x}$, where F is a nonsingular matrix and $\underline{\hat{x}}^T = [\underline{\hat{x}}_1{}^T, \underline{\hat{x}}_2{}^T]$, which brings the feedforward block to the form

$$\dot{\underline{\hat{x}}}_1 = A_{11}\underline{\hat{x}}_1 + A_{12}\underline{\hat{x}}_2 + B_1\underline{u} \quad \text{(C-19)}$$

$$\dot{\underline{\hat{x}}}_2 = A_{22}\underline{\hat{x}}_2 \quad \text{(C-20)}$$

$$\underline{v} = [C_1, C_2]\begin{vmatrix} \underline{\hat{x}}_1 \\ \underline{\hat{x}}_2 \end{vmatrix} + J\underline{u} \quad \text{(C-21)}$$

(C_1, A_{11}) is *completely observable* [8]. For this case one has the following result [5, 9, 11].

THEOREM C-5. The necessary and sufficient conditions for the feedback system described by Eqs. (C-19), (C-20), (C-21), (C-6), and (C-2) or (C-3) to be *asymptotically hyperstable* are as follows:

1. The matrix A_{22} is a Hurwitz matrix (all its eigenvalues have negative real parts).
2. The transfer matrix of the completely controllable/completely observable part of the feedforward block,

$$H_1(s) = H(s) = C_1(sI - A_{11})^{-1}B_1 + J \quad \text{(C-22)}$$

is *strictly positive real.*

C-4 PROPERTIES OF COMBINATIONS OF THE HYPERSTABLE BLOCKS

LEMMA C-6 [4]. Any block obtained by the *parallel* combination of *two hyperstable blocks* [see Fig. C-2(a)] is *hyperstable.*

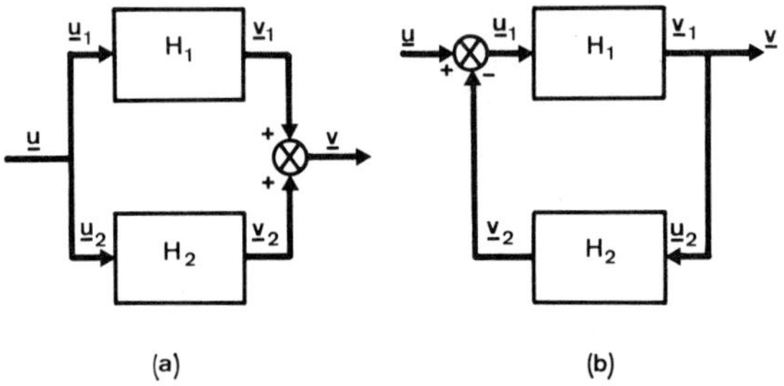

Fig. C-2 (a) Parallel and (b) feedback combination of two hyperstable blocks.

LEMMA C-7 [4]. Any block obtained from the *feedback* combination of *two hyperstable blocks* [see Fig. C-2(b)] is *hyperstable*.

LEMMA C-8. The *parallel* combination of two blocks which are *hyperstable in the large* forms a block which is *hyperstable in the large*.

LEMMA C-9. The *feedback* combination of two blocks which are *hyperstable in the large* forms a block which is *hyperstable in the large*.

The proofs of Lemmas C-6 and C-7 can be found in Ref. 4. The proofs of Lemmas C-8 and C-9 proceed along the same lines. These proofs are based on the observation that for both combinations shown in Fig. C-2 one has

$$\int_{t_0}^{t_i} \underline{v}^T \underline{u}\, dt = \int_{t_0}^{t_i} \underline{v}_1^T \underline{u}_1\, dt + \int_{t_0}^{t_i} \underline{v}_2^T \underline{u}_2\, dt$$

LEMMA C-10. Any *feedback system* obtained by the feedback combination of a *hyperstable* block and of a block which is *hyperstable in the large* is *globally stable*.

Lemma C-10 is a direct consequence of the various hyperstability theorems (Theorems C-1 to C-4) combined with the definition of hyperstability in the large (in the above-mentioned theorems the feedback blocks are, in fact, hyperstable in the large).

Note that the hyperstability properties are conserved through parallel and feedback combinations but not through series combinations (except particular cases).

REFERENCES

1. V. M. Popov, Criterii de stabilitate pentru sistemele automate continînd elemente neunivoce, Probleme de Automatizare, 143-151, Publishing House of the Rumanian Academy, Bucharest, 1960.
2. V. M. Popov, Solution of a new stability problem for controlled systems, Autom. Remote Control, *24*, no. 1, 1-23 (1963) (trans.).
3. V. M. Popov, Hyperstability of automatic systems with several nonlinear elements, Rev. Roum. Sci. Tech. Ser. Electrotech. Energ., *9*, no. 4, 33-45 (1964).
4. V. M. Popov, *Hyperstability of Automatic Control Systems*, Springer, New York, 1973 (trans.).
5. V. M. Popov, *Hiperstabilitatea Sistemelor Automate*, Editura Academiei Republicü Socialiste România, Bucharest, 1966.
6. V. M. Popov, Hyperstability of control systems with variable coefficients, Rev. Roum. Sci. Tech. Ser. Electrotech. Energ., *10*, no. 2, 285-298 (1965).
7. I. D. Landau, Hyperstability concepts and their application to discrete control systems, in *Proceedings of the Joint Automatic Control Conference*, Stanford University, Stanford, August 1972.
8. C. T. Chen, *Introduction to Linear Systems Theory*, Holt, Rinehart and Winston, New York, 1970.
9. V. M. Popov, Incompletely controllable positive systems and applications to optimization and stability of automatic control systems, Rev. Roum. Sci. Tech. Ser. Electrotech. Energ., *12*, no. 3, 337-357 (1967).
10. T. Ionescu, Hyperstability of linear time varying discrete systems, IEEE Trans. Autom. Control, *AC-15*, no. 6, 645-647 (1970).
11. V. M. Popov, Hyperstability and optimality of automatic control system with several control functions, Rev. Roum. Sci. Tech. Ser. Electrotech. Energ., *9*, 629-690 (1964).

Appendix D

SOLUTIONS FOR AN INTEGRAL INEQUALITY

D.1 CONTINUOUS CASE

Problem Formulation

To achieve the synthesis of model reference adaptive systems (MRAS's) using the hyperstability approach one must solve the following problem [1]: Consider a block with the input $\underline{v}$ and the output $\underline{w}$ where $\underline{v}$ and $\underline{w}$ are vectors of the same dimension (n). The output of this block is defined by

$$\underline{w}(t) = [\int_0^t \Phi_1(\underline{v}, t, \tau)\, d\tau + \Phi_2(\underline{v}, t) + A_0]\underline{y}(t) \qquad \text{(D.1-1)}$$

where $\Phi_1(\underline{v}, t, \tau)$ is a matrix functional of $\underline{v}$ [(n × q)-dimensional] [i.e., it depends on $v(\tau)$, $0 \le \tau \le t$], $\Phi_2(\underline{v}, t)$ is a matrix function of $\underline{v}(t)$ [(n × q)-dimensional], $\underline{v}$ is a piecewise continuous vector function (n-dimensional), A_0 is a constant matrix [(n × q)-dimensional], and $\underline{y}(t)$ is a piecewise continuous vector function (q-dimensional).

The Problem

Find the most general solutions for $\Phi_1(\underline{v}, t, \tau)$ and $\Phi_2(\underline{v}, t)$ in Eq. (D.1-1) such that the following inequality holds:

$$\eta(0, t_1) \triangleq \int_0^{t_1} \underline{v}^T\underline{w}\, dt \ge -\gamma_0^2 \quad \text{for all } t_1 \ge 0 \qquad \text{(D.1-2)}$$

where γ_0^2 is an arbitrary finite positive constant.

In addition to this, in most of the cases one asks that $\Phi_1(\underline{v}, t, \tau)$ is such that

$$\lim_{t\to\infty} \underline{v}(t) = 0 \Longrightarrow \lim_{t\to\infty} \int_0^t \Phi_1(\underline{v}, t, \tau)\, d\tau = \text{constant} \neq 0 \qquad \text{(D.1-3)}$$

This of course implies that the adaptation mechanism will have memory (see Sec. 2.4).

Introducing the expression for $\underline{w}(t)$ given by Eq. (D.1-1) into Eq. (D.1-2), one obtains

$$\eta(0, t_1) \triangleq \int_0^{t_1} \underline{v}^T \left[\int_0^t \Phi_1(\underline{v}, t, \tau)\, d\tau + A_0\right]\underline{y}\, dt + \int_0^{t_1} \underline{v}^T \Phi_2(\underline{v}, t)\underline{y}\, dt \geq -\gamma_0^2 \quad \text{for all } t_1 \geq 0 \qquad \text{(D.1-4)}$$

A simple example of this type of problem is solved in Sec. 3.3, but here we shall focus our attention on finding general solutions for Φ_1 and Φ_2 allowing us to satisfy Eq. (D.1-4).

The inequality of Eq. (D.1-4) [and of course the inequality of Eq. (D.1-2)] is satisfied if each of the two terms on the left-hand side satisfies an inequality of the same type (this is a *sufficient* condition). Therefore the original problem can be replaced by the following two subproblems, whose solutions imply that the inequality of Eq. (D.1-2) holds. Note that in many cases each of these two subproblems corresponds to the synthesis of a distinct component of the adaptation law, namely one with memory and another one without memory.

SUBPROBLEM D.1-1. Find the most general solutions for $\Phi_1(\underline{v}, t, \tau)$ such that the following inequality holds:

$$\eta_1(0, t_1) \triangleq \int_0^{t_1} \underline{v}^T \underline{w}_1\, dt = \int_0^{t_1} \underline{v}^T \left[\int_0^t \Phi_1(\underline{v}, t, \tau)\, d\tau + A_0\right]\underline{y}\, dt \geq -\gamma_1^2 \quad \text{for all } t_1 \geq 0 \qquad \text{(D.1-5)}$$

where γ_1^2 is a finite positive constant. In addition to the condition (D.1-5), $\Phi_1(\underline{v}, t, \tau)$ should also satisfy the condition of Eq. (D.1-3) (when the adaptation mechanism should have memory).

SUBPROBLEM D.1-2. Find the most general solutions for $\Phi_2(\underline{v}, t)$ such that the following inequality holds:

$$\eta_2(0, t_1) \triangleq \int_0^t \underline{v}^T \underline{w}_2 \, dt = \int_0^{t_1} \underline{v}^T \Phi_2(\underline{v}, t)\underline{y} \, dt \geq -\gamma_2^2$$

$$\text{for all } t_1 \geq 0 \qquad \text{(D.1-6)}$$

where γ_2^2 is an arbitrary finite positive constant.

Solutions for Subproblem D.1-1

Before stating the main results obtained for Subproblem D.1-1 we shall discuss some reformulations of Subproblem D.1-1 which will help us to understand the meaning of the results. First, the matrix functional $\Phi_1(\underline{v}, t, \tau)$ can be decomposed into column vectors:

$$\Phi_1 = [\underline{\phi}_1^{\ 1} \cdots \underline{\phi}_1^{\ j} \cdots \underline{\phi}_1^{\ q}] \qquad \text{(D.1-7)}$$

Using this decomposition, the inequality of Eq. (D.1-5) can be expressed as

$$\eta_1(0, t_1) \triangleq \sum_{j=1}^{q} \int_0^{t_1} y_j \underline{v}^T [\int_0^t \underline{\phi}_1^{\ j}(\underline{v}, t, \tau) \, dt + \underline{a}_0^{\ j}] \, dt \geq -\gamma_1^2$$

$$\text{for all } t_1 \geq 0 \qquad \text{(D.1-8)}$$

where the $\underline{a}_0^{\ j}$ are the column vectors of the matrix A_0:

$$A_0 = [\underline{a}_0^{\ 1} \cdots \underline{a}_0^{\ j} \cdots \underline{a}_0^{\ q}] \qquad \text{(D.1-9)}$$

For the inequality of Eq. (D.1-8) to be satisfied *it is sufficient* that each of the q terms on the left-hand side of Eq. (D.1-8) satisfy this type of inequality. Therefore Subproblem D.1-1 is reduced to the search for the vector functional $\underline{\phi}_1^{\ j}(\underline{v}, t, \tau)$ which satisfies the inequality

$$\int_0^{t_1} y_j \underline{v}^T [\int_0^t \underline{\phi}_1^{\ j}(\underline{v}, t, \tau) \, d\tau + \underline{a}_0^{\ j}] \, dt \geq -\gamma_{1j}^2$$

$$\text{for all } t_1 \geq 0 \text{ and } j = 1, \ldots, q \qquad \text{(D.1-10)}$$

where γ_{1j}^2 is an arbitrary finite positive constant and the condition (when the adaptation mechanism should have memory)

$$\lim_{t\to\infty} \underline{v}(t) = 0 \Longrightarrow \lim_{t\to\infty} \int_0^t \underline{\phi}_1^{\ j}(\underline{v}, t, \tau) \, d\tau = \text{constant} \neq 0 \qquad \text{(D.1-11)}$$

But the matrix functional $\Phi_1(\underline{v}, t, \tau)$ can also be decomposed into row vectors as

$$\Phi_1^T(\underline{v}, t, \tau) \triangleq [\underline{\phi}_1^{1T} \cdots \underline{\phi}_1^{iT} \cdots \underline{\phi}_1^{nT}] \tag{D.1-12}$$

By using this decomposition the inequality of Eq. (D.1-5) becomes

$$\eta_1(0, t_1) \triangleq \sum_{i=1}^{n} \int_0^{t_1} v_i \underline{y}^T [\int_0^t \underline{\phi}_1^{iT}(\underline{v}, t, \tau)\, d\tau + \underline{a}_0^{iT}]\, dt \geq -\gamma_1^2$$
$$\text{for all } t_1 \geq 0 \tag{D.1-13}$$

where the $\underline{a}_0^i$ are the row vectors of the matrix A_0:

$$A_0^T = [\underline{a}_0^{1T} \cdots \underline{a}_0^{iT} \cdots \underline{a}_0^{nT}] \tag{D.1-14}$$

As for the previous decomposition of Φ_1, for the inequality of Eq. (D.1-13) to be satisfied, it is sufficient that

$$\int_0^{t_1} v_i \underline{y}^T [\int_0^t \underline{\phi}_1^{iT}(\underline{v}, t, \tau)\, d\tau + \underline{a}_0^{iT}]\, dt \geq -\gamma_{1i}^2$$
$$\text{for all } t_1 \geq 0 \text{ and } i = 1, \ldots, n \tag{D.1-15}$$

where γ_{1i}^2 is a finite arbitrary positive constant and in addition to this $\underline{\phi}_1^{iT}(\underline{v}, t, \tau)$ should verify a relation of the form of Eq. (D.1-11) when the adaptation mechanism should have memory.

At this point one can make the following observations:

1. The integral on the left-hand side of the inequality in Eq. (D.1-10) can be interpretated as the integral of the input/output inner product for a block with

$$\text{input:} \quad y_j(t)\underline{v}(t) = \underline{f}(t) \tag{D.1-16}$$

$$\text{output:} \quad \underline{a}^j(t) = \int_0^t \underline{\phi}_1^j(\underline{v}, t, \tau)\, d\tau + \underline{a}_0^j \tag{D.1-17}$$

or for a block with

$$\text{input:} \quad \underline{v}(t) \tag{D.1-18}$$

$$\text{output:} \quad [\underline{w}_1(t)]_j = y_j[\int_0^t \underline{\phi}_1^j(\underline{v}, t, \tau)\, d\tau + \underline{a}_0^j] \tag{D.1-19}$$

where

$$\sum_{j=1}^{q} [\underline{w}_1]_j = \underline{w}_1 \tag{D.1-20}$$

With these notations the inequality of Eq. (D.1-10) can be expressed as

$$\int_0^{t_1} y_j(t)\underline{v}^T(t)\underline{a}^j(t)\ dt = \int_0^{t_1} \underline{f}^T(t)\underline{a}^j(t)\ dt$$

$$= \int_0^{t_1} \underline{v}^T(t)[\underline{w}_1(t)]_j\ dt \geq -\gamma_{1j}^{\ 2}$$

$$\text{for all } t_1 \geq 0 \tag{D.1-21}$$

2. A similar interpretation can be established for the left-hand side of the inequality of Eq. (D.1-15).

One can now state some of the main results for Subproblem D.1-1.

LEMMA D.1-1 [2]. The inequality of Eq. (D.1-5) [together with the condition of Eq. (D.1-3)] is satisfied by

$$\underline{\phi}_1^{\ j}(\underline{v}, t, \tau) = F^j(t - \tau)\underline{v}(\tau)y_j(\tau), \quad j = 1, \ldots, q \tag{D.1-22}$$

where the $\underline{\phi}_1^{\ j}$ are the column vectors of $\Phi_1(\underline{v}, t, \tau)$ and $F^j(t - \tau)$ is a *positive definite square matrix kernel* whose Laplace transform is a *positive real transfer matrix with a pole at* $s = 0$.[1]

LEMMA D.1-2 [1]. The inequality of Eq. (D.1-5) [together with the condition of Eq. (D.1-3)] is satisfied by

$$\Phi_1(v, t, \tau) = F(t - \tau)\underline{v}(\tau)[G\underline{y}(\tau)]^T \tag{D.1-23}$$

where $F(t - \tau)$ is a *positive definite square matrix kernel* whose Laplace transform is a *positive real transfer matrix with a pole at* $s = 0$ and G is a *positive definite matrix.*

LEMMA D.1-3 [1]. The inequality of Eq. (D.1-5) [together with the condition of Eq. (D.1-3)] is satisfied by

$$\underline{\phi}_i^{\ iT}(\underline{v}, t, \tau) = G^i(t - \tau)\underline{y}(\tau)v_i(\tau), \quad i = 1, \ldots, n \tag{D.1-24}$$

---

[1]For the definitions of positive real transfer matrices and positive definite kernels, see Appendix B, Sec. B.1, Definitions B.1-5 and B.1-7.

where the $\underline{\phi}_1^{\,i}$ are the row vectors of $\Phi_1(\underline{v}, t, \tau)$ and $G^i(t - \tau)$ is a *positive definite square matrix kernel* whose Laplace transform is a *positive real transfer matrix with a pole at* $s = 0$.

LEMMA D.1-4. The inequality of Eq. (D.1-5) [together with the condition of Eq. (D.1-3)] is satisfied by

$$\phi_1^{ij}(\underline{v}, t, \tau) = k^{ij}(t - \tau)v_i(\tau)y_j(\tau), \qquad i = 1, \ldots, q, \quad j = 1, \ldots, n \tag{D.1-25}$$

where the ϕ_1^{ij} are the elements of $\Phi_1(\underline{v}, t, \tau)$ and $k^{ij}(t - \tau)$ is a *positive definite scalar kernel* whose Laplace transform is a *positive real transfer function with a pole at* $s = 0$.

We shall start by giving the proof of Lemma D.1-1, from which some of the other lemmas are derived.

Proof of Lemma D.1-1: To prove that the inequality of Eq. (D.1-5) holds it is sufficient to show that the inequality of Eq. (D.1-10) holds; i.e.,

$$\eta^j(0, t_1) = \int_0^{t_1} y_j\underline{v}^T\left[\int_0^t F^j(t - \tau)\underline{v}(\tau)y_j(\tau)\, d\tau + \underline{a}_0^{\,j}\right] dt \geq -\gamma_{1j}^{\,2} \quad \text{for all } t_1 \geq 0 \tag{D.1-26}$$

If one replaces $y_j\underline{v}$ by the vector $\underline{f}$ as in Eq. (D.1-16), the inequality of Eq. (D.1-26) becomes

$$\int_0^{t_1} \underline{f}^T(t)\left[\int_0^t F^j(t - \tau)\underline{f}(\tau)\, d\tau + \underline{a}_0^{\,j}\right] dt \geq -\gamma_{1j}^{\,2} \tag{D.1-27}$$

If $\underline{a}_0^{\,j} = 0$, one can note that

$$\int_0^{t_1} \underline{f}(t)^T\left[\int_0^t F^j(t - \tau)\underline{f}(\tau)\, d\tau\right] dt \geq 0 \quad \text{for all } t_1 \geq 0 \tag{D.1-28}$$

because $F^j(t - \tau)$ is a positive definite kernel (see Definition B.1-7 and Lemma B.1-2). For $\underline{a}_0^{\,j} \neq 0$, one can consider that $\underline{a}_0^{\,j}$ is the output of the system characterized by the matrix kernel $F^j(t - \tau)$ at $t = 0$, as a consequence of the inputs applied at $t < 0$ (see Ref. 3). Because the Laplace transform of $F^j(t - \tau)$ has a pole at $s = 0$, this means that the operator acting on $\underline{f}(t)$ has a pure integrator.

Therefore, for any $t \geq 0$, there exist a time $t_0 < \infty$ and a constant finite vector $\underline{f}(-t_0)$ such that

$$\underline{a}_0^j = \int_{-t_0}^{0} F^j(t - \tau)\underline{f}(-t_0)\, d\tau, \quad t_0 < \infty, \quad ||f(-t_0)|| < \infty \tag{D.1-29}$$

Using the notation

$$\underline{f}'(t) = \begin{cases} \underline{f}(t) & t \geq 0 \\ \underline{f}(-t_0), & -t_0 \leq t < 0 \end{cases} \tag{D.1-30}$$

the inequality of Eq. (D.1-27) becomes

$$\int_{-t_0}^{t_1} \underline{f}'(t)^T [\int_{-t_0}^{t} F^j(t - \tau)\underline{f}'(\tau)\, d\tau]\, dt - \int_{-t_0}^{0} \underline{f}(-t_0)^T \cdot [\int_{-t_0}^{t} F^j(t - \tau)\underline{f}(-t_0)\, d\tau]\, dt \geq -\gamma_{1j}^2 \tag{D.1-31}$$

The first term on the left-hand side of the inequality in Eq. (D.1-31) is greater than or equal to zero because $F^j(t - \tau)$ is a positive definite kernel. To estimate the second term one uses the property that the elements of $F^j(t - \tau)$ are bounded because its Laplace transform is a stable transfer matrix (a necessary condition for a transfer matrix to be positive real; see Definition B.1-5). Also using the fact that $F^j(t - \tau)$ is a positive definite kernel, one obtains

$$0 \leq \int_{-t_0}^{0} \underline{f}(-t_0)^T [\int_{-t_0}^{t} F^j(t - \tau)\underline{f}(-t_0)\, d\tau]\, dt$$

$$\leq \int_{-t_0}^{0} \int_{-t_0}^{0} m_j ||\underline{f}(-t_0)||^2\, d\tau\, dt = m_j ||\underline{f}(-t_0)||^2 \frac{t_0^2}{2},$$

$$0 \leq m_j < \infty \tag{D.1-32}$$

Using this result in Eq. (D.1-26), one obtains

$$\eta^j(0, t_1) \geq -m_j ||\underline{f}(-t_0)||^2 \frac{t_0^2}{2} = -\gamma_{1j}^2 \tag{D.1-33}$$

and the proof of Lemma D.1-1 is achieved since the pole at $s = 0$ (pure integration) implies verification of the condition specified by Eq. (D.1-11) [or by Eq. (D.1-3)].

Proof of Lemma D.1-2: Consider that in Lemma D.1-1 all the $F^j(t - \tau)$ are identical:

$$F^j(t - \tau) = F(t - \tau) \quad \text{for all } j = 1, \ldots, q \tag{D.1-34}$$

Therefore one can write

$$\begin{aligned}\Phi_1(\underline{v}, t, \tau) &= [F(t - \tau)\underline{v}y_1, \ldots, F(t - \tau)\underline{v}y_q] \\ &= F(t - \tau)\underline{v}[y_1, y_2, \ldots, y_q] \\ &= F(t - \tau)\underline{v}(\tau)\underline{y}^T(\tau)\end{aligned} \tag{D.1-35}$$

and Lemma D.1-2 is proved for the case G = I. For the case G ≠ I but positive definite, one has

$$G = G_1 G_1^T \tag{D.1-36}$$

where G_1 is a nonsingular matrix and the inequality of Eq. (D.1-5) can be written using Eq. (D.1-23) as

$$\begin{aligned}&\int_0^{t_1} \underline{v}^T[\int_0^t F(t - \tau)\underline{v}(\tau)\underline{y}^T(\tau)G^T \, d\tau + A_0]\underline{y} \, dt \\ &= \int_0^{t_1} \underline{v}^T[\int_0^t F(t - \tau)\underline{v}(\tau)\underline{y}(\tau)^T G_1^T \, d\tau + A_0 G_1^{-1}]G_1\underline{y} \, dt \\ &= \int_0^{t_1} \underline{v}^T[\int_0^t F(t - \tau)\underline{v}(\tau)\underline{y}'^T(\tau) \, d\tau + A_0']\underline{y}' \, dt \geq -\gamma_1^2 \quad \text{for all } t_1 \geq 0\end{aligned} \tag{D.1-37}$$

where $\underline{y}' = G_1\underline{y}$ and $A' = AG_1^{-1}$, and the problem becomes similar to the case G = I.

Proof of Lemma D.1-3: This proof is similar to the proof of Lemma D.1-1 by considering the inequality of Eq. (D.1-15) instead of the inequality of Eq. (D.1-10).

Proof of Lemma D.1-4: The inequality of Eq. (D.1-5) can also be written as

$$\begin{aligned}&\eta_1(0, t_1) \\ &= \sum_{i=1}^{n} \sum_{j=1}^{q} \int_0^{t_1} v_i y_j [\int_0^t \phi_1^{ij}(\underline{v}, t, \tau) \, dt + a_0^{ij}] \, dt \\ &= \sum_{i=1}^{n} \sum_{j=1}^{q} \int_0^{t_1} v_i y_j [\int_0^t k^{ij}(t - \tau)v_i(\tau)y_j(\tau) \, d\tau + a_0^{ij}] \, dt \geq -\gamma_1^2\end{aligned} \tag{D.1-38}$$

To show that $\phi_1^{ij}(\underline{v}, t, \tau)$ given by Eq. (D.1-25) satisfies the inequality (D.1-38) it is sufficient that each of the n × q terms on the left-hand side of the inequality in Eq. (D.1-38) satisfies this type of inequality. From this point the proof is a particular case of the proof for Lemma D.1-1 (scalar case).

Solutions for Subproblem D.1-2

For Subproblem D.1-2 one has the following results.

LEMMA D.1-5 [1]. The inequality of Eq. (D.1-6) is satisfied by

$$\Phi_2(\underline{v}, t) = F'(t)\underline{v}(t)[G'(t)\underline{y}(t)]^T \tag{D.1-39}$$

where F'(t) and G'(t) are time-varying positive semidefinite matrices for all $t \geq 0$.

Proof: Introducing Eq. (D.1-39) into Eq. (D.1-6), one obtains

$$\eta_2(0, t_1) = \int_0^{t_1} [\underline{v}^T F'(t)\underline{v}][\underline{y}^T G'\underline{y}]\, dt \geq 0 \quad \text{for all } t_1 \geq 0 \tag{D.1-40}$$

since the products $\underline{v}^T F'\underline{v}$ and $\underline{y}^T G'\underline{y}$ will be greater than or equal to zero for any $t \geq 0$.

LEMMA D.1-6. The inequality of Eq. (D.1-6) is satisfied by

$$\phi_2^{ij}(v, t) = \alpha_{ij}(t)v_i(t)y_j(t), \quad i = 1, \ldots, n,\ j = 1, \ldots, q \tag{D.1-41}$$

where the ϕ_2^{ij} are the elements of $\Phi_2(v, t)$ and $\alpha_{ij}(t)$ are positive or nonnegative functions for all $t \geq 0$.

Proof: The proof is obvious after the decomposition of the inequality of Eq. (D.1-6) into a sum of n × q inequalities.

D.2 DISCRETE CASE

Problem Formulation

For the synthesis of discrete MRAS's using the hyperstability approach one must solve the following problem [1, 3, 4]: Consider a block with input $\underline{v}$ and output $\underline{w}$, where $\underline{v}$ and $\underline{w}$ are n-dimensional vectors. The output at instant k + 1 is expressed by

$$\underline{w}_{k+1} = \left[\sum_{\ell=0}^{k} \Phi_1(\underline{v}, k, \ell) + \Phi_2(\underline{v}, k) + A_0\right]\underline{y}_k \tag{D.2-1}$$

where $\Phi_1(\underline{v}, k, \ell)$ is a discrete matrix functional of $\underline{v}$ (i.e., it depends on $\underline{v}_\ell$, $0 \le \ell \le k$), $\Phi_2(\underline{v}, k)$ is a matrix function of $\underline{v}_k$ [both $(n \times q)$-dimensional], $\underline{y}$ is a finite q-dimensional vector, and A_0 is a constant $(n \times q)$-dimensional matrix.[2]

The Problem

Find the most general solutions for $\Phi_1(\underline{v}, k, \ell)$ and $\Phi_2(\underline{v}, k)$ in Eq. (D.2-1) such that the following inequality holds:

$$\eta(0, k_1) \triangleq \sum_{k=0}^{k_1} \underline{v}_{k+1}{}^T \underline{w}_{k+1} \ge -\gamma_0{}^2 \quad \text{for all } k_1 \ge 0 \tag{D.2-2}$$

where $\gamma_0{}^2$ is an arbitrary positive finite constant. In addition, one requires in most of the cases (in order to assure the memory of the adaptation mechanism) that $\Phi_1(\underline{v}, k, \ell)$ is such that

$$\lim_{k\to\infty} \underline{v}_k = 0 \Longrightarrow \lim_{k\to\infty} \sum_{\ell=0}^{k} \Phi_1(\underline{v}, k, \ell) = \text{constant} \neq 0 \tag{D.2-3}$$

Using Eq. (D.2-1), the inequality of Eq. (D.2-2) can be written as:

$$\eta(0, k_1) \triangleq \sum_{k=0}^{k_1} \underline{v}_{k+1}{}^T \left[\sum_{\ell=0}^{k} \Phi_1(\underline{v}, k, \ell) + A_0\right]\underline{y}_k$$
$$+ \sum_{k=0}^{k_1} \underline{v}_{k+1}{}^T \Phi_2(\underline{v}, k)\underline{y}_k \geq -\gamma_0{}^2 \quad \text{for all } k_1 \geq 0 \tag{D.2-4}$$

A simple example of this type of problem is solved in Sec. 5.2. As in the continuous case, in order that inequality of Eq. (D.2-4) be satisfied it is sufficient that each of the two terms on the left-hand side of Eq. (D.2-4) satisfy such a type of inequality. Therefore we will have to solve the following two subproblems.

---

[2]Note that in the continuous-time case [Eq. (D.1-1)] the output is $\underline{w}(t)$ and the integral is considered up to the time t. In the discrete-time case the output $\underline{w}$ is given at instant k + 1 and the summation on the right-hand side of Eq. (D.2-1) is made only up to instant k.

Subproblem D.2-1. Find the most general solutions for $\Phi_1(\underline{v}, k, \ell)$ such that the following inequality holds:

$$\eta_1(0, k_1) \triangleq \sum_{k=0}^{k_1} \underline{v}_{k+1}^T \underline{w}_{1,k+1}$$

$$= \sum_{k=0}^{k_1} \underline{v}_{k+1}^T [\sum_{\ell=0}^{k} \Phi_1(\underline{v}, k, \ell) + A_0]\underline{y}_k \geq -\gamma_1^2$$

$$\text{for all } k_1 \geq 0 \qquad \text{(D.2-5)}$$

where γ_1^2 is a finite positive constant. In addition to the condition (D.2-5), $\Phi_1(\underline{v}, k, \ell)$ should also verify in most of the cases the condition of Eq. (D.2-3) (if the adaptation mechanism should have memory).

Subproblem D.2-2. Find the most general solutions for $\Phi_2(\underline{v}, k)$ such that the following inequality holds:

$$\eta_2(0, k_1) \triangleq \sum_{k=0}^{k_1} \underline{v}_{k+1}^T \underline{w}_{2,k+1}$$

$$= \sum_{k=0}^{k_1} \underline{v}_{k+1}^T \Phi_2(\underline{v}, k)\underline{y}_k \geq -\gamma_2^2 \quad \text{for all } k_1 \geq 0 \qquad \text{(D.2-6)}$$

where γ_2^2 is a finite positive constant.

The solutions to the above problems are direct counterparts of the results given for the continuous case. The only difference is that the variables $\underline{y}$ and $\underline{v}$ appearing in the solutions will be indexed at instant k for $\underline{y}$ and at instant k + 1 for $\underline{v}$. We mention below two results which are used in Sec. 5.4.

LEMMA D.2-1: The inequality of Eq. (D.2-5) [together with the condition of Eq. (D.2-3)] is satisfied by:

$$\Phi_1(\underline{v}, k, \ell) = F(k - \ell)\underline{v}_{\ell+1}[G\underline{y}_\ell]^T \qquad \text{(D.2-7)}$$

where $F(k - \ell)$ is a *positive definite discrete matrix kernel* whose z-transform is a *positive real discrete transfer matrix* with a pole at z = 1, and G is a *positive definite matrix*.

LEMMA D.2-2: The inequality of Eq. (D.2-6) is satisfied by:

$$\Phi_2(\underline{v}, k) = F_k' \underline{v}_{k+1} [G_k' \underline{y}_k]^T \tag{D.2-8}$$

where F_k' and G_k' are positive semidefinite time varying matrices for all $k \geq 0$.

For other results and proofs see Refs. 1, 4, 5, and 6.

REFERENCES

1. I. D. Landau, Sur une méthode de synthèse des systèmes adaptatifs avec modèle utilisés pour la commande et l'identification d'une classe de procédés physiques, Thèse Docteur ès Sciences, No. C.N.R.S.: A.0.8495, University of Grenoble, Grenoble, June 1973.
2. I. D. Landau, A generalization of the hyperstability conditions for model reference adaptive systems, IEEE Trans. Autom. Control, *AC-17*, no. 2, 246-247 (1972).
3. V. M. Popov, *Hyperstability of Automatic Control Systems,* Springer, New York, 1973.
4. I. D. Landau, Algorithmes d'adaptation hyperstables pour une classe de systèmes adaptatifs échantillonnés, C. R. Acad. Sci. Ser. A, *T. 275*, 1391-1394 (Dec. 18, 1972).
5. I. D. Landau, Design of discrete model reference adaptive systems using the positivity concepts, in *Proceedings of the Third Symposium on Sensitivity, Adaptivity and Optimality*, Ischia, 1973, pp. 307-314.
6. I. D. Landau and G. Bethoux, Algorithms for discrete time model reference adaptive systems, in *Proceedings of the Sixth IFAC Congress,* Part I-D, Boston, 1975, pp. 58-4.1, 58-4.11. (Intern. Federation of Automatic Control, Instruments Society of America, Pittsburgh, 1975.

INDEX